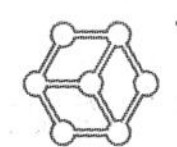

普通高等教育“十三五”规划教材

河南省“十二五”普通高等教育规划教材

双语版

高等有机化学

Advanced Organic Chemistry 2nd

第二版

谢普会 胡思前 徐翠莲 主编
Puhui Xie Siqian Hu Cuilian Xu Edit-in-Chief
彭望明 鲍峰玉 金 秋 副主编

化学工业出版社
·北京·

《高等有机化学（双语版）第二版（Advanced Organic Chemistry 2nd）》第 1 章介绍了分子结构和化学键与反应活性的定性关系。第 2 章为立体化学，主要介绍了立体异构、消旋化合物的拆分、立体选择性和立体特异性反应、不对称合成等。第 3 章主要介绍有机反应的类型，通过举例来解释反应机理，并结合机理讨论反应中间体的结构、稳定性和化学性质。第 4 章讨论亲核取代反应的机理、立体化学和影响因素。第 5 章讨论加成和消除反应的机理和立体化学。第 6 章主要介绍羰基化合物的亲核加成反应。第 7 章介绍重排反应。第 8 章讨论芳香亲核和亲电取代反应，同时介绍定位效应。第 9 章介绍各类氧化和还原反应及歧化反应。第 10 章介绍周环反应。每章后附有相关习题、专业术语词汇表。为方便学生理解，每章后有中文列出的知识要点。

《高等有机化学（双语版）第二版（Advanced Organic Chemistry 2nd）》可供化学类专业高年级本科生和研究生使用，亦可供从事高等有机化学教学的教师参考。

图书在版编目 (CIP) 数据

高等有机化学：双语版：汉文、英文/谢普会，胡思前，徐翠莲主编. —2 版. —北京：化学工业出版社，2019.11
ISBN 978-7-122-35555-3

Ⅰ.①高… Ⅱ.①谢… ②胡… ③徐… Ⅲ.①有机化学-高等学校-教材-汉、英 Ⅳ.①O62

中国版本图书馆 CIP 数据核字（2019）第 249571 号

责任编辑：李 琰 宋林青　　装帧设计：王晓宇
责任校对：宋 玮

出版发行：化学工业出版社（北京市东城区青年湖南街 13 号 邮政编码 100011）
印 装：三河市双峰印刷装订有限公司
787mm×1092mm 1/16 印张 14¾ 字数 392 千字 2020 年 2 月北京第 2 版第 1 次印刷

购书咨询：010-64518888 售后服务：010-64518899
网 址：http://www.cip.com.cn
凡购买本书，如有缺损质量问题，本社销售中心负责调换。

定 价：48.00 元

前　言

Advanced Organic Chemistry 是河南省高等学校双语教学示范课程，目前市场上还没有一本适合该门课程课时及内容难易适中的双语教材。由于国外原版教材内容多、难度大、价格高，因此编写一本适合国内化学相关专业高年级本科生或研究生用的双语教材势在必行。这样可使学生融会贯通地深入学习有机化学的理论与反应机理，同时提高英语阅读水平和应用能力。

《高等有机化学（双语版）第二版（Advanced Organic Chemistry 2nd）》共分10章，前3章主要讨论有机化合物结构、立体化学与有机反应活性的关系，简单概述了反应机理的基础知识。后7章具体介绍了各类重要的有机反应机理，探讨了不同影响因素对这些反应机理、反应活性及产物的立体化学的影响。每章后附有相关习题、专业术语词汇表，专业术语词汇表中的词汇在文中以斜体表示以方便查阅。最后附有三套高等有机化学测试题及答案。

《高等有机化学（双语版）第二版（Advanced Organic Chemistry 2nd）》修改了第一版中的错误，将各级标题补充了中文，并在主要内容的章节后面用中文知识点加以概括以帮助理解。第1章、第3章由杨国玉编写，第2章由徐翠莲编写，第4章、第6章由潘振良编写，第5章由金秋编写，第7章、第9章由鲍峰玉编写，第8章由胡思前、彭望明、王亮编写，第10章由谢普会编写，附录由高光芹编写。全书由谢普会统稿、定稿。

《高等有机化学（双语版）第二版（Advanced Organic Chemistry 2nd）》得到河南省省级教学质量工程《高等有机化学》双语教学示范课程（豫教高〔2011〕173号）、河南农业大学校级教学质量工程《高等有机化学》双语教学示范课程、河南农业大学校级教学改革研究项目、江汉大学高等有机化学研究生教材立项的经费资助，并经河南省普通高等教育教材建设指导委员会审定。感谢河南省“十二五”规划教材评审专家对本教材编写提出的宝贵意见！

由于笔者水平有限，书中疏漏之处在所难免，诚恳希望读者批评指正。

编　者

2019年4月于河南农业大学

Contents

Chapter 1

Chemical Bonding and Molecular Structure （化学键与分子结构）

Information about molecular structure and ideas about bonds can be used to interpret and predict physical properties and chemical reactivities. Molecular structure specifies the relative position of all atoms (bond lengths and bond angles) in a molecule in quantitative terms. Structural information and interpretation can also be provided by computational chemistry. In this chapter, molecular orbital theory is applied to the interpretation of molecular structure and properties.

1.1 Localized chemical bonding（定域化学键）

Valence bond theory was the first structural theory applied to the empirical information about organic chemistry. In this theory, carbon almost always formed four bonds, nitrogen three, oxygen two, and the halogen one. Kekule's structure for benzene, published in 1865, was a highlight of this period. However, the structural formulas were developed without understanding of the nature of the chemical bond that is represented by the lines in the formulas. The key advance in understanding the concept of origin of chemical bonds was the introducing of the concept of electron-pair and the octet rule proposed by G. N. Lewis in 1916. The concept of bonds as electron pairs gave a fuller meaning to the traditional structural formulas, since the lines then specifically represent single, double, and triple bonds. However, Lewis structures convey relatively little information about molecular structure. The hybridization concept developed by Linus Pauling provided an approximate molecular geometry. The Pauling hybridization scheme provides an effective structural framework for most molecules. However, a particular hybridization scheme does not provide a unique description of molecular structure. Valence bond theory is neither a unique nor a complete description of *electron density*. Qualitative information about electron distribution can be deduced by applying the concepts of polarity and resonance.

Molecular orbital (MO) theory is an alternative way of describing molecular structure and electron density. In the molecular-orbital method, bonding is considered to arise from the overlap of atomic orbitals. When atomic orbitals overlap in a molecule, they combine to form new molecular orbitals with an equal number of atomic orbitals. Molecular orbitals (MOs) differ from atomic orbitals in that there are electron clouds that surround the nuclei of two or more atoms, rather than just one atom. MOs that involve only two atoms are called *localized molecular orbitals*. In localized bonding, the number of atomic orbitals that overlap is two (each containing one electron), so that two molecular orbitals are generated. One of them, called a *bonding orbital*, has lower energy than the original atomic orbitals, and the other, called an *antibonding orbital*, has higher energy than the original atomic orbitals. Bonding orbitals are filled first. Since the two original atomic orbitals each held one electron, both of these electrons can now go into the new molecular bonding orbital, since any orbital can hold two electrons. The antibonding orbital remains empty in the *ground state*. The greater the overlap is, the stronger the bond is. Fig 1.1 shows the bonding and antibonding orbitals that arise by the overlap of two 1s orbitals. Since the antibonding orbital has a node between the nuclei, there is practically no electron density in that area, so that this orbital cannot bond very well.

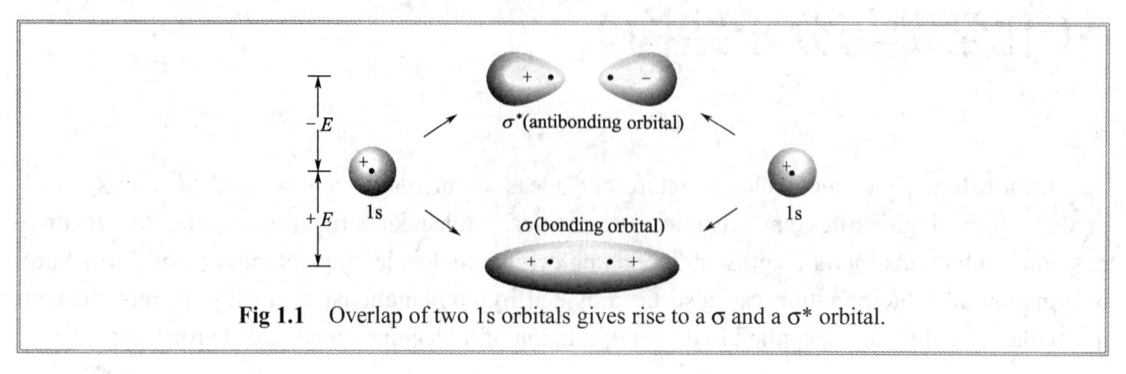

Fig 1.1 Overlap of two 1s orbitals gives rise to a σ and a σ* orbital.

Molecular orbitals formed by the overlap of two atomic orbitals are called σ (sigma) orbitals when most of its electron density is centered along the line connecting the two nuclei, and the bonds are called σ bonds. Corresponding antibonding orbitals are designated as σ*. Sigma orbitals are formed not only by the overlap of two s orbitals, but also by the overlap of any two atomic orbitals of s, p, d, or f. But the two *lobes* of the two atomic orbitals that can overlap must have the same signs: a positive s orbital can form a bond only by overlapping with another positive s orbital or with a positive lobe of a p, d, or f orbital. Any σ orbital may be represented as approximately ellipsoidal in shape.

A pi bond (π bond) results from overlap between two p orbitals oriented perpendicular to the line connecting the nuclei. These parallel orbitals overlap sideways, with most of the electron density centered above and below the line connecting the nuclei. This overlap is parallel, not linear (as in a sigma bond), so a π molecular orbital is not cylindrically symmetrical.

When ethylene forms, the combination of a σ bond and a π bond between the two carbons is involved. The two p orbitals of the two carbon atoms overlap and form a π bonding and a π antibonding (π*) molecular orbital, as shown in Fig 1.2. The π bonding molecular orbital of ethylene forms when the p atomic orbitals overlap with the same signs of the *wave function* in the bonding region between the two nuclei. This type of overlap is called a *constructive overlap*. Formation of the π antibonding molecular orbital occurs when the p atomic orbitals with opposite signs for the wave functions overlap in the bonding region. This type of overlap is called a *destructive overlap*. In the ground state of ethylene, two electrons are in the bonding MO, but the antibonding MO is vacant.

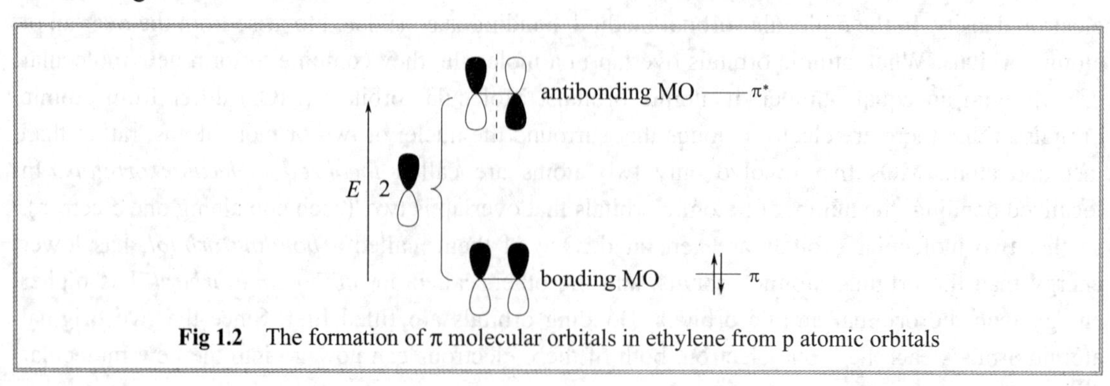

Fig 1.2 The formation of π molecular orbitals in ethylene from p atomic orbitals

The distances between atoms in a molecule (bond lengths) are characteristic properties of the molecule and can give us information if we compare the same bond in different molecules. There is a correlation of bond strengths with bond lengths. In general, shorter bonds are stronger bonds. Bonds become weaker as we move down the periodic table. Compare C—O and C—S, or the carbon–halogen bonds C—F, C—Cl, C—Br, C—I, since bond distances must increase as we go down the periodic table because the number of inner electrons increases. Double bonds are both shorter and stronger than the corresponding single bonds, but not twice as strong, because π overlap is weaker than σ overlap. This means that a σ bond is stronger than a π bond.

原子轨道在线性组合成分子轨道时（即两个波函数相加得到的分子轨道），得到的分子轨道的能量比组合前的原子轨道能量低的分子轨道叫成键轨道。组合得到的分子轨道的能量比组合前的原子轨道能量高的分子轨道叫作反键轨道。成键轨道与反键轨道总是成对出现。成键轨道是由两个原子轨道符号相同的部分相加重叠而成。成键轨道中，核间的电子的概率密度大。电子在成键轨道中可以使两个原子核结合在一起，形成稳定分子。成键轨道有 σ 成键轨道和 π 成键轨道(以符号 σ 和 π 标记)。反键轨道有 σ*反键轨道和 π*反键轨道(以符号 σ*和 π*标记)。

定域共价键只存在于两个原子之间。只包含定域键的多原子分子可以看成由相对独立的两个原子之间的化学键把原子连接起来形成的，而忽略相邻化学键的影响，把描述双原子分子中化学键的方法用到多原子分子的定域键上。如：乙烯中有一个 C—C、四个 C—H σ 键和一个 C—Cπ 键。定域键具有比较恒定的键性质。例如：一定类型定域键的键长、键偶极矩、键极化度、键力常数、键能等在不同分子中近似保持不变。

1.2 Dipole moments（偶极矩）

A bond with the electrons shared equally between two atoms is a nonpolar covalent bond. In most bonds between two different elements, the bonding electrons are attracted more strongly to one of the two nuclei. An unequally shared pair of bonding electrons is a polar covalent bond. The bond polarity is measured by dipole moment (μ), defined to be the product of charge separations (δ^+ and δ^-) and the bond length. The total dipole moment of the molecule is the vector sum of the individual bond moments. This vector sum reflects both the magnitude and the direction of each individual bond dipole moment.

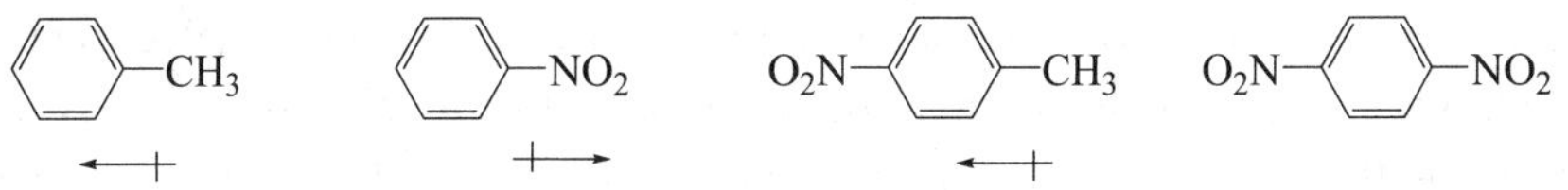

C_6H_5—CH_3 μ=0.43D C_6H_5—NO_2 μ=3.93D O_2N—C_6H_4—CH_3 μ=4.39D O_2N—C_6H_4—NO_2 μ=0D

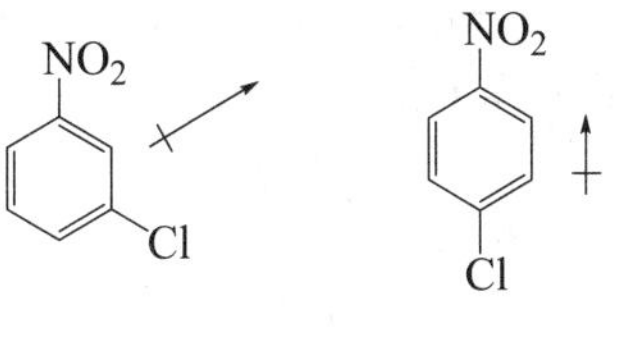

Lone pairs of electrons contribute to the dipole moments of bonds and molecules. Each lone pair corresponds to a charge separation, with the nucleus having a partial positive charge balanced

by the negative charge of lone pair.

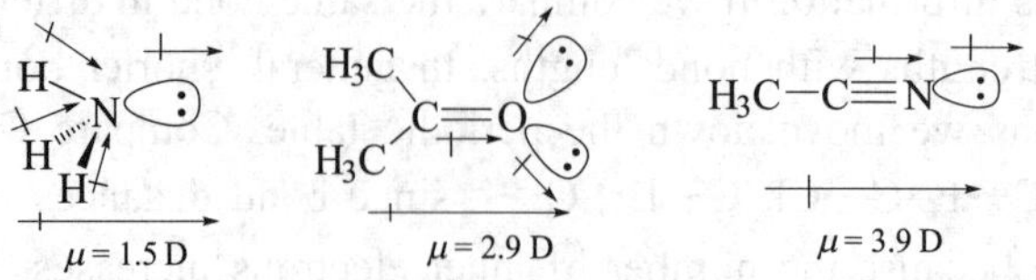

$\mu = 1.5$ D　　$\mu = 2.9$ D　　$\mu = 3.9$ D

由不同原子形成的共价键，由于成键原子电负性不同，成键电子云偏向电负性较大的原子，该原子带上部分负电荷，而电负性较小的原子带部分正电荷。这种共价键具有极性。共价键的极性主要决定于成键原子的相对电负性大小。电负性差别越大，键的极性越大。还受相邻键和不相邻原子或基团的影响。共价键的偶极矩表示键的极性大小。偶极矩 $\mu=qd$，方向：正电中心指向负电中心。分子的偶极矩为所有共价键偶极矩的矢量和。

1.3 Inductive effects and field effects（诱导效应和场效应）

The C—C bond in ethane has no polarity because it connects two equivalent carbon atoms. However, the C—C bond in chloroethane is polarized by the presence of the more *electronegative* chlorine atom. Electronegativity differences are the origin of polar bonds. The C1 atom in chloroethane, having been deprived of some of its electron density by Cl with the greater electronegativity, is partially compensated by drawing the C—C electrons closer to itself, resulting in a polarization of this bond and a slightly positive charge on the C1 atom. This polarization of a bond caused by the polarization of an adjacent bond is called the *inductive effect*. The effect is greatest for adjacent bonds but may also be felt farther away; thus the polarization of the C—C bond causes a (slight) polarization of the three methyl C—H bonds.

$$\overset{\delta^{++}}{\underset{2}{CH_3}} \rightarrow \overset{\delta^{+}}{\underset{1}{CH_2}} \rightarrow \rightarrow \overset{\delta^{-}}{Cl}$$

Functional groups can be classified as *electron-withdrawing* (–*I*) or *electron-donating* (+*I*) groups relative to hydrogen. The most common –*I* groups include $—NR_3^+$, $—SR_2^+$, $—NH_3^+$, $—NO_2$, $—SO_2R$, —CN, $—SO_2Ar$, —COOH, —F, —Cl, —Br, —I, —OAr, —COOR, —OR, —COR, —SH, —SR, —OH, —C≡CR, —Ar and $—C{=}CR_2$. The most common +*I* groups include $—O^-$, $—COO^-$, $—CR_3$, $—CHR_2$, $—CH_2R$, $—CH_3$ and D. The groups are listed approximately in order of decreasing strength for both –*I* and +*I* groups. It can be seen that most groups are electron withdrawing. The +*I* groups are groups with a formal negative charge, or atoms with low electronegativity (Si, Mg, etc., and alkyl groups). Deuterium is electrondonating with respect to hydrogen. Atoms with sp hybridization orbital generally have a greater electron-withdrawing power than those with sp^2 hybridization orbital, which in turn have more electron-withdrawing power than those with sp^3 hybridization orbital. Inductive effects always decrease with increasing distance, and in most cases (except when a very powerful +*I* or –*I* group is involved), cause very little difference in four bonds away or more.

Typical order of –*I* groups: —F > —OH > $—NH_2$ > $—CH_3$ (central atom in the same period), F > Cl > Br > I (in the same group), —C≡CR > $—C{=}CR_2$ > $—CR_2CR_3$.

Typical order of +*I* groups: $—CR_3$ > $—CHR_2$ > $—CH_2R$ > $—CH_3$

The *field effect* operates not through bonds, but directly through space or solvent molecules. The field effect depends on the geometry of the molecule. But the inductive effect depends only on the nature of the bonds. For example, in isomers **1** and **2**, the inductive effect of the chlorine atoms on the position of the electrons in the COOH group should be the same since the bonds intervene is the same. But the field effect is different because the chloro groups are closer in space to the COOH in **1** than they are in **2**. Thus a comparison of the acidity of **1** and **2** should reveal whether a field effect is truly operating. The *–I* effect of the chloro group in **3** should cause a stronger acidity than **4.** The result is reverse because of the field effect between Cl and —COOH, leading to less ability of releasing H^+ in **3** than in **4**.

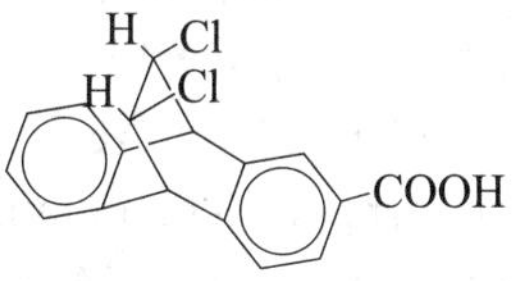

pK_a=6.07 **1**　　pK_a=5.67 **2**　　pK_a 6.25 **3**　　pK_a 6.04 **4**

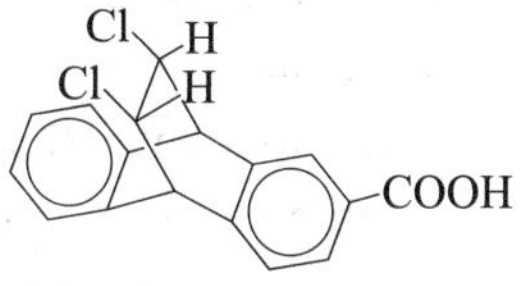

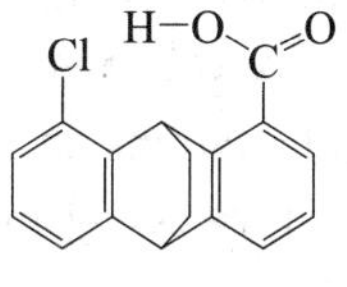

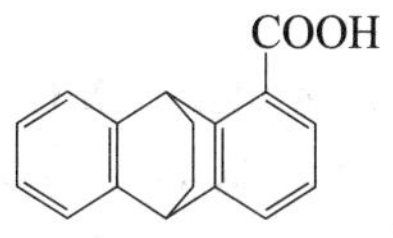

因分子中某一原子或基团的电负性不同引起分子中 σ 电子云向一个方向传递的效应叫诱导效应。通常以 H 的电负性作为衡量标准。凡是比 H 电负性大的原子或基团叫吸电子基，其诱导效应叫吸电子诱导效应（*–I*），反之叫供电子基和供电子诱导效应（+*I*）。

诱导效应的特点如下：

① 诱导效应以静电诱导方式沿键（包括 σ 和 π 键）进行传递，只涉及电子云分布状况的改变和键极性的改变。一般不引起整个电荷的转移和价态的变化；

② 诱导效应沿键迅速减小，其影响一般在三个原子内起作用；

③ 传递方向具有单一性。

诱导效应的相对强度如下：

电负性越大的原子或基团，其*–I* 越大。电负性越小，原子或基团的+*I* 效应越大。同主族元素，从上到下，电负性降低，*–I* 作用降低（+*I* 增加）。*–I* 效应：—F>—Cl>—Br>—I。

同周期元素的基团或原子，从左到右，电负性增加，*–I* 增加（+*I* 降低）。–I 效应：—F>—OH>—NH_2>—CH_3。

当烷基与不饱和碳原子或电负性原子(基团)相连时为+*I* 效应。+*I* 效应：—CR_3>—CHR_2>—CH_2R>—CH_3。

原子或基团带正电荷时具有相对较强的*–I* 效应，原子或基团带负电荷时具有相对较强的+*I* 效应。

诱导效应对物质性质的影响如下：

（1）对反应活性中间体稳定性的影响

中心碳原子上连的烷基越多的碳正离子和碳自由基的稳定性越大，而碳负离子的稳定性正相反。

（2）对羧酸酸性的影响

羧酸的酸性主要取决于 O—H 键离解的倾向以及共轭碱的稳定性，而诱导效应对两者均有影响。凡是烃基上带吸电基时将增加羧酸的酸性，带供电基时减小其酸性。

场效应：分子中的极性基团通过空间电场的传递的静电作用叫场效应。场效应依赖分子的几何构型。场效应也会对物质的酸碱性产生影响。

1.4 Delocalized chemical bonding（离域共价键）

Some compounds contain one or more bonding orbitals that are not restricted to two atoms, but spread out over three or more atoms. Such bonding is said to be delocalized. Delocalized bonding can not be adequately described by a single Lewis structure.

For planar unsaturated and aromatic molecules, many molecular-orbital calculations (MO calculations) have been made by treating σ and π electrons separately. It is assumed that the σ orbitals can be treated as localized bonds and the calculations involve only the π electrons. Hückel molecular-orbital (HMO) calculations is often used to treat delocalized bonding. According to HMO, a molecular orbital is formed by linear combinations of atomic orbitals. The number of π molecular orbitals is always the same as the number of p atomic orbitals used to form π MOs. These MOs have energies that are symmetrically distributed above and below the energy of the starting p orbitals. Half of the MOs are bonding MOs. Half of the MOs are antibonding MOs.

The construction of the π molecular orbitals for a conjugated diene requires that four p atomic orbitals from the four carbon atoms overlap. The overlap of these orbitals leads to four π MOs, each MO encompasses all four atoms that contributed the p orbitals. In the molecular orbital picture (Fig. 1.3), the overlap of four p orbitals gives two bonding orbitals that contain the four electrons and two vacant antibonding orbitals. It can be seen that each orbital has one more *node* than the one of next lower energy. The nodes in a molecular orbital are always symmetrically distributed. For example, the π_3 molecular orbital has two nodes equidistant from the center and the two ends of the orbital. Each of the two lower energy π MOs is filled with two electrons, whereas the two higher energy π MOs remain empty.

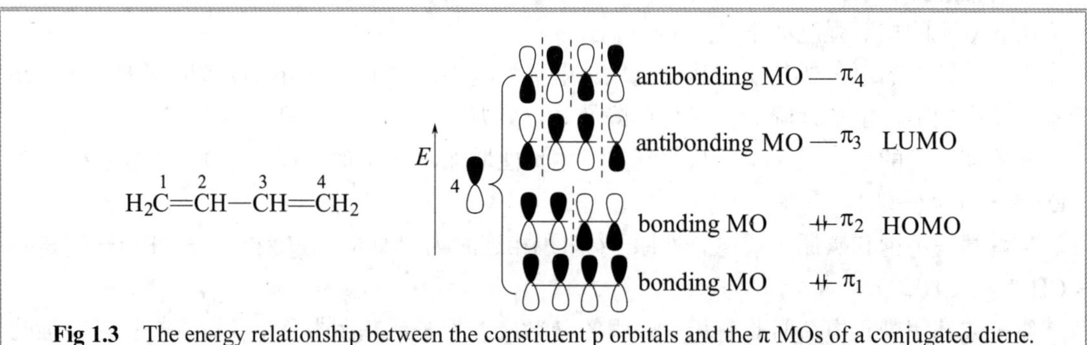

Fig 1.3 The energy relationship between the constituent p orbitals and the π MOs of a conjugated diene.

As shown in Fig 1.3, The π_1 molecular orbital consists only of bonding overlaps between the constituent atomic orbitals of 1,3-butadiene. As a result, the π_1 molecular orbital is very stable. The $\pi_{1\text{-}4}$ MOs of a conjugated diene are called *delocalized molecular orbitals* because they extend over more than two atoms.

The π_1 molecular orbital of a conjugated diene is lower in energy than the π_1 molecular orbital of an unconjugated diene or alkene. The π_1 molecular orbital of a conjugated system also exhibits the other two characteristics of a conjugated system: planar conformation and a shorter single bond between C2—C3 than typical single bond. This MO has the lowest energy among the four MOs formed in a conjugated system, so it fills with electrons before any other levels.

The π_2 MO involves bonding overlap between C1—C2 and C3—C4 along with an antibonding overlap (node) between C2 and C3. The π_2 MO is higher in energy than the π_1 MO because of the node between C2 and C3. After the π_1 MO fills with electrons, the π_2 MO then fills. The π_2 MO is *the highest occupied molecular orbital* (abbreviated HOMO) of 1,3-butadiene.

The π_3 MO involves a bonding overlap between C2—C3 and an antibonding overlap between C1—C2 and C3—C4. The π_3 molecular orbital is *the lowest unoccupied molecular orbital* (LUMO) of 1,3-butadiene. The π_4 MO involves all antibonding overlaps between adjacent carbons. The π_4 MO is the highest unoccupied molecular orbital.

Where a p orbital is on an atom adjacent to a double bond, there are three parallel p orbitals that overlap. As previously noted, it is a general rule that the overlap of *n* atomic orbitals creates *n* molecular orbitals, so overlap of a p orbital with an adjacent double bond gives rise to three new orbitals, as shown in Fig 1.4. The middle orbital is a *nonbonding molecular orbital* of zero bonding energy. The central carbon atom does not participate in the nonbonding orbital. Electrons in this orbital have the same energy as an isolated p orbital.

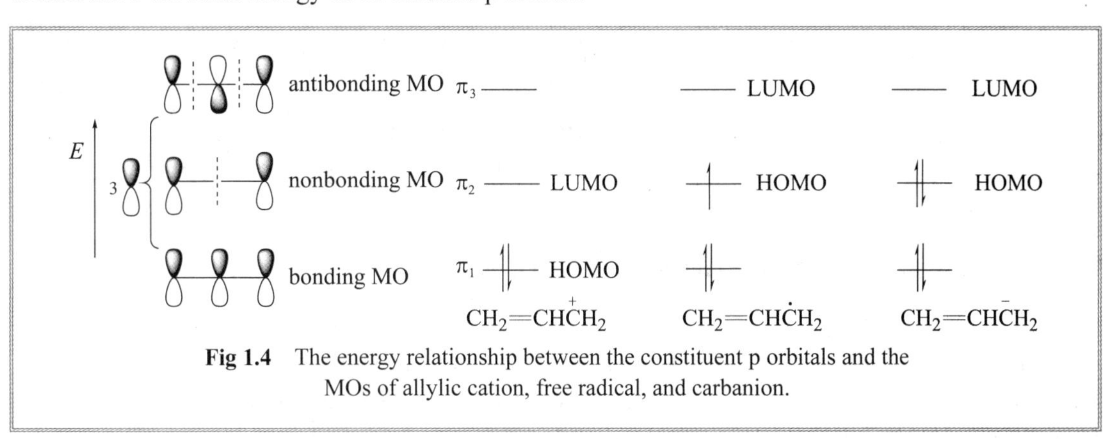

Fig 1.4 The energy relationship between the constituent p orbitals and the MOs of allylic cation, free radical, and carbanion.

There are three cases: the original p orbital may have contained no, one, or two electrons. Since the original double bond contributes two electrons, the total number of electrons accommodated by the new molecular orbitals is two, three, or four respectively. For example, the orbital structures of the allylic cation, free radical, and carbanion differ from each other, therefore, only in that the nonbonding orbital is empty, half-filled, or filled. The electrons in the nonbonding orbital do not contribute to the bonding energy, positively or negatively. The two cases, where the original p orbital contains only one or no electron, are generally found only in free radicals and cations, respectively. Allylic free radicals have one electron in the nonbonding orbital. In allylic cations, this orbital is vacant and only the bonding orbital is occupied. An allylic carbocation is more stable than an ordinary carbocation. The case where the original p orbital contains two electrons are found in allylic carbanions as well as any system containing an atom that has an unshared pair and that is directly attached to a multiple-bond atom, e.g. CH_2═CH—Cl, CH_2═CH—Cl, CH_2═CH—OCH_3, etc.

分子轨道理论：描述共价键形成的一种理论。包括以下要点：

① 分子中任何电子都可看成在所有核和其余电子所构成的势场中运动，描述分子中电子运动状态的波函数称为分子轨道。

② 分子轨道由能量相同/相近的原子轨道的线性组合而构成。对于离域 π 键，π 分子轨道是由原子的 P_z 原子轨道线性组合而成。

$$\Psi = C_1\phi_1 + C_2\phi_2 + \cdots + C_n\phi_n \text{。}$$

式中，C_1，C_2，C_n 为原子轨道系数；ϕ_1, $\phi_2 \cdots \phi_n$ 为原子轨道，Ψ 为分子轨道。

③ 每个分子轨道的能量：$E = \int \Psi H \Psi \mathrm{d}\tau$ 分子的总能量等于被电子占据的分子轨道能量

的总和。

④ 分子中电子的填充：Pauli 不相容原理和 Hund 规则。

每个分子轨道最多能容纳两个电子，且电子自旋反平行。对于能量相等的分子轨道，电子将尽可能分占不同的轨道，且自旋平行。

分子轨道的基本特点如下：

① 分子轨道数目等于参与线性组合的原子轨道数目。

② 离域 π 键分子轨道分为成键轨道、非键轨道和反键轨道。

能量低于原子轨道的分子轨道叫成键轨道（有利于共价键的形成），能量等于原子轨道的分子轨道叫非键轨道（对共价键的形成没有任何作用），能量高于原子轨道的分子轨道叫反键轨道（不利于共价键的形成）。

前线轨道理论：日本诺贝尔化学奖获得者福井谦一提出了前线轨道理论。其要点如下：有电子占据的能量最高的分子轨道叫最高已占分子轨道（HOMO）。没有电子占据的能量最低的分子轨道叫最低未占分子轨道（LUMO）。HOMO 与 LUMO 统称为前线轨道（FMO）。前线轨道中的电子是化学反应中最活泼的电子，是有机化学反应的核心。

1.5 Resonance structures（共振结构）

A resonance structure may be drawn for any conjugated structure by moving electrons from one place to another, so long as the rules of Lewis structures are followed. The electron delocalization described by resonance enhances the stability of the molecules, and compounds or ions composed of such molecules often show exceptional stability. Resonance structures are not different compounds, but different ways of drawing the same compound. The actual molecule is said to be a resonance hybrid of its resonance forms.

In drawing resonance structures, we try to draw structures that are as low in energy as possible. The best candidates are those that have the maximum number of octets and the maximum number of bonds. Only electrons can be delocalized. Nuclei cannot be delocalized. They must remain in the same places, with the same bond lengths and angles in all the resonance contributors. The following general rules will help us to draw realistic resonance structures:

(1) All the resonance structures must be valid Lewis structures for the compound.

(2) Only the placement of electrons may be shifted from one structure to another (electrons in double bonds and lone pairs are the ones that are most commonly shifted.). Nuclei cannot be moved, and the bond angles must remain the same.

(3) All the *canonical forms* must have the same number of paired and unpaired electrons. Most stable compounds have no unpaired electrons, and all the electrons must remain paired in all the resonance structures.

(4) The major resonance contributor is the one with the lowest energy. Good contributors generally have all octets satisfied, as many bonds as possible, and as little charge separation as possible. Positive charge is best accommodated on atoms of low electronegativity, and negative charge on high electronegative atoms.

(5) Resonance stabilization is most important when it serves to delocalize a charge over two or more atoms.

$$CH_2{=}CH{-}\bar{C}H_2 \leftrightarrow \bar{C}H_2{-}CH{=}CH_2$$

$$CH_2{=}CH{-}\dot{C}H_2 \leftrightarrow \dot{C}H_2{-}CH{=}CH_2$$

$$CH_2{=}CH{-}\overset{+}{C}H_2 \leftrightarrow \overset{+}{C}H_2{-}CH{=}CH_2$$

$$HO^{+}{=}CH{-}CH{=}CH{-}\ddot{O}H \leftrightarrow H\ddot{O}{-}CH{=}CH{-}CH{=}OH^{+}$$

$$H_2\overset{+}{C}{-}CH{=}CH{-}CH{=}CH{-}CH_3 \leftrightarrow H_2C{=}CH{-}CH{=}CH{-}\overset{+}{C}H{-}CH_3 \leftrightarrow H_2C{=}CH{-}\overset{+}{C}H{-}CH{=}CH{-}CH_3$$

$$H_3C{-}\ddot{S}{-}CH_2^{+} \leftrightarrow H_3C{-}\overset{+}{S}{=}CH_2$$

$$H_3C{-}C(=\!O){-}\ddot{C}H_2^{-} \leftrightarrow H_3C{-}C(-O^{-}){=}CH_2$$

当一个分子、离子或自由基的结构不能用路易斯结构式正确地描述时，可以用多个路易斯式表示，这些路易斯式称为共振结构（又称极限式或正则结构）。在共振结构之间用双箭头“↔”联系，以表示它们的共振关系。要正确写出共振结构式，应符合下列几条规则：

① 共振结构式之间只允许键和电子的移动，而不允许原子核位置的改变。

② 所有的共振结构式必须符合 Lewis 结构式。

③ 所有的共振结构式必须具有相同数目的未成对电子。

④ 共振结构式中所有的原子都具有完整的价电子层，是较为稳定的。

⑤ 有电荷分离的共振结构式稳定性较低。

⑥ 负电荷在电负性较大的原子上的共振结构式较稳定。

1.6 Conjugative effect（共轭效应）

Multiple bonds that alternate with single bonds are said to be conjugated. In the conjugated systems, many p orbitals overlap to form a more stable π bond where the π electrons are delocalized than similar systems with isolated double bonds.

The interaction effect of atoms on the electronic density in conjugated systems is called conjugative effect (C). Conjugative effect from the overlap and interaction of p orbitals whose axis are parallel to each other is a characteristic of conjugative systems. Compared with inductive effect, conjugative effect only exists in the conjugative systems. The intensity of conjugative effect is not weakened along with the increase of the distance. Conjugative effect can be divided into electron withdrawing conjugative effect (–*C*) and electron donating conjugative effect (+*C*).

$$\overset{\delta-}{C}H_2{=}\overset{\delta+}{C}H{-}\ddot{C}l \quad (+C) \qquad \overset{\delta+}{C}H_2{=}CH{-}\overset{\delta+}{C}H{=}\overset{\delta-}{O} \quad (-C)$$

Common +*C* and –*C* groups:

+*C* groups: O^-, S^{2-}, —NR_2, —NHR, —NH_2, —NHCOR, —OR, —OCOR, —SR, —SH, —F, —Cl, —Br, —I, —R, —Ar.

–*C* groups: —NO_2, —CN, —COOH, —COOR, —$CONH_2$, —CONHR, —$CONR_2$, —CHO, —COR, —SO_3R, —SO_2R, —NO, —Ar.

+*C* groups decrease in order: —F > —Cl > —Br > —I, —OR > —SR > —SeR, —O^- > —OR> —O^+R_2.

In π-π conjugative systems, –*C* groups decrease in order: —C═O > —C═N > —C═C, —C═N^+HR_2 > —C═NR.

In p-π conjugative systems, +*C* decreases in the order: —NR_2> —OR > —F.

The conjugative effect can affect the chemical properties of some compounds. The conjugative effect can affect on the acidity of compounds. For example, carboxylic acids are more acidic than alcohols owing to the conjugative effect of hydroxy group in carboxylic acids. Let's look at the stabilities of *carboxylate anions*, the negative charge in the carboxylate is shared by both oxygen atoms. In other words, a carboxylate anion is a resonance *hybrid* of two equivalent structures. Since a carboxylate anion is more stable than an *alkoxide anion*, a carboxylate anion is lower in energy and is present in greater amount at equilibrium than an alkoxide anion.

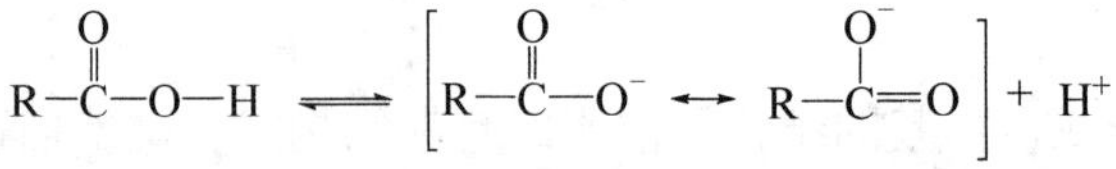

$$R-\overset{\overset{\displaystyle O}{\|}}{C}-O-H \rightleftharpoons \left[R-\overset{\overset{\displaystyle O}{\|}}{C}-O^- \longleftrightarrow R-\overset{\overset{\displaystyle O^-}{|}}{C}=O\right] + H^+$$

The conjugative effect can also affect the basicity of compounds. Arylamines, such as aniline, are weaker bases than alkylamines by a factor of about 10^6. The nitrogen lone-pair electrons in an *arylamine* are shared by orbital overlap with the π orbitals of the aromatic ring, and they are therefore less available for bonding to an acid. In contrast to amines, amides ($RCONH_2$) are nearly neutral. The main reason for the decreased basicity of amides relative to amines is that the lone-pair electrons on nitrogen atom of amides are shared by orbital overlap with the adjacent carbonyl group π orbital. The electrons are therefore much less available for bonding to an acid.

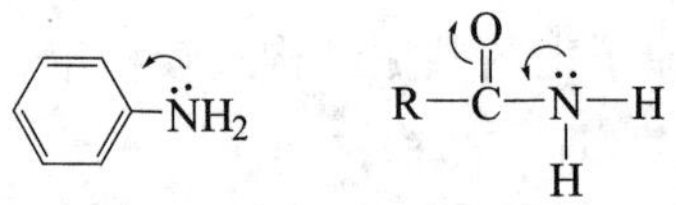

$$C_6H_5-\ddot{N}H_2 \qquad R-\overset{\overset{\displaystyle O}{\|}}{C}-\underset{\underset{\displaystyle H}{|}}{\ddot{N}}-H$$

The conjugative effect can affect the type of some addition reactions. The reaction of ethylene with HCl is an *electrophilic* addition because the electrons in the π bond are accessible to electrophiles. In *acrylaldehyde*, however, the π electrons in the C═C bond are accessible to *nucleophiles* because the electronegative oxygen atom of α, β-unsaturated carbonyl compound withdraws electrons from the β carbon, thereby making it more electron-poor and more electrophilic than a typical C═C bond in an alkene.

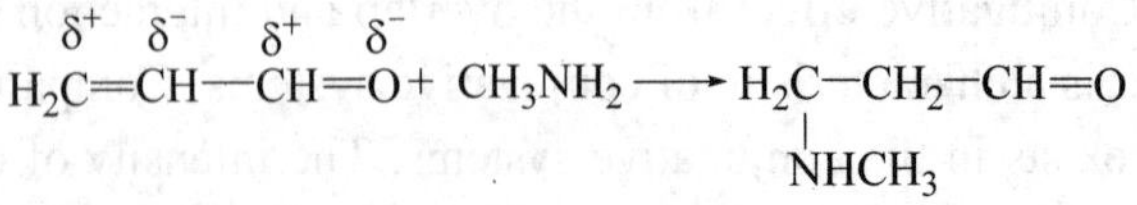

$$H_2\overset{\delta^+}{C}=\overset{\delta^-}{C}H-\overset{\delta^+}{C}H=\overset{\delta^-}{O} + CH_3NH_2 \longrightarrow H_2\underset{\underset{\displaystyle NHCH_3}{|}}{C}-CH_2-CH=O$$

包含三个或三个以上原子的 π 键叫共轭 π 键，这样的体系叫共轭体系。共轭体系中的电子云是离域的而不是定域的，使得所涉及的化学键平均化，这种体现在共轭体系中原子之间的相互影响的电子效应叫共轭效应。

基本特点如下：共轭效应只存在于共轭体系中，只沿共轭体系传递；无论共轭体系有多大，共轭效应都能传递到共轭体系末端，而且不因传递距离增加而减弱，在共轭体系中的原子依次出现电荷呈正负(疏密)交替分布。

基本体现如下：共轭分子存在共轭能，分子能量降低，氢化热降低，稳定性增加。共轭分子的共价键的键长平均化，源于电子云的平均化。

共轭效应的分类如下：

① 静态共轭效应（C_s，分子固有的）与动态共轭效应（C_d，反应瞬间）；

② π-π 共轭与 p-π 共轭；

③ 吸电子共轭效应（$-C$）与供电子共轭效应（$+C$）。

共轭效应的相对强度如下：

① 同一主族元素，随着原子序数增大，原子半径增大，与碳的 $2p_z$ 轨道重叠程度逐渐减小，$+C$ 降低。

$+C$：$-F > -Cl > -Br > -I$, $-OR > -SR > -SeR$, $-O^- > -OR > -O^+R_2$

② 同周期元素

π-π 共轭体系中，随原子序数增大，电负性增大，$-C$ 增大。$-C$：—C═O>—C═N>—C═C, —C═N^+HR_2>—C═NR.

p-π 共轭体系中，原子序数增大，电负性增大，给电子能力下降，$+C$ 下降。$+C$：$-NR_2$>−OR>−F。

共轭效应对化学性质的影响如下：

① 影响羧酸的酸性强弱；

② 加成反应方向。

1.7 Hyperconjugative effect（超共轭效应）

Resonance and conjugation involves π electrons, while hyperconjugation involves σ electrons. Hyperconjugation is the interaction of the electrons in a σ bond (usually C—H or C—C) with an adjacent empty (or partially filled) non-bonding p orbital, antibonding π orbital, or filled π orbital, to give an extended molecular orbital that increases the stability of the system.

Hyperconjugation affects several properties.

(1) Bond length: Hyperconjugation is a key factor in shortening of σ bonds. For example, the single C—C bond methylacetylene is approximately 1.46 Å in length, much less than the value of around 1.54 Å found in saturated hydrocarbons. This can be explained as hyperconjugation between the alkyl and alkynyl parts.

(2) *Dipole moments*: The large increase in dipole moment of 1,1,1-trichloroethane as compared with chloroform can be attributed to hyperconjugated structures.

(3) The heat of formation of molecules with hyperconjugation is greater than sum of their bond energies. And the heat of hydrogenation per double bond is less than the heat of hydrogenation of ethylene.

(4) Stability of carbocations and radicals:

$(CH_3)_3\overset{+}{C} > (CH_3)_2\overset{+}{C}H > CH_3\overset{+}{C}H_2 > \overset{+}{C}H_3$, $(CH_3)_3\overset{\bullet}{C} > (CH_3)_2\overset{\bullet}{C}H > CH_3\overset{\bullet}{C}H_2 > \overset{\bullet}{C}H_3$

As the C—C σ bond adjacent to the cation is free to rotate, the three C—H σ bonds of the methyl group in turn undergoes the stabilization interaction. The more adjacent C—H bonds in hyperconjugated structures are, the larger hyperconjugation stabilization is.

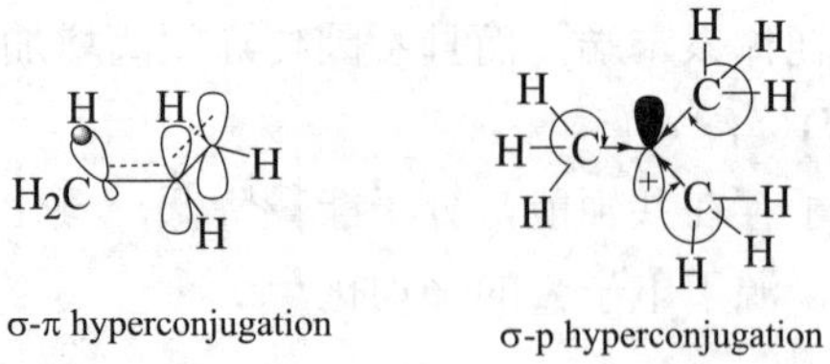

σ-π hyperconjugation σ-p hyperconjugation

Hyperconjugative effect results in the electrons in the C—H bond are closer to the carbon than they would be if the molecule did not contribute at all. In alkenes, the electron-releasing effect of the alkyl substituents can be represented by hyperconjugated resonance structures. The implication of these resonance structures is that some electron density is transferred from the C—H σ bond to the empty π^* orbital.

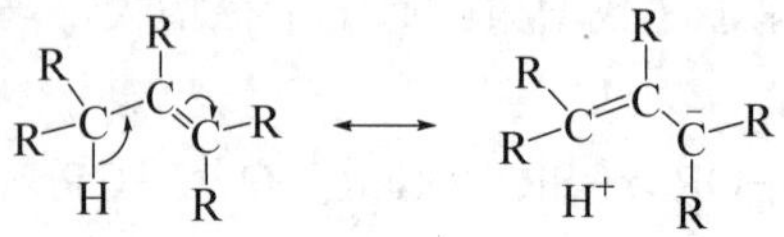

There is also hyperconjugation across the double bond. This interaction may be even stronger because the double bond is shorter than a corresponding single bond, permitting better orbital overlap. Because these resonance structures show equivalent compensating charge transfer, there is no net charge separation, but structural features such as bond lengths and spectroscopic properties are affected.

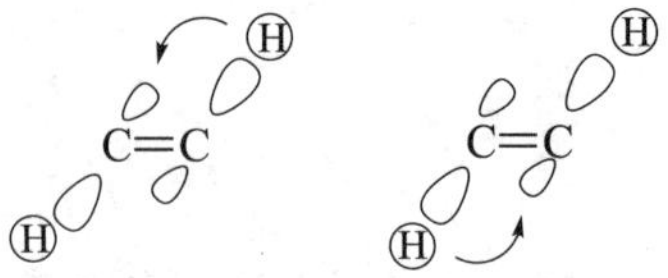

Heteroatoms with unshared electron pairs can also interact with adjacent σ* bonds. For example, oxygen and nitrogen substituents substantially weaken an adjacent (geminal) C—H bond.

凡是 α —C—H σ 键与 π 键或 p 轨道的共轭叫超共轭，所产生的电子效应叫超共轭效应。但超共轭效应的强度不如共轭效应强，是一种弱电子效应。结果使 C—C 变短，α —C—H 键的数目越多，超共轭效应越强，相应的烯烃就越稳定。烷基自由基和正离子的稳定性顺序为：

$(CH_3)_3C^+ > (CH_3)_2CH^+ > CH_3CH_2^+ > CH_3^+$

$(CH_3)_3C\cdot > (CH_3)_2CH\cdot > CH_3CH_2\cdot > CH_3\cdot$

取代基的电子效应影响有机分子的反应活性。

1.8 Steric effect（立体效应）

Steric effects arise from the fact that each atom within a molecule occupies a certain amount of space. If atoms are brought too close together, there is an associated cost in energy due to overlapping electron clouds (Pauli or Born repulsion), and this may affect the molecule's preferred conformation and reactivity. Steric effects are often contrasted and complemented by electronic effects implying the influence of effects such as induction, conjunction, orbital symmetry etc.

It occasionally happens that a reaction proceeds much faster or much slower than expected on the basis of electrical effects alone. In these cases it can often be shown that steric effects are influencing the rate. For example, relative rates for the S_N2 *ethanolysis* of certain alkyl halides are affected by steric effects (Table 1.1).

Table 1.1 Relative rates of reaction of RBr with ethanol

R	Relative rate	R	Relative rate
CH_3	17.6	$(CH_3)_2CH\ CH_2$	0.030
CH_3CH_2	1	$(CH_3)_3CCH_2$	4.2×10^{-6}
$CH_3CH_2CH_2$	0.28		

These compounds are primary bromides; the branching is on the second carbon, so that field-effect differences should be small. As Table 1.1 shows, the rate decreases with increasing β branching and reaches a very low value for neopentyl bromide. This reaction is known to involve an attack by the nucleophile from a position opposite to that of the bromine. The great decrease in rates can be attributed to steric hindrance, a sheer physical blockage to the attack of the nucleophile.

Not all steric effects decrease reaction rates. In the hydrolysis of RCl by an S_N1 mechanism, the first step, which is a *rate-determining step*, involves ionization of the alkyl chloride to a carbocation:

$$R_3C{-}Cl \longrightarrow R_3C^+ + Cl^-$$

If the halide is tertiary and the three alkyl groups are large enough, they will be pushed together by the enforced *tetrahedral* angle, resulting in strain. This type of strain is called B strain, and it can be relieved by ionization to the carbocation. The rate of ionization of a molecule in which there is B strain is therefore expected to be larger than in cases where B strain is not present. Table 1.2 shows that this is so. Substitution of ethyl groups for the methyl groups of *tert*-butyl chloride does not cause B strain; the increase in rate is relatively small, and the rate smoothly rises with the increasing number of ethyl groups. The rise is caused by normal field and resonance (hyperconjugation) effects. Substitution by one isopropyl group is not greatly different. But with the second isopropyl group the crowding is now great enough to cause B strain, and the rate is increased 10-fold. Substitution of a third isopropyl group increases the rate still more.

Table 1.2 Rates of hydrolysis of tertiary alkyl chlorides at 25℃ in 80% aqueous ethanol

Halide	Rate	Halide	Rate
Me_3CCl	0.033	Et_3CCl	0.099
Me_2EtCCl	0.055	$Me_2(i\text{-}Pr)CCl$	0.029
$MeEt_2CCl$	0.086	$Me(i\text{-}Pr)_2CCl$	0.45

Another type of strain, that can affect rates of cyclic compounds, is called Ⅰ strain (internal strain). Ⅰ strain results from changes in ring strain in going from a tetrahedral to a *trigonal* carbon or vice versa. For example, S_N1 *solvolysis* of an alkyl halide involves a change in the bond angle of the central carbon from ~109.5 ° to ~120 °. This change is highly favored in 1-chloro-1-methylcyclopentane because it relieves eclipsing strain. Thus this compound undergoes solvolysis in 80% ethanol at 25℃, 43.7 times faster than the reference compound *tert*-butyl chloride. In the corresponding cyclohexyl compound, this factor is absent because the substrate does not have eclipsing strain.

	t-BuCl	1-chloro-1-methylcyclopentane (Me, Cl)	1-chloro-1-methylcyclohexane (Me, Cl)
Relative solvolysis rates	1.0	43.7	0.35

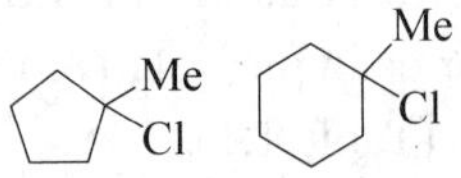

Understanding steric effects is critical to chemistry, biochemistry, and pharmacology. In organic chemistry, steric effects are nearly universal and affect the rates and activation energies of most chemical reactions. Steric effects often dictate reaction pathways in organic synthesis due to the fact that there are fewer configurations in which molecules can collide and successfully react. In biochemistry, steric effects are often exploited in naturally occurring molecules such as enzymes, where the catalytic site may be buried within a large protein structure. In pharmacology, steric effects determine how and at what rate a drug will interact with its target bio-molecules.

立体效应指因分子中靠近反应中心的原子或基团占有一定的空间位置而影响分子的反应活性的现象。降低分子反应活性的立体效应称立体阻碍。例如，邻位双取代的苯甲酸的酯化反应要比没有取代的苯甲酸困难得多。同样，邻位双取代的苯甲酸酯也较难水解。这是由于邻位上的基团占据了较大的空间位置，阻碍了试剂（水、醇等）对羧基碳原子的进攻。相反，反应物转变为活性中间体的过程中，如降低反应物的空间拥挤程度，则能提高反应速度。这种立体效应称立体助效。例如，叔丁基正离子比甲基正离子容易形成，这是因为在形成叔丁基正离子的反应中，空间拥挤程度降低得多一些，而在形成甲基正离子的反应中，空间拥挤程度相对降低得少一些。空间效应是影响有机反应历程的重要因素。空间位阻效应又称立体效应，主要是指分子中某些原子或基团彼此接近而引起的空间阻碍和偏离正常键角，从而引起的分子内的张力立体效应，影响反应速度。

1.9 Aromaticity and Hückel's rule（芳香性与休克尔规则）

1.9.1 *Annulenes*（轮烯）

Aromatic compounds undergo distinctive reactions which set them apart from other functional groups. They are highly unsaturated compounds, but unlike alkenes and alkynes. They are relatively unreactive and will tend to undergo reactions which involve retention of their unsaturation. Benzene is a six-membered ring structure with three double bonds. The six π electrons are delocalized around the ring which results in an increased stability, so benzene undergoes reactions where the aromatic ring system is retained. All six carbon atoms in benzene are sp^2 hybridized, and the molecule itself is cyclic and planar. *Aromaticity* is a chemical property describing the way in which a conjugated ring of unsaturated bonds, lone pairs, or empty orbitals exhibits a kind of stabilization stronger than expected by the stabilization of conjugation alone. Aromaticity can be defined as the ability to sustain an induced ring current. A compound with this ability is called *diatropic*. Aromatic compounds are characterized by a special stability and they undergo substitution reactions more easily than addition reactions.

A molecular orbital description of benzene provides a more satisfying and more general treatment of aromaticity. Benzene has a planar hexagonal structure and all the carbon-carbon bonds are equal in length. As shown in Fig 1.5, the cyclic array of six p-orbitals of carbon atoms overlap to generate six molecular orbitals, three bonding orbitals and three antibonding orbitals. The first MO (π_1) is lowest in energy with no nodes, with six p orbitals having the greatest overlap to form a continuously bonding ring of electron density. The degenerate bonding orbitals (π_2 and π_3) are higher in energy than π_1. Both π_2 and π_3 have one *nodal plane*. Six electrons then occupy these three molecular orbitals in pairs, resulting in a fully occupied set of bonding molecular orbitals. It is this completely filled set of bonding orbitals, or closed shell, that gives the benzene ring its thermodynamic and chemical stability, just as a filled valence shell octet confers stability on the inert gases. The degenerate pair π_4* and π_5* orbitals with two nodal planes in each are antibonding, yet not as high in energy as the all-antibonding π_6* orbital with three nodal planes.

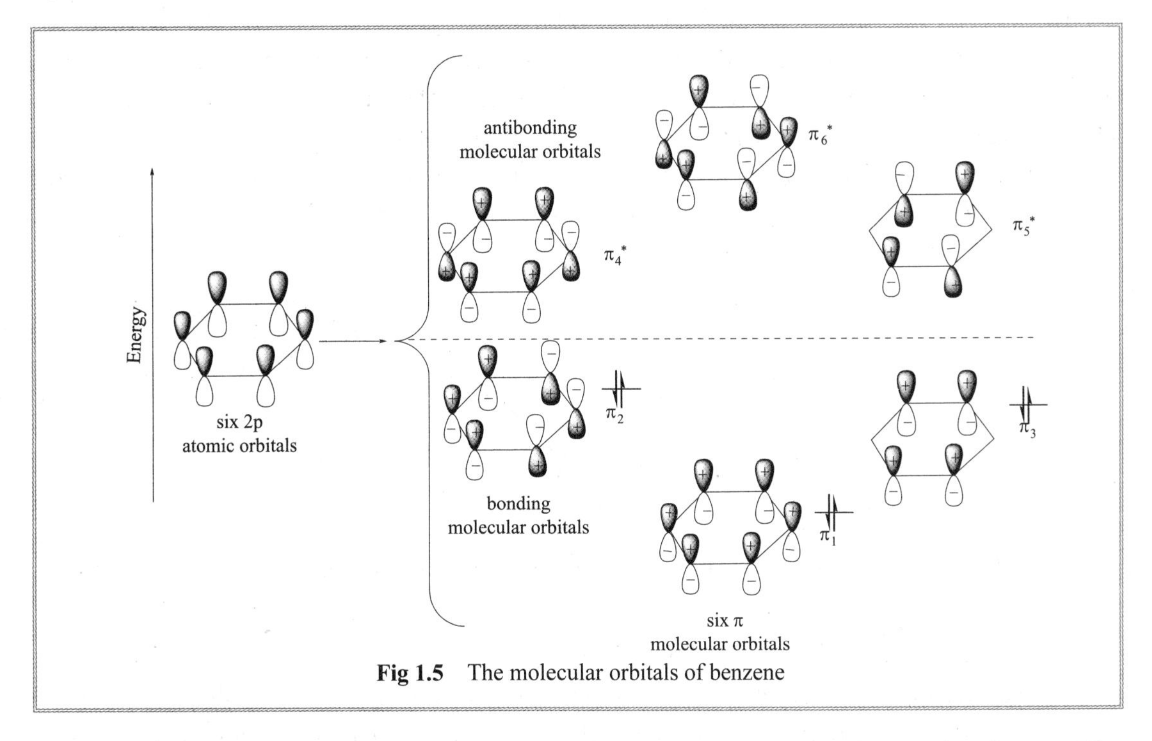

Fig 1.5 The molecular orbitals of benzene

Cyclobutadiene is of much low stability and immediately polymerize into its dimmer. The bond lengths of C—C and C═C of this compound are 150.6 and 137.6 pm, respectively. Its instability is explained by the molecular orbital description. As shown in Fig 1.6, four sp^2 hybridized carbons form the cyclobutadiene ring. And their four p orbitals overlap to form four molecular orbitals. The lowest energy MO is π_1, the all-bonding MO with no nodes. Two electrons fill π_1. The π_2 and π_3 orbitals are degenerate, each having one symmetrically situated nodal plane. Each MO has two bonding interactions and two antibonding interactions. The net bonding order is zero, so these two MOs are nonbonding. The final MO, π_4^* has two nodal planes and is entirely antibonding. The remaining two electrons occupy the degenerate π_2 and π_3 orbitals, with unpaired *spinning* electrons (Hund's rule). The compound exists as a *diradical* (two unpaired electrons in nonbonding orbitals) in its ground state. So cyclobutadiene is very reactive.

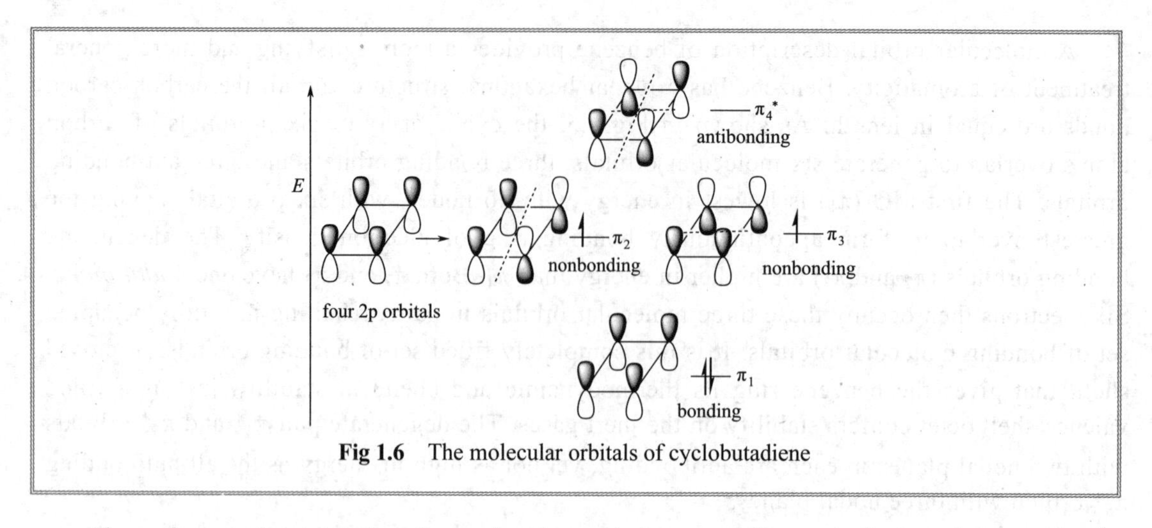

Fig 1.6 The molecular orbitals of cyclobutadiene

The *polygon* rule states that the molecular orbital energy diagram of a regular, completely conjugated cyclic system has the same polygonal shape as the compound, with one vertex (the all-bonding MO) at the bottom. The nonbonding line cuts horizontally through the center of the polygon. Fig 1.7 shows how the polygon rule predicts the MO energy diagrams for benzene, cyclobutadiene, cyclooctatetraene. The π electrons are filled into the orbitals in accordance with *the aufbau principle* (lowest energy orbitals are filled first) and Hund's rule.

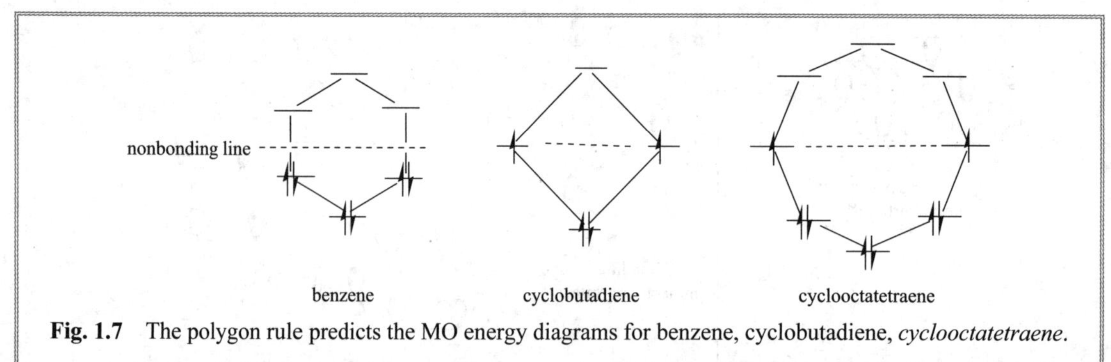

Fig. 1.7 The polygon rule predicts the MO energy diagrams for benzene, cyclobutadiene, *cyclooctatetraene.*

Aromatic compounds are those that meet the following criteria:

(1) The structure must be cyclic, containing some number of conjugated π bonds.

(2) Each atom in the ring must have an unhybridized p orbital. The ring atoms are usually sp^2 hybridized or occasionally sp hybridized.

(3) The unhybridized p orbital must overlap to form a continuous ring of parallel orbitals. In most cases, the structure must be planar (or nearly planar) for effective overlap to occur.

(4) Delocalization of the π electrons over the ring must lower the electronic energy.

Aromatic structures are more stable than their open-chain counterparts. For example, benzene is more stable than 1,3,5-hexatriene.

An *antiaromatic* compound is one that meets the first three criteria, but delocalization of the π electrons over the ring increases the electronic energy. Cyclobutadiene meets the first three criteria, but delocalization of the π electrons increases the electronic energy. Cyclobutadiene is less stable than its open-chain counterpart (1,3-butadiene), and it is antiaromatic.

A cyclic compound that does not have a continuous, overlapping ring of p orbitals can't be aromatic or antiaromatic, it is *nonaromatic* or aliphatic. Its electronic energy is similar to its open-chain counterpart. For example, 1,3-cyclohexadiene is about as stable as *cis,cis*-2,4-hexadiene.

Erich Hückel developed a shortcut for predicting which of the annulenes and related compounds are aromatic and which are antiaromatic. Hückel's rule: if the number of electrons in the cyclic system is ($4n$+2), the system is aromatic; if the number of electrons in the cyclic system is ($4n$), the system is antiaromatic; where n is zero or any positive integer. The compound must have a continuous ring of overlapping p orbitals, usually in planar conformation.

Benzene is [6] annulene, with a ring of overlapping p orbitals. It's a $4n$ +2 system, with n=1. Hückel's rule predicts benzene to be aromatic.

Like benzene, cyclobutadiene ([4] annulene) has ring of overlapping p orbitals, but it has four π electrons. Hückel's rule predicts cyclobutadiene to be antiaromatic.

Cyclooctatetraene is [8] annulene, with eight π electrons. It is a $4n$ system, with n=2. If Hückel's rule were applied to cyclooctatetraene, it would predict antiaromaticity. However, cyclooctatetraene is a stable hydrocarbon. It doesn't show the high reactivity associated with antiaromaticity. Although all the carbon atoms in the ring are sp^2 hybridized, it is not planar, the π electrons are not delocalized and the molecule consists of alternating single and double bonds. Cyclooctatetraene is more flexible than cyclobutadiene, and it assumes a nonplannar "tub" conformation that avoids most of the overlap between adjacent π electrons. Hückel's rule simply does not apply. Its reactions are typical of alkenes. So it is nonaromatic.

Like cyclooctatetraene, larger [$4n$] annulenes such as [12] annulene, [16] annulene and [20] annulene, do not show antiaromaticity because they have the flexibility to adopt nonplannar conformations. They all react as partially conjugated polyenes.

In the case of all-*cis*-[10]-annulene, there is a contradiction arised from the angle requirements by the geometry requirements and sp^2 hybridization. Therefore, all-*cis*-[10]-annulene is not very stable and highly reactive due to the angular strain. As for *cis,trans,cis,cis,trans*-[10] annulene and [14] annulene, They are both destabilized by the *stereorepulsion* among the interior directed hydrogens and cannot adopt a planar comformation. Thus, [14] annulene is not aromatic except under a temperature lower than −60℃. Even though they have ($4n$+2) π electrons, with n = 2, 3 respectively. The *transannular* hydrogen crowding that destabilizes *cis,trans,cis,cis,trans*-[10] annulene may be eliminated by replacing the interior hydrogens with a bond or a short bridge. As expected, the resulting 10π-electron annulene derivatives exhibit aromatic stability and reactivity as well as characteristic ring current anisotropy in the NMR. Naphthalene and *azulene* are [10]annulene analogs stabilized by a transannular bond. Although the CH_2 bridged structure to the right of naphthalene in the diagram is not exactly planar, the conjugated 10π-electron ring is sufficiently close to planarity to achieve aromatic stabilization. The bridged [14] annulene compound on the far right, also has aromatic properties.

As the ring is large enough to be planar allowing no stereorepulsion among the hydrogen atoms directed toward the interior of ring, [18] annulene, however, is a planar molecule with aromatic properties and a delocalized π system.

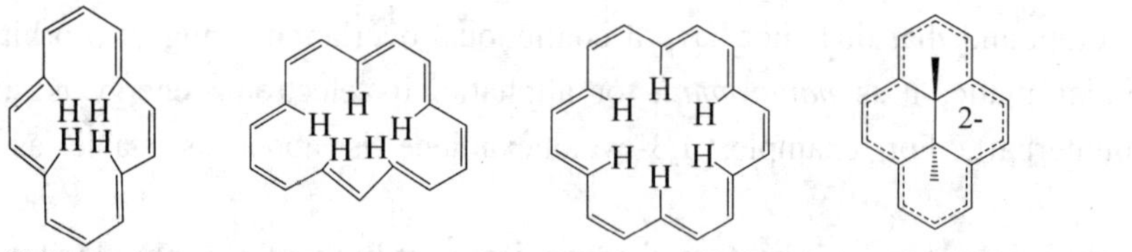

[14] Annulene nonaromatic

[16] Annulene antiaromatic

[18] Annulene aromatic

[14] Annulene dianion antiaromatic

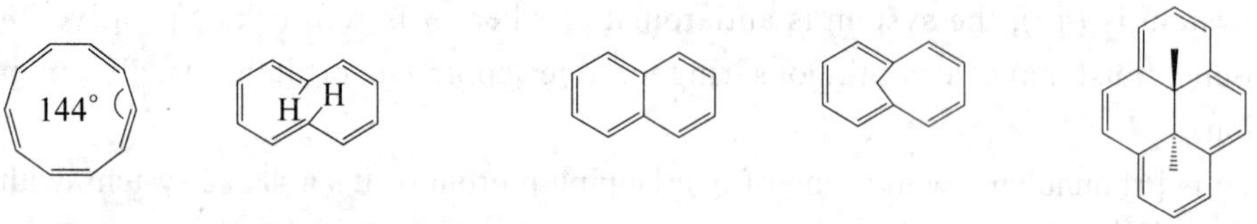

all *cis* nonaromatic

cis, trans, cis, cis, trans nonaromatic

[10] Annulenes

Bridged [10] Annulenes

Bridged [14] Annulene

stable aromatic hydrocarbons

1.9.2 Aromatic ions（芳香性离子）

It is also possible to get aromatic ions. The *cyclopentadienyl* anion and the *cycloheptatrienyl* cation are both aromatic. Both are cyclic and planar, containing six π electrons and all the atoms in the ring are sp^2 hybridized. Cyclopentadiene contains a sp^3 hybrid (—CH_2—) carbon without an unhybridized p orbital, so there can be no continuous ring of p orbital. *Deprotonation* of the —CH_2— group leaves an orbital occupied by a pair of electrons. This orbital can rehybridize to a p orbital, completing a ring of p orbitals containing six π electrons. The cyclopentadienyl anion can be formed by abstracting a proton from cyclopentadiene, which is unusually acidic for an alkene. Cyclopentadiene is nearly as acidic as H_2O and more acidic than many alcohols. It is entirely ionized by potassium *t*-butoxide:

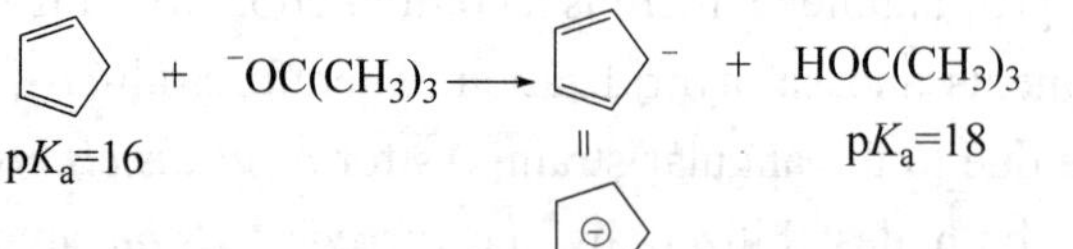

$$\text{C}_5\text{H}_6\ (\text{p}K_a=16) + {}^{-}\text{OC(CH}_3)_3 \longrightarrow \text{C}_5\text{H}_5^{-} + \text{HOC(CH}_3)_3\ (\text{p}K_a=18)$$

cyclopentadienyl anion, aromatic

Hückel's rule predicts that cyclopentadiene cation, with four π electrons is antiaromatic. Cyclopentadiene cation is not easily formed. Protonated *2,4-cyclopentadienol* does not lose water to give cyclopentadiene cation even in concentrated sulfuric acid. The antiaromatic cation is too unstable.

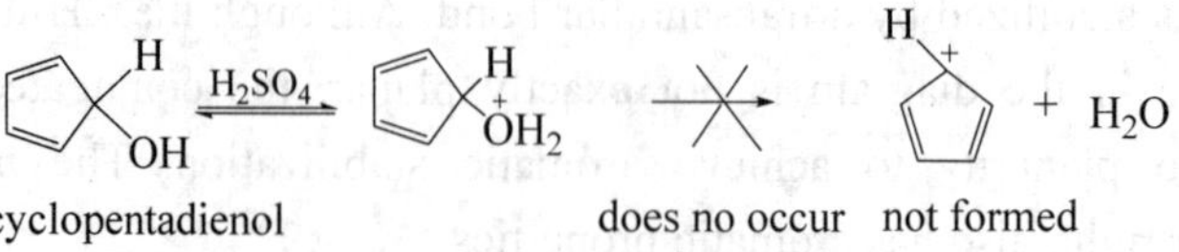

$$\text{2,4-cyclopentadienol} \xrightarrow{H_2SO_4} \text{C}_5\text{H}_5\text{-}\overset{+}{\text{O}}\text{H}_2 \not\longrightarrow \text{C}_5\text{H}_5^{+} + H_2O$$

2,4-cyclopentadienol   does no occur   not formed

The cycloheptatrienyl cation (*tropylium* ion) is easily formed by treating the corresponding alcohol with dilute sulfuric acid. However, the cycloheptatrienyl anion is difficult to form because it is antiaromatic.

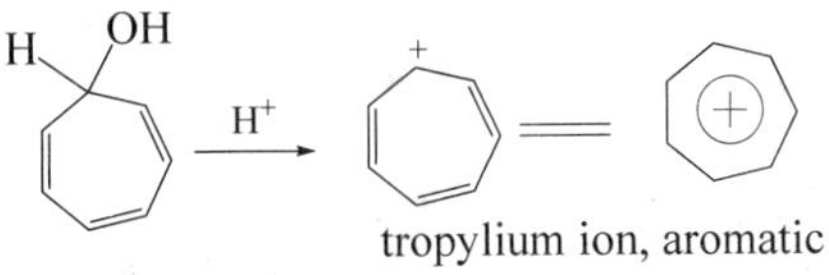

tropylium ion, aromatic

Dianions of hydrocarbons are rare and are difficult to form. However, cyclooctatetraence reacts with potassium metal to form an aromatic cyclooctatetraence dianion with ten π electrons, which has a planar, regular octagonal structure, with the C—C bond length close to benzene.

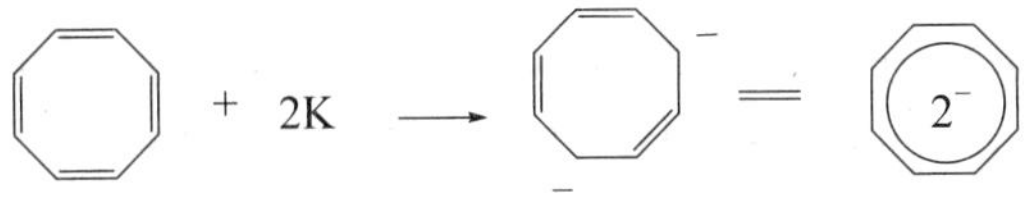

cyclooctatetraence dianion, aromatic

An additional "class" of aromatic molecules includes homoaromatic compounds. *Homoaromaticity* in organic chemistry refers to a special case of aromaticity in which conjugation is interrupted by a single sp^3 hybridized carbon atom. Although this sp^3 center disrupts the continuous overlap of p-orbitals, traditionally thought to be a requirement for aromaticity, considerable thermodynamic stability and many of the spectroscopic, magnetic, and chemical properties associated with aromatic compounds are still observed for such compounds. This formal discontinuity is apparently bridged by p-orbital overlap, maintaining a continuous cycle of π electrons that is responsible for this preserved chemical stability. To date, homoaromatic compounds are known to exist as cationic and anionic species, and some studies support the existence of neutral homoaromatic molecules, though these are less common. The "homotropylium" cation ($C_8H_9^+$) is the best studied example of a homoaromatic compound.

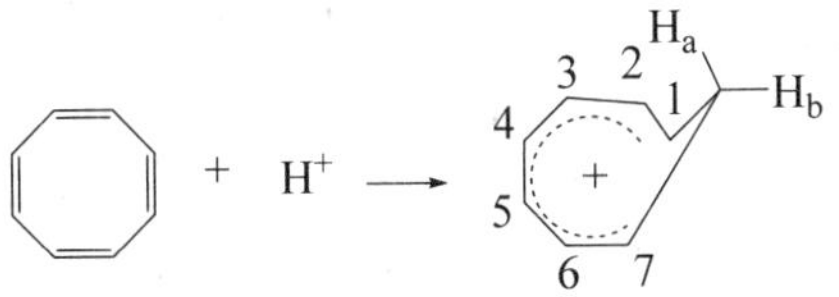

cyclooctatetraene　　homotropylium ion, aromatic

When cyclooctatetraene is dissolved in concentrated H_2SO_4, a proton is added to one of the double bonds to form the homotropylium ion. In this species, an aromatic *sextet* is spread over seven carbons, as in the tropylium ion. The eighth carbon is an sp^3 carbon and so cannot take part in the aromaticity. The NMR spectra show the presence of a diatropic ring current: H_b is found at $\delta = -0.3$; H_a at 5.1; H_1 and H_7 at 6.4; H_2~H_6 at 8.5. This ion is an example of a homoaromatic compound, containing one or more sp^3-hybridized carbon atoms in a conjugated cycle.

Using a resonance theory, we might incorrectly expect both the cyclopentadienyl anions and the cyclopentadienyl cations to be stable. The resonance structures shown below both spread the negative charge of anion and the positive of the cation over all five carbon atoms of the ring. With conjugated cyclic systems such as these, the resonance approach is a poor predictor of stability. The Hückel's rule based on molecular orbital theory is a much better predictor of stability for these aromatic and antiaromatic systems.

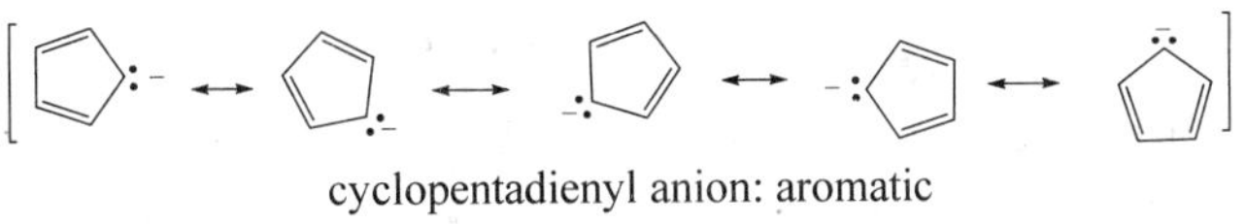

cyclopentadienyl anion: aromatic

cyclopentadienyl cation: antiaromatic
The resonance picture gives a misleading suggestion of stability

Simple resonance theory predicts that *pentalene*, azulene, and *heptalene* should be aromatic, Molecular-orbital calculations show that azulene should be stable but not the other two. Heptalene reacts readily with oxygen, acids, and bromine, is easily hydrogenated, and polymerizes on standing. Analysis of its NMR spectrum shows that it is not planar. Pentalene has not been prepared. Azulene, a blue solid, is quite stable and many of its derivatives are known. Azulene readily undergoes aromatic substitution.

heptalene, 12 π electrons, antiaromatic if planar　　pentalene, 8 π electrons, antiaromatic if planar

azulene, 10 π electrons, aromatic　　indole, aromatic

1.9.3 Heterocyclic aromatic compounds（杂环芳香族化合物）

Heterocyclic compounds of $(4n+2)$ π electrons, with rings containing sp^2 hybridized atoms of other elements can be aromatic. If a five membered ring has two double bonds, and the fifth atom possesses an unshared pair of electrons, the ring has five p orbitals that can overlap to create five new orbitals: three bonding orbitals and two antibonding orbitals. There are six π electrons for these orbitals: the four p orbitals of the double bonds each contribute one and the filled orbital contributes the other two. The six electrons occupy the bonding orbitals and constitute an aromatic sextet. As shown in Fig 1.8, the heterocyclic compounds pyrrole, thiophene, and furan are the most important examples of this kind of aromaticity.

Pyridine is an aromatic nitrogen analogue of benzene. It has a six membered ring with six π electrons. The nonbonding pair of electrons on nitrogen is in a sp^2 hybridized orbital in the plane of the ring. They are perpendicular to the π system and do not overlap with it. Pyridine shows all the characteristics of aromatic compounds. Pyridine protonates to give pyridinium ion, it is still aromatic because the additional proton has no effect on the electrons of the aromatic sextet. Pyridine derivatives such as *N*-oxides are still aromatic. Pyran is not aromatic, but the pyrylium ion is.

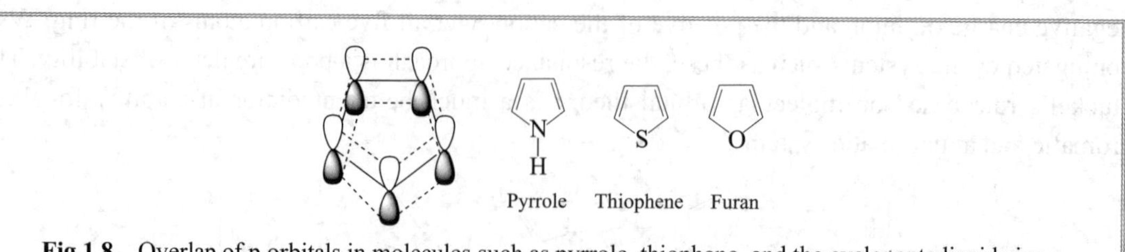

Fig 1.8 Overlap of p orbitals in molecules such as pyrrole, thiophene, and the cyclopentadienide ion

pyridine, aromatic   pyridinium ion, aromatic

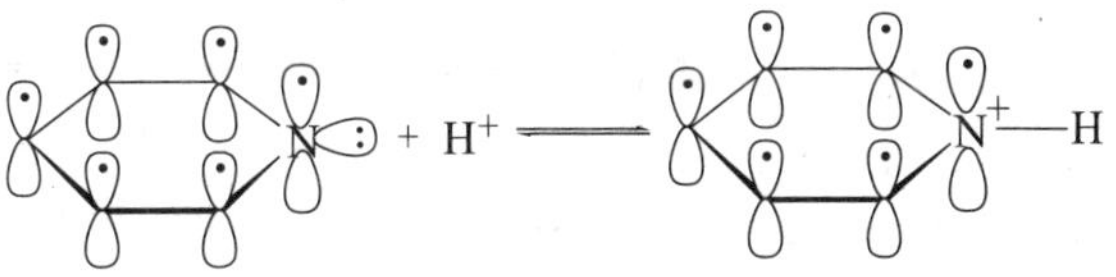

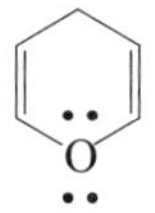

Pyran, nonaromatic 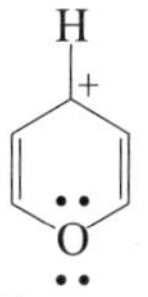   Pyrylium, aromatic

*Tropon*es and *tropolones* would have an aromatic sextet if the two C═O electrons stayed away from the ring and resided near the electronegative oxygen atom. In fact, tropones are stable compounds, and tropolones are found in nature. In sharp contrast to tropones, *cyclopentadienone* has been isolated only in an argon matrix < 38K. Above this temperature it dimerizes. As in tropolones, the electronegative oxygen atom draws electron to itself, but in this case it leaves only four electrons and the molecule is unstable.

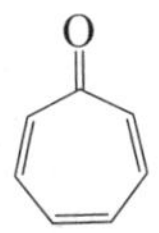

Tropone, aromatic 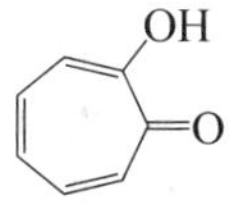   tropolone, aromatic 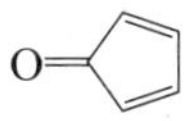   cyclopentadienone, antiaromatic

The dianion of *squaric acid* is an aromatic system. The stability of this system is illustrated by the fact that the pK_1 of squaric acid is ~1.5 and the pK_2 is ~3.5, which means that even the second proton is given up much more readily than the proton of acetic acid.

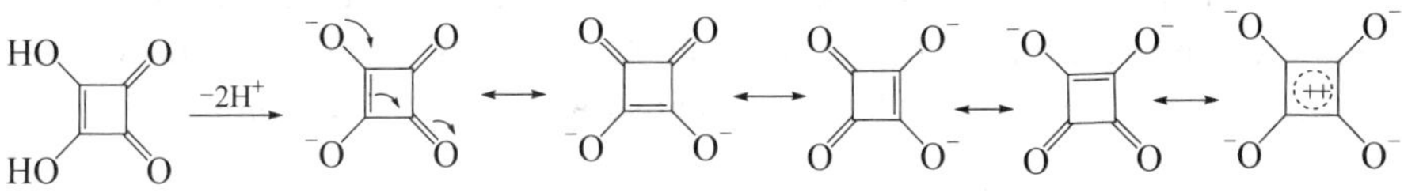

Applications of Hückel's rule to a variety of cyclic π systems are summarized. The 2, 6, 10 π-electron systems are aromatic, while the 4π and 8π-electron systems are antiaromatic if they are planar. *Cyclopropenyl cation* (2π-electron system), benzene, cyclopentadienyl anion, cycloheptatrienyl cation, pyrrole, thiophene, furan, and pyridine (6π-electron systems), azulene, naphthalene, cyclononatetraenyl anion, cyclooctatetraenyl dianion and indole (10π-electron systems) are aromatic compounds or ions. Cyclobutadiene, cyclopropenyl anion, cyclopentadienyl cation (4π-electron systems) are antiaromatic compounds or ions. Pentalene, cycloheptatrienyl anion, *cyclononatetraenyl cation* (8π-electron systems), and heptalene (12π-electron system) are planar and antiaromatic compounds or ions. Cyclooctatetraene (8π-electron system) and [12]annulene (12 π-electron system) are not planar, they are *nonaromatic*.

1.9.4 Fused aromatic hydrocarbons（稠合芳烃）

Fused aromatic compounds such as naphthalene, anthracene, phenanthrene, *pyrene*, *benzo* [*α*] pyrene and *coronene* are polynuclear aromatic hydrocarbons. Unlike benzene, all the C—C bond lengths in these fused ring aromatics are not the same, and there is some localization of the π-electrons. Therefore they are not as strongly stabilized as benzene.

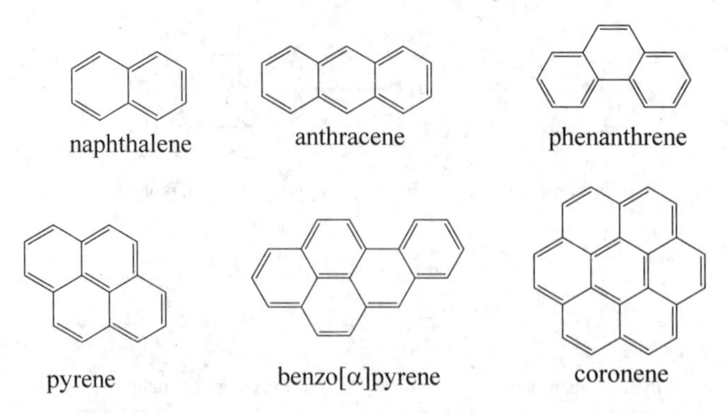

As these extended aromatic compounds become larger, the ratio of hydrogen to carbon decreases. For example, the symmetrical hexacyclic compound coronene has an H/C ratio =1/2, compared with 1 for benzene. If we were to imagine fused ring systems of this kind to be further extended in space, the H/C ratio would approach zero, and the resulting compound would be a form of carbon. Amorphous carbon, Diamond and graphite are known to exist as carbon allotropes. Diamond is an extended array of sp^3 hybridized carbon atoms; whereas, graphite consists of overlapping sheets of sp^2 hybridized carbon atoms arranged in a hexagonal pattern. *Buckminsterfullerene* (C_{60}) represents a forth class of carbon allotropes, a soccer ball-like representation of the 12 five and 20 six-membered rings composes its surface. Although C_{60} is composed of fused benzene rings its chemical reactivity resembles that of the cycloalkenes more than benzene. Most of the reactions thus far reported for C_{60} involve addition to, rather than substitution of, the core structure. Strain introduced by the curvature of the surface may be responsible for the enhanced reactivity of C_{60}. Larger fullerenes, such as C_{70}, C_{76}, C_{82} and C_{84} have ellipsoidal or distorted spherical structures, and fullerene-like assemblies up to C_{240} have been detected. A fascinating aspect of these structures is that the space within the carbon cage may hold atoms, ions or small molecules. *Nanotubes* may be viewed as rolled up segments of graphite. The chief structural components are six-membered rings, but changes in tube diameter, branching into side tubes and the capping of tube ends is accomplished by fusion with five and seven-membered rings. Many interesting applications of these unusual structures have been proposed.

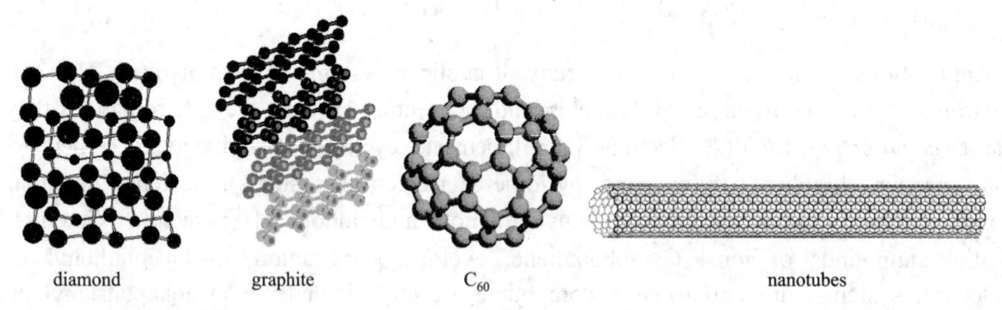

（1）轮烯的芳香性

环状闭合共轭体系，π 电子高度离域，具有较高的离域能，体系能量低，分子较稳定。在化学性质上表现为易发生亲电取代反应，不易发生加成反应和氧化反应的性质称为芳香性。

休克尔规则：一个单环化合物只要具有平面离域体系，它的 π 电子数为 $4n+2$（n=0,1,2,3…整数），就有芳香性（当 $n>7$ 时，有例外）。其中 n 相当于简并的成键轨道和非键轨道的组数。

苯有六个 π 电子，符合 $4n+2$ 规则，六个碳原子在同一平面内，故苯有芳香性。

而环丁二烯、环辛四烯的 π 电子数不符合 $4n+2$ 规则，故无芳香性。

只有当单环多烯的 π 电子全部进入能量较低的成键分子轨道和非键分子轨道且全满时，

才能使体系稳定，具有芳香性。凡符合休克尔规则的单环多烯具有芳香性。双键完全共轭的单环多烯烃类叫轮烯（C_nH_n）。当环碳原子共平面，环内氢原子没有或很少有空间排斥作用，π 电子数目符合 4*n*+2 规则时，该轮烯具有芳香性，属非苯芳烃。[18]轮烯有芳香性;[10]轮烯的 π 电子数为 10，但因环内氢排斥力大为非平面的，无芳香性; [12]轮烯不服从休克尔规则无芳香性，但其双负离子具有芳香性。

（2）芳香性离子

环丙烯正离子具有芳香性。环戊二烯基负离子具有芳香性，五个氢等同。杯烯为偶极分子，具有芳香稳定性。环庚三烯正离子具有芳香性。稳定的甘菊蓝烃（azulene）为一偶极分子，具有芳香性。非平面的环辛四烯没有芳香性，但与金属钠反应得到环辛四烯双负离子（平面，10π 电子），具有芳香性。

同芳香性：共轭双键的环被一个或两个亚甲基所分隔开，这个亚甲基在环平面之外，使环上的 π 电子构成芳香体系。例如环辛三烯正离子具有芳香性。

环状多烯和相应的开链烃相比稳定性差不多的性质叫非芳香性。分子非平面，无共轭能。例如环辛四烯为非芳香性的。

平面环状共轭多烯的稳定性小于相应的开链共轭烯烃的性质叫反芳香性。π 电子数为 4*n* 的环状共轭多烯为反芳香性。

方酸电离出 2 个质子，因有 4 个等价的共振结构和一个非苯芳烃结构而稳定。

（3）芳香性杂环化合物

具有（4*n*+2）π 电子，成环原子为 sp^2 杂化的杂环化合物具有芳香性。例如吡咯、呋喃、噻吩、吡啶、吲哚、喹啉等。

（4）稠合芳香烃

萘、蒽、菲、芘、苯并芘、六苯并苯都具有芳香性，但其 π 电子并非完全离域，因此它们都没有苯稳定。

Problems

1. The following molecules and ions are grouped by similar structures. Classify each as aromatic, antiaromatic, or nonaromatic. For the aromatic and antiaromatic species, give the number of π electrons in the ring.

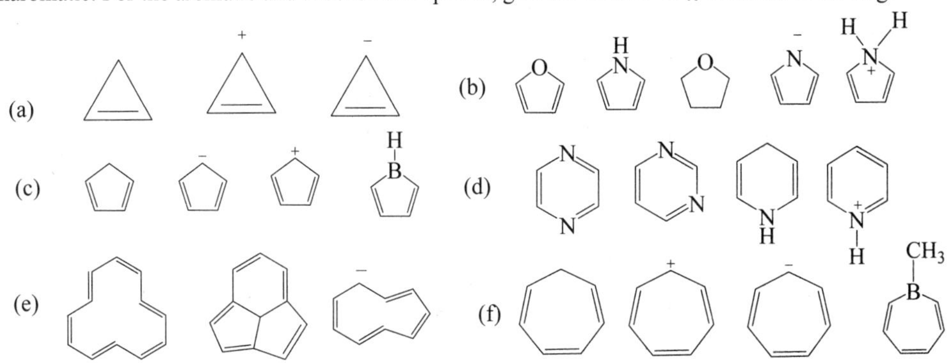

2. The polarization of a carbonyl group can be represented by a pair of resonance structures:

$$\left[\ \rangle C=\ddot{O} \longleftrightarrow \rangle C^{+}-\ddot{O}^{-}\ \right]$$

Cyclopropenone and cycloheptatrienone are more stable than anticipated. Cyclopentadienone is relatively unstable and rapidly undergoes a Diels-Alder dimerization. Explain.

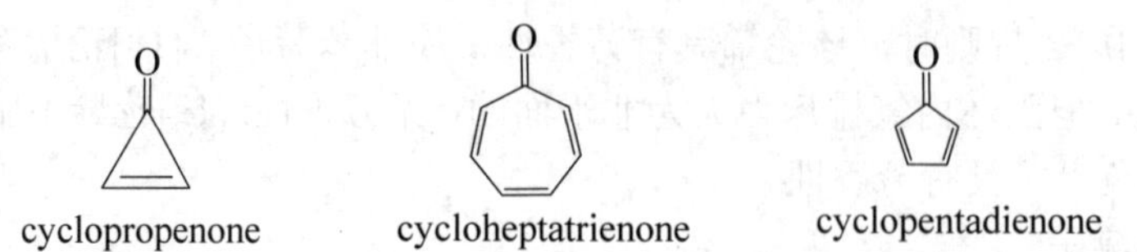

3. The following hydrocarbon has an unusually large dipole moment. Explain how a large dipole moment might arise.

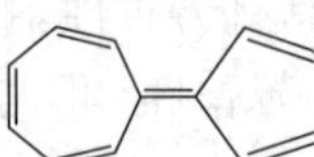

4. (1) From what you remember of electronegativities, show the direction of the dipole moments of the following bonds. (2) In each case, predict whether the dipole moment is relatively large or small.

(a) C—Cl (b) C—H (c) C—Li (d) C—N (e) C—O

(f) C—B (g) C—Mg (h) N—H (i) O—H (j) C—Br

5. Show by a crossed arrow for the direction of dipole moment of the following molecules:

(a) CH_3OCH_3 (b) CH_3Br (c) $C_2H_5NH_2$ (d)SO_2 (e) CH_3NO_2

6. For each pair of ions, determine which ion is more stable. Use resonance forms to explain your answers.

(a) $H_3C-\overset{H}{\underset{+}{C}}-CH_3$ or $H_3C-\overset{H}{\underset{+}{C}}-OCH_3$

(b) $H_2C{=}\underset{H}{C}-\overset{H}{\underset{+}{C}}-CH_3$ or $H_2C{=}\underset{H}{C}-\overset{H_2}{C}-\overset{+}{C}H_2$

(c) $H_2\bar{C}-CH_3$ or $H_2\bar{C}-CN$

(d) $\overset{+}{C}H_2$ or $\overset{+}{C}H_2$

7.

(a) *p*-Nitroaniline has a dipole moment (6.10 D) greater than the sum of the dipole moments of nitrobenzene (3.95 D) and aniline (1.53 D), Explain.

(b) Explain the difference in the dipole moment of the following compounds: CH_3CHO (μ=2.65D) and $CH_3CH{=}CH_2$ (μ=0.35D)

(c) Which of the following two compounds will have a larger dipole moment and why? CH_3NO_2 or $C_6H_5NO_2$

8. Which ones are aromatic?

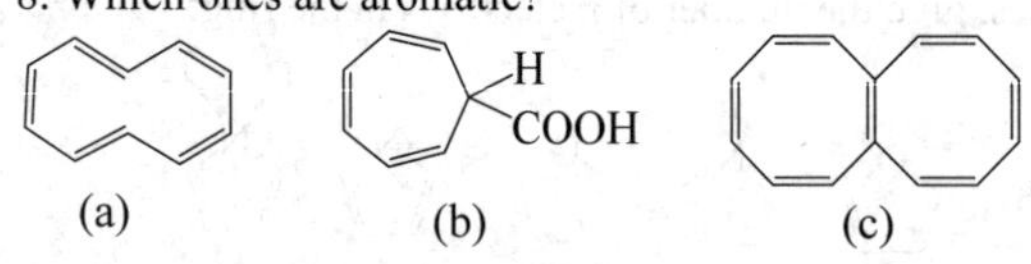

9. Put in order of each groups.

(1) $-I$

$-SO_3^-$, $-SR$, $-SO_2R$

(2) $+C$

$-\underset{R}{N}-COR$ $\quad -\underset{R}{N}-\overset{NR}{\overset{\|}{C}}R$ $\quad -\underset{R}{N}-CH_2R$ $\quad -\underset{R}{N}-\overset{\overset{+}{N}R_2}{\overset{\|}{C}}-R$

(3) $-C$

$-CONHR$ $\quad -\overset{NR}{\overset{\|}{C}}NR_2$ $\quad -\overset{\overset{+}{N}R_2}{\overset{\|}{C}}NR_2$

Vocabulary

electron density 电子密度
localized molecular orbitals 定域分子轨道
bonding orbital 成键轨道
antibonding orbital 反键轨道
ground state 基态
lobe [ləʊb] 圆形突出部；耳垂；裂片
wave function 波函数
constructive overlap 构建性重叠
destructive overlap 非构建性重叠
electronegative [ɪlektrəʊ'negətɪv] 负电的，带负电的
inductive effect 诱导效应
electron-withdrawing 吸电子的
electron-donating 供电子的
field effect 场效应
node [nəʊd] 节点
delocalized molecular orbitals 离域分子轨道
the highest occupied molecular orbital 最高占有分子轨道
the lowest unoccupied molecular orbital 最低未占分子轨道
nonbonding molecular orbital 非键轨道
resonance structure 共振结构
conjugated structure 共轭结构
canonical form 规范格式
conjugative effect 共轭效应
carboxylate anion 羧酸负离子
hybrid ['haɪbrɪd] 杂化
alkoxide anion 烷氧负离子
arylamine [æ'raɪlæmaɪn] 芳基胺
electrophilic [ɪlektrəʊ'fɪlɪk] 亲电的
acrylaldehyde [ækrɪ'lældəhaɪd] 丙烯醛
nucleophile['njuːklɪəfaɪl] 亲核试剂
hyperconjugative effect 超共轭效应
dipole moment 偶极矩
steric effect 位阻效应
ethanolysis [eθə'nɒləsɪs] 乙醇解
rate-determining step 决速步骤
tetrahedral 四面体的
trigonal ['trɪgənəl] 三角形的
solvolysis [sɒl'vɒlɪsɪs] 溶剂解
aromaticity [ærəʊmæ'tɪsɪtɪ] 芳香性
annulene [ə'nʌliːn] 轮烯
diatropic [daɪə'trɒpɪk]横向性的、抗磁性的
nodal plane 波节面
spin [spɪn] 自旋
diradical [daɪ'rædɪkəl] 双自由基
polygon ['pɔliː gɔn] 多边形
the aufbau principle 构造原理
cyclooctatetraene [saɪkluːktætet'riːn] 环辛四烯
antiaromatic 反芳香性的
nonaromatic 非芳香性的
stereorepulsion 立体排斥
transannular [ˌtræn'sænjʊlə] 跨环的，
azulene [æʒ'juːliːn] 甘菊环
cyclopentadienyl [saɪkləʊpen'teɪdɪənɪl] anion 环戊二烯负离子
cycloheptatrienyl [saɪkləʊhep'teɪtrɪənɪl] catin 环庚三烯正离子
deprotonation [deprəʊtɒ'neɪʃn] 去质子化
2,4-cyclopentadienol 2,4-环二烯醇
tropylium [trə'piliəm 'aɪən] ion 草鎓离子
dianion [daɪ'ænaɪən] 二价阴离子
homoaromaticity 同芳香性
sextet 六隅体
pentalene 并环戊二烯
heptalene 庚间三烯并庚间三烯
tropone ['trəʊpəʊn] 环庚三烯酮
tropolone ['trəʊpələʊn] 环庚三烯酚酮
cyclopentadienone 环戊二烯酮
squaric acid 芳酸
cyclopropenyl cation 环丙烯正离子
cyclononatetraenyl cation 环壬四烯正离子
pyrene ['paɪəriːn] 芘
benzo[α]pyrene 苯并芘
coronene ['kərəniːn] 六苯并苯
Buckminsterfullerene [ˌbʌkmɪnstə'fʊləriːn]勃克明斯特富勒烯
nanotube 纳米管

Chapter 2

Stereochemistry

（立体化学）

Stereochemistry refers to the 3-dimensional properties and reactions of molecules, including the static and dynamic aspects of the three-dimensional shapes of molecules. It is a foundation for understanding the structure and reactivity of molecules. It has its own language and terms that need to be learned in order to fully communicate and understand the concepts.

Because an understanding of *isomer*s is integral to the discussion of stereochemistry, let's begin with an overview of isomers. Isomers are different compounds with the same molecular formula. Isomers are molecules that contain the same number of atoms and also the same kind of atoms. However, they have different bonding arrangements. There are two major classes of isomers: *constitutional* isomers and *stereoisomer*s.

Constitutional (or structural) isomers differ in how the atoms are arranged and connected. Constitutional isomers have: ①different IUPAC names; ②the same or different functional groups; ③different physical properties, so they are separable by physical techniques such as distillation; and ④different chemical properties. They behave differently or give different products in chemical reactions. The classification is as follows:

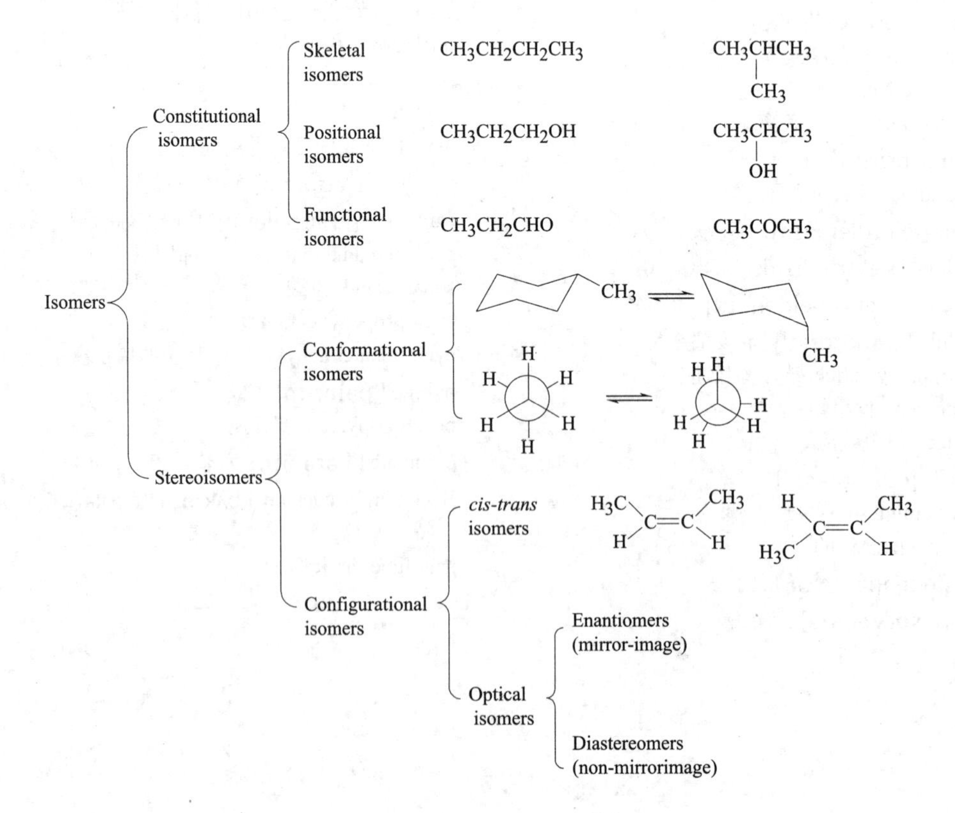

Stereoisomers include *conformational* isomers and configurational isomers. Conformational isomers are compounds that posses the same molecular formula and atomic connectivity but differ in a rotation about a bound. In other words, conformational isomers can be interconverted by rotation about single bonds. They are not separable at room temperature. There are different kinds of conformational isomers. They are eclipsed, staggered, anti, and gauche conformations. Configurational isomers are those isomers which can only be interconverted by breaking bonds. Stereoisomers have identical IUPAC names (except for a prefix like *cis* or *trans*).

Isomers can be recognized by steps in Fig 2.1.

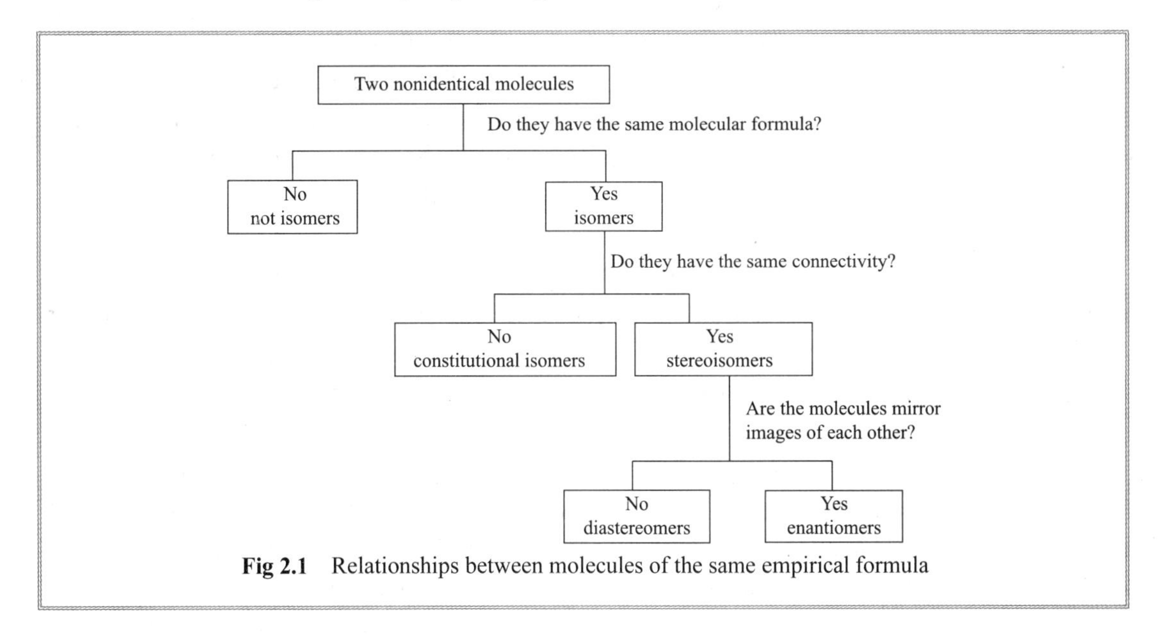

Fig 2.1 Relationships between molecules of the same empirical formula

2.1 Enantiomers（对映异构体）

2.1.1 Chirality and molecular structure（手性与分子结构）

（1）Optical activity and *chirality*（光学活性与手性）

Any material that rotates the plane of polarized light is said to be optically active. If a pure compound is optically active, the molecule is nonsuperimposable on its mirror image. If a molecule is superimposable on its mirror image, the compound does not rotate the plane of polarized light; it is optically inactive. The property of nonsuperimposability of an object on its mirror image is called chirality. If a molecule is not superimposable on its mirror image, it is chiral. If it is superimposable on its mirror image, it is *achiral*. The relationship between optical activity and chirality is absolute. This is both a necessary and a sufficient condition.

（2）Enantiomers（对映体）

If a molecule is chiral, the mirror image must be a different molecule, since superimposability is the same as identity. In each case of optical activity of a pure compound there are two and only two isomers, called *enantiomers*, and thus these two forms are said to be enatiomeric with each other. They differ in structure only in the left and right handedness of their orientations (Fig 2.2). Enantiomers differ at the configuration of every stereocenter.

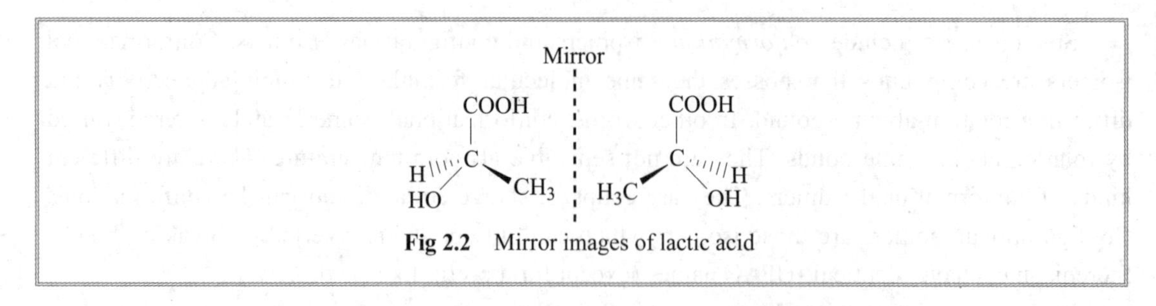

Fig 2.2 Mirror images of lactic acid

Enantiomers have identical physical and chemical properties except in two important respects.

① They rotate the plane of polarized light in opposite directions, although in equal amounts. The isomer that rotates the plane to the left (counterclockwise) is called the levo isomer and is designated (–), for example **(–)-*epinephrine*,** while the one that rotates the plane to the right (clockwise) is called the dextro isomer and is designated (+), for example (+)-epinephrine. Because they differ in this property they are often called optical antipodes. This property is called optical activity or optical rotation.

② They react at different rates with other chiral compounds. These rates may be so close together that the distinction is practically useless, or they may be so far apart that one enantiomer undergoes the reaction at a convenient rate while the other does not react at all. This is the reason that many compounds are biologically active while their enantiomers are not. Enantiomers react at the same rate with achiral compounds.

Therefore, (+)- and (–)-epinephrine have the same melting point, solubility, chromatographic retention time, infrared spectroscopy (IR) and nuclear magnetic resonance (NMR) spectra, but different biological activity. Because enantiomers have identical physical properties, they cannot be separated by common physical techniques like distillation or crystallization.

Biological systems commonly distinguish between enantiomers, and two enantiomers may have totally different biological properties. Enzymes in living systems are chiral, and they are capable of distinguishing between enantiomers. Usually, only one enantiomer of a pair fits properly into the chiral active site of an enzyme (Fig 2.3). Biological systems are capable of distinguishing between enantiomers of many different chiral compounds. In general, just one of the enantiomers produces the characteristic effect; the other either produces no effect or has a different effect. Some typical different behaviors of enantiomers are shown in the following table.

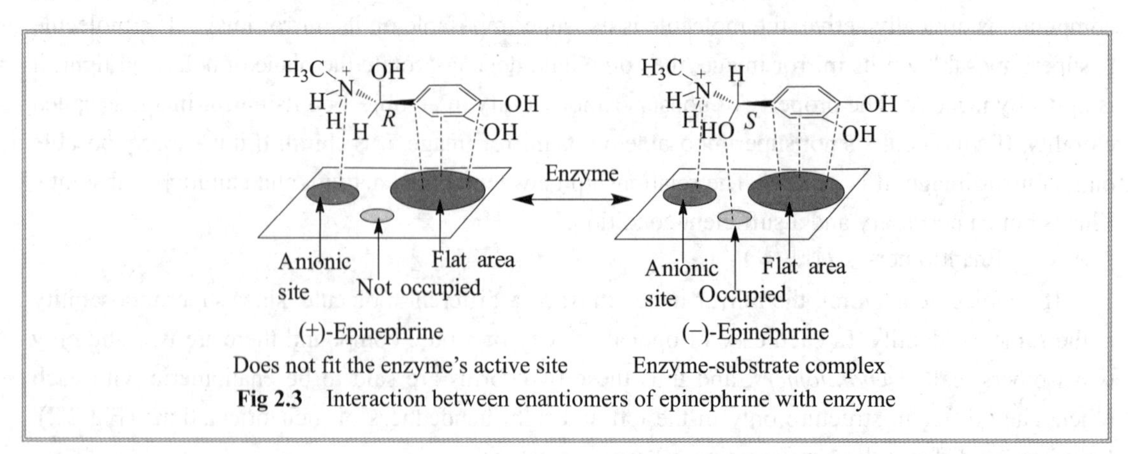

Fig 2.3 Interaction between enantiomers of epinephrine with enzyme

Examples of the different behaviors of enatiomers

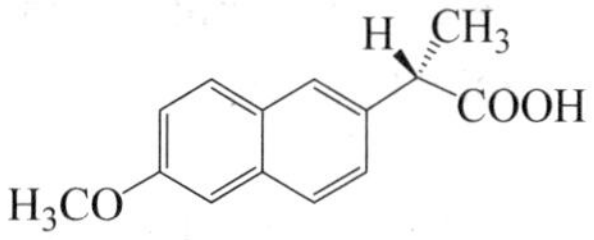

(*S*)-Naproxen (anti-inflammatory agent)

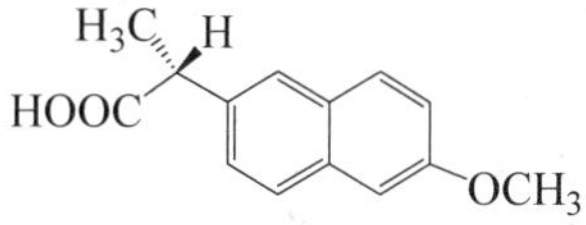

(*R*)-Naproxen (liver toxin)

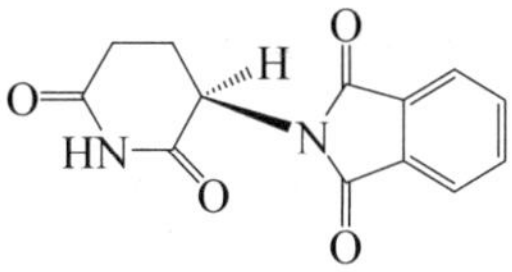

(*S*)-Thalidomider (teratogen)

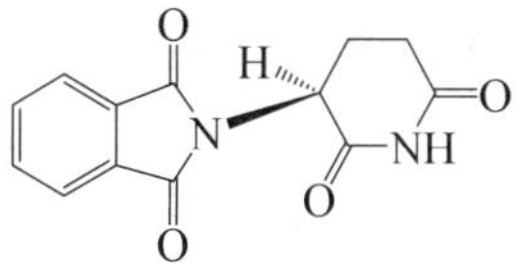

(*R*)-Thalidomider(-sedative)

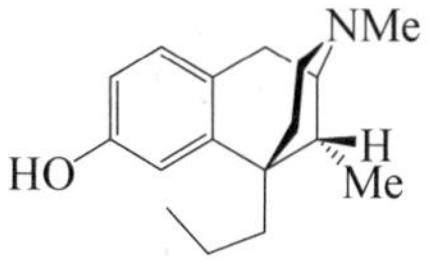

(−) Benzomorphia

(eases pain, unhabituational)

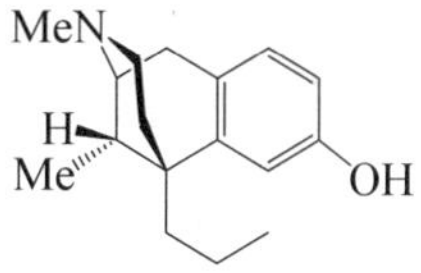

(+) Benzomorphia

(faintly pain-easing, habituational),

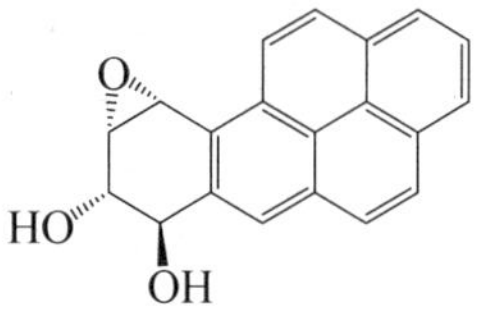

(−)-Benzopyryldiol

(strong carcinogenicity)

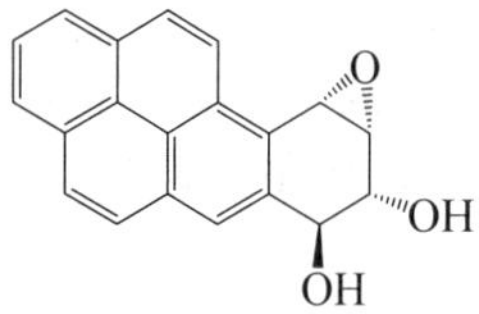

(+)-Benzopyryldiol

(no carcinogenicity)

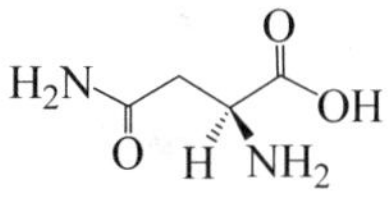

L-Asparagine (bitter)

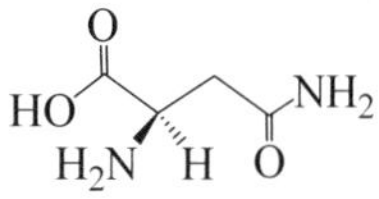

D-Asparagine (sweet)

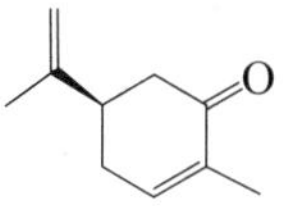

(*R*)-Carvone (spearmint)

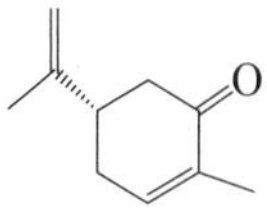

(*S*)-Carvone (caraway odor)

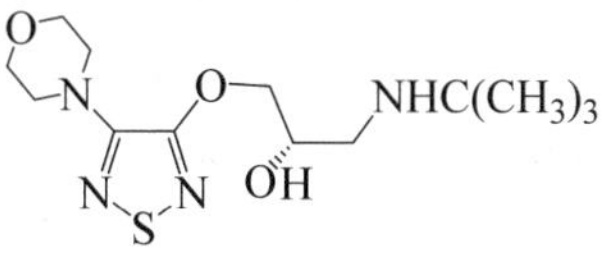

(*R*)-Timolol (adrenergic blocker)

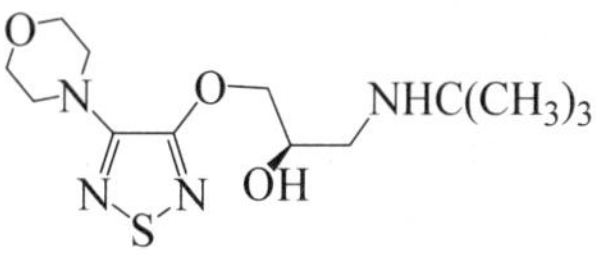

(*S*)- Timolol (ineffective)

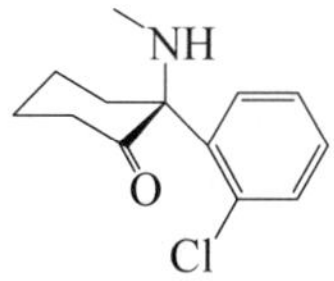

(*S*)-Ketamine (anaesthetin)

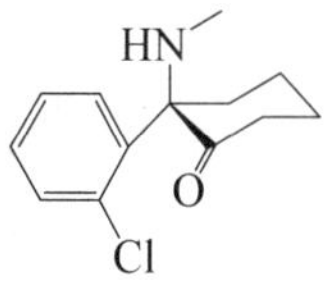

(*R*)- Ketamine (hallucinogen)

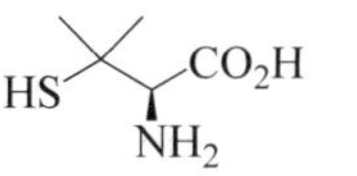

(*S*)-Penicillamine (antiiarthritic)

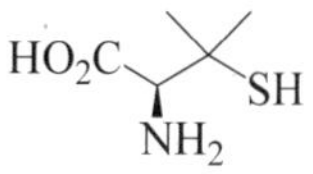

(*R*)-Penicillamine (mutagen)

(3) What kinds of molecules display optical activity?（什么类型的分子具有光学活性？）

Molecules that possess certain elements of symmetry are not chiral, if the element of symmetry ensures that the mirror image forms are superimposable. Four elements of symmetry are discussed here.

A plane of symmetry (σ) is a plane passing through an object such that the part on one side of the plane is the exact reflection of the part on the other side (the plane acting as a mirror). Compounds possessing such a plane are always optically inactive (Fig 2.4).

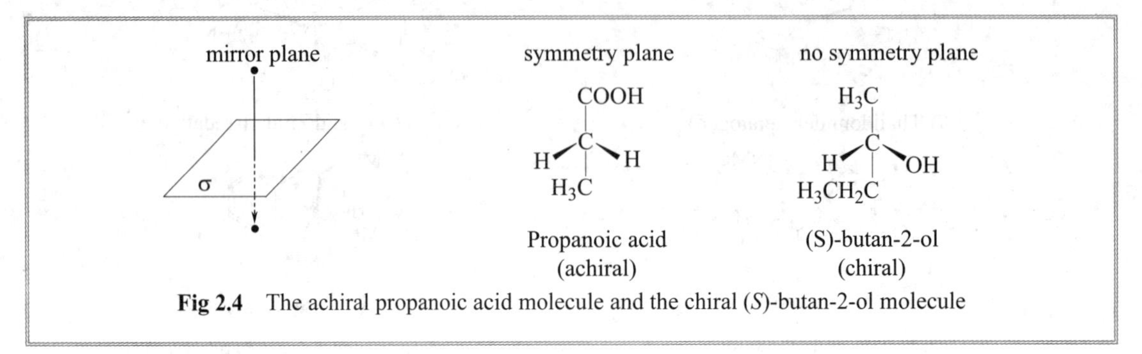

Fig 2.4 The achiral propanoic acid molecule and the chiral (*S*)-butan-2-ol molecule

A center of symmetry (i) is a point within an object such that a straight line drawn from any part or element of the object to the center and extended an equal distance on the other side encounters an equal part or element. Compounds possessing such a center are also optically inactive (Fig 2.5).

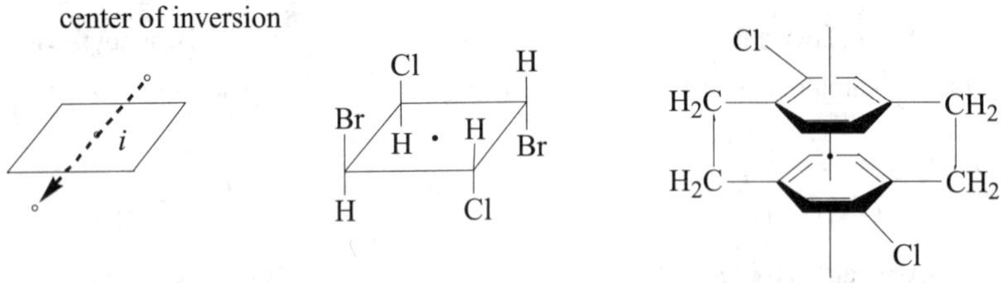

Fig 2.5 The symmetry centre of cyclic molecules

A rotational axis of symmetry (C_n) of order n is a line in space about which an object may be rotated by $360°/n$ such that its initial and final positions are indistinguishable. As illustrated below, if an object can be rotated about an axis and repeats itself every 90° of rotation then it is said to have an axis of 4-fold rotational symmetry. The axis along which the rotation is performed is an element of symmetry referred to as a rotation axis. A molecule with rotational axes may or may not be a chiral molecule (Fig 2.6).

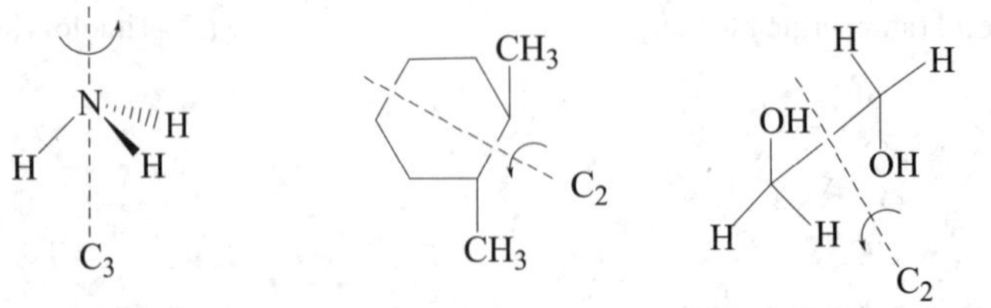

Fig 2.6 Molecules with rotational axis of symmetry

An *alternating axis of symmetry* (S_n) of order n is an axis such that when an object containing such an axis is rotated by $360°/n$ about the axis and then reflection is performed across a plane at right angles to the axis, a new object is obtained that is indistinguishable from the original one. It is also called an improper rotation axis or a rotary-reflection axis. An S_n axis is composed of two successive transformations, first a rotation through $360°/n$, followed by a reflection through a plane. An S_1 axis is identical to a simple mirror plane, and an S_2 axis is equivalent to a center of inversion i. Compounds that lack an alternating axis of symmetry are always chiral.

Molecules which do not possess a plane of symmetry σ or a center of inversion i, but an S_4 axis are not very common, the examples given below are 1,3,5,7-tetrabromo-2,4,6,8-tetramethyl-cyclooctane and 2,3,7,8-tetramethyl-spiro[4.4] nonane (both S_4 symmetry only). Combinations of S_4 axis with other symmetry elements are common, e.g. methane CH_4 possess three S_4 axes along with six mirror planes σ, four C_3 and three C_2 axes (Fig 2.7).

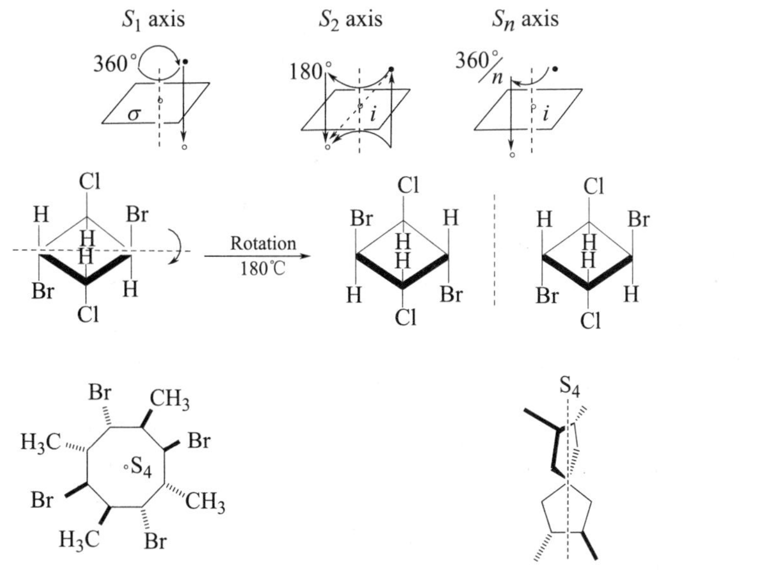

1,3,5,7-tetrabromo-2,4,6,8-tetramethyl-cyclooctane   2,3,7,8-tetramethyl-spiro[4.4]nonane

Fig. 2.7 Molecules with an alternating axis of symmetry

On a molecular level, many biomolecules fundamental to life are chiral. On a macroscopic level, many naturally occurring objects possess handedness. For example, the human body is chiral, and hands, feet, and ears are all chiral. Most of the biological macromolecules of living systems occur in nature in one enantiomeric form only. A biologically active chiral compounds interacts with its acceptor site in a chiral manner, and enantiomers may be discriminated by the receptor in very different ways. Thus it is not surprising that the two enantiomers of a drug may interact differently with the receptor, leading to different effects. Therefore, the two enantiomers of a drug have unequal degrees or different kinds of activity. One may be therapeutically effective, while the other may be ineffective or even toxic. For example, the levorotatory form of epinephrine is one of the principal hormones secreted by the adrenal medulla. When synthetic epinephrine is given to a patient, the (–) form has the same stimulating effect as the natural hormone. The (+) form lacks this effect and is mildly toxic.

2.1.2 The asymmetric molecular with a stereogenic center（具有不对称中心的不对称分子）

A *stereogenic center* or *stereocenter* is an atom bearing groups such that an interchanging of any two groups leads to a stereoisomer. A stereogenic center means an atom which is bonded to four different groups in a spatial arrangement which is not superimposable on its mirror image. In organic chemistry a chiral center usually refers to a carbon, phosphorus, or sulfur atom, though it is also possible for other atoms to be chiral centers in organic and inorganic chemistry.

(1) Compounds with stereogenic centers on carbon atoms（不对称中心位于碳原子的化合物）

A chiral carbon or asymmetric carbon is a carbon atom which is asymmetric. Having a chiral carbon is usually a prerequisite for a molecule to have chirality, though the presence of a chiral carbon does not necessarily make a molecule chiral (see *meso* compound). A chiral carbon is often denoted by C*.

For example, L-alanine has one stereogenic center, (1*R*, 2*S*)-ephedrine has two stereogenic centers, and D-*glucose* has four stereogenic centers.

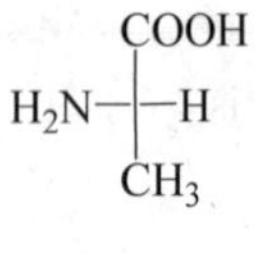

L-alanine 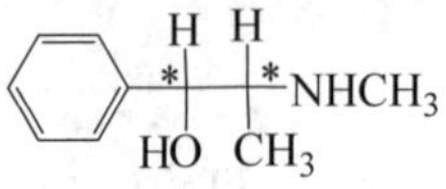 (1*R*, 2*S*)-ephedrine 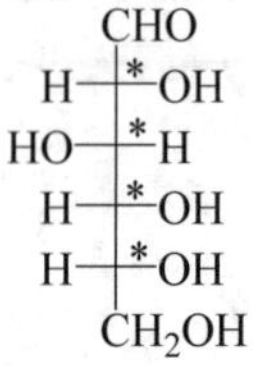 D-glucose

The situation is more complex for compounds with two stereogenic centers, because more stereoisomers are possible. Moreover, a molecule with two stereogenic centers may or may not be chiral.

For *n* stereogenic centers, the maximum number of stereoisomers is 2^n.

When $n = 1$, $2^n = 2$. With one stereogenic center there are always two stereoisomers and they are enantiomers.

When $n=2$, $2^n=4$. With two unidentical stereogenic centers, the maximum number of stereoisomers is four, although sometimes there are fewer than four. For example, there are four stereoisomers for ephedrine: enantiomers A and B, and enantiomers C and D. What is the relationship between two stereoisomers like A and C ? A and C are *diastereomer*s. Diastereomers are stereoisomers that are not mirror images of each other. A and B are diastereomers of C and D, and vice versa. Molecules that are not mirror images but differ in spatial arrangements of atoms are diastereomers. Stereoisomers that differ at some stereocenters but not at others are not mirror images, so they are not enantiomers. Instead, they are diastereomers.

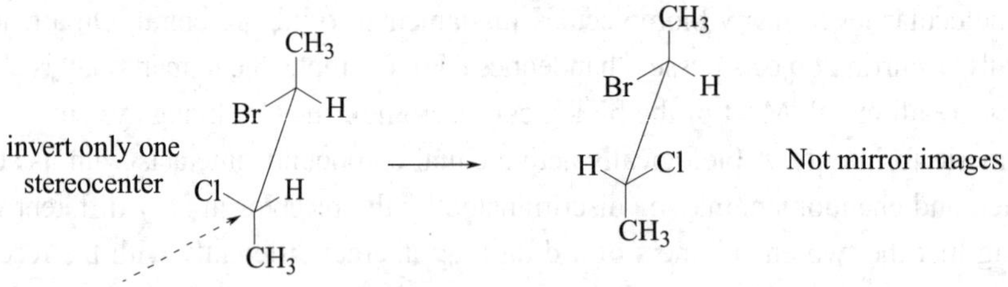

invert only one stereocenter → Not mirror images

diastereomers

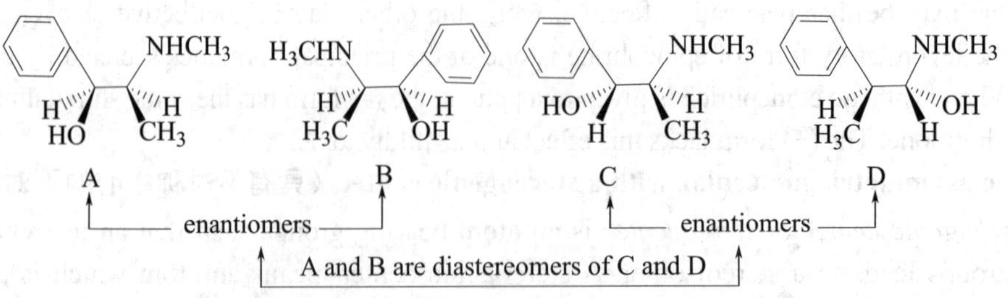

A B C D

enantiomers enantiomers

A and B are diastereomers of C and D

If there are identically substituted stereocenters, the number of steroisomers will be lower than 2^n.

There are two stereocenters in 2,3-dibromobutane, we might expect four distinct *RR*, *SS*, *RS*, *SR*. These could then be organized into enantiomeric pairs: *RR* | *SS* and *RS* | *SR*. But in fact, 2,3-dibromobutane exists as three stereoisomers only: a pair of enantiomers and an achiral meso diastereomer.

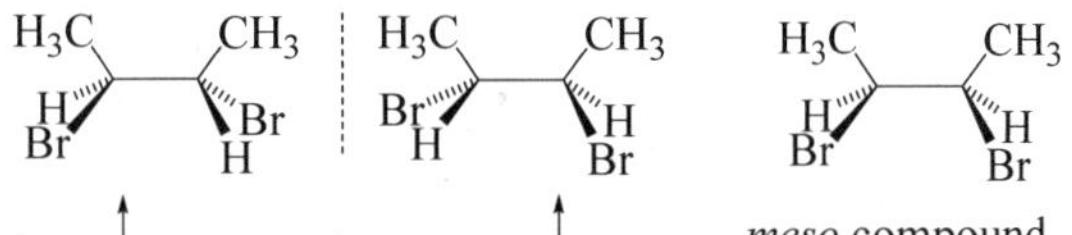

enantiomers *meso* compound

In 2,3-dibromobutane, the *RS* | *SR* pair of molecules are superimposable and therefore identical. They are *meso* compounds, a compound containing two or more stereocenters that is superimposable with its mirror image.

Take note, first assign (*R*) or (*S*) to each chiral carbon if you need to determine the relationship (enantiomer? diastereomer? *meso*?) between the molecules with stereocenters, then according to the following basic points:

① Enantiomers have opposite configurations at each corresponding chiral carbon.

② Diastereomers have some matching, some opposite configurations.

③ Meso compounds have internal mirror plane.

(2) Compounds with stereogenic centers in cyclic compounds

Stereogenic centers may also occur at carbon atoms that are part of a ring. To find stereogenic centers on ring carbons you can always draw the rings as flat polygons, and look for tetrahedral carbons that are bonded to four different groups, as usual. Each ring carbon is bonded to two other atoms in the ring, as well as two substituents attached to the ring. When the two substituents on the ring are different, we must compare the ring atoms equidistant from the atom in question. The following molecules have one, two, and ten stereogenic centers, respectively.

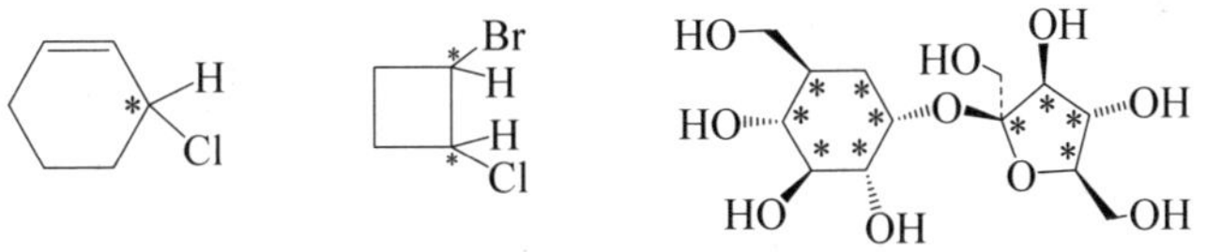

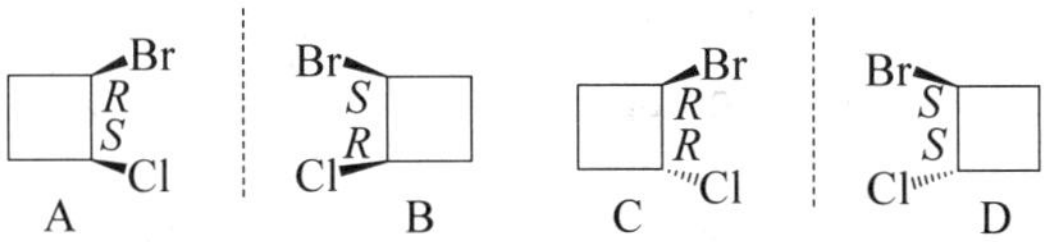

A B C D

Take 2-bromo-1-chlorocyclobutane for example, there are four *RR*, *SS*, *RS* and *SR*. The two *cis* isomers A and B are the pair of enantiomers; so are the two *trans* isomers C and D. Their configurations are shown above. A, B and C, D are enantiomers of one another, respectively. Compounds A and B are each diasteromers of C and D. We can think of *cis* and *trans*-2-bromo-1-chlorocyclobutane as diastereomers of each other.

(3) Compounds with stereogenic centers other than carbon（不对称中心位于除碳以外的化合物）

Although asymmetrically substituted carbon atoms are by far the most common type of stereogenic center in organic compounds, several other kinds of stereogenic centers are encountered, such as stereogenic nitrogen, stereogenic phosphorus, and stereogenic sulfur centers. Tetravalent nitrogen (ammonium) and phosphorus (phosphonium) ions are obvious extensions. Phosphine oxides are also tetrahedral and are chiral if all three substituents (in addition to the oxygen) are different. Not quite so evident are the cases of trivalent sulfur and phosphorus compounds, including sulfonium salts, sulfoxides, and phosphines. The heteroatom in these structures is approximately tetrahedral, with an electron pair occupying one of the tetrahedral positions. Because there is a relatively high energy barrier to inversion of these tetrahedral molecules, they can be obtained as pure enantiomers.

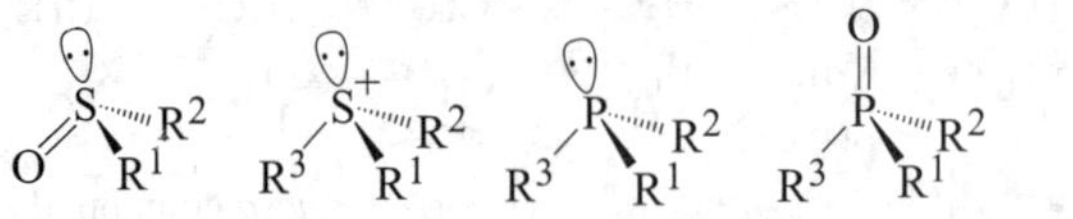

sulfoxide sulfonium phosphine phosphine oxi de

Amines with three different substituents are potentially chiral because of the pseudotetrahedral arrangement of the three groups and the lone-pair electrons. The unshared electron pair in the sp^3 orbital is like a fourth "group". Under normal conditions, however, these enantiomers are not separable and are not considered chiral compounds because of the rapid inversion at the nitrogen center (about 10 times per second for NH_3). This inversion allows an amine to rapidly change into its non-superimposable mirror image. As soon as the lone-pair electrons are fixed by the formation of quaternary ammonium salts, tertiary amide *N*-oxide, or any other fixed bonding, the inversion is prohibited, and consequently the enantiomers of chiral nitrogen compounds can be separated.

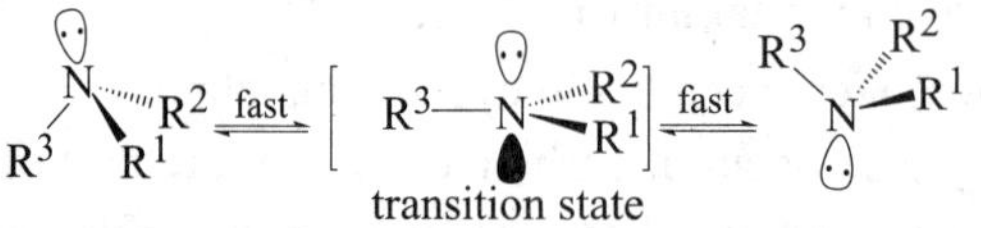

transition state

In contrast to the amines, inversion of configuration for phosphines is generally negligibly slow at ambient temperature. So, phosphines with three different substituents are widely used as chiral ligands in transition metal-catalyzed *asymmetric synthesis*.

Similarly, the configuration of organosulfur species is pyramidal due to the presence of lone-pair electrons, and the pyramidal reversion is slow at ambient temperature. Thus two enantiomers of chiral sulfoxides are possible and separable.

2.1.3 Asymmetric compounds without stereogenic centers（不具有不对称中心的不对称化合物）

While they have no chiral atoms like *allene*s and biphenyls, some molecules are chiral because they are so bulky or highly strained that they cannot easily convert from one chiral conformation to the mirror-image conformation. The mirror image of A cannot be *superimpose*d on B no matter how it is oriented in space. So it is a different compound that we label as B.

Crowed derivatives of biphenyls are chiral, because they cannot achieve the most symmetric conformation (being "locked" into a conformation) due to their steric hindrance or ring strain.

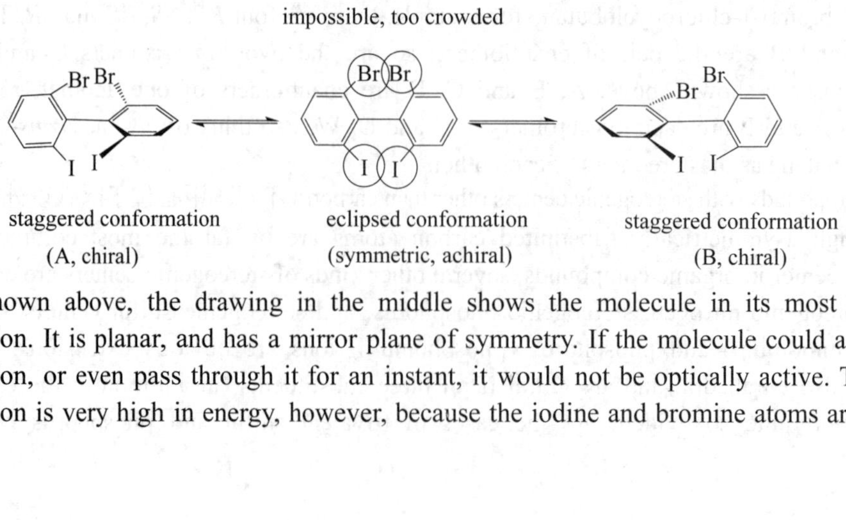

impossible, too crowded

staggered conformation (A, chiral) eclipsed conformation (symmetric, achiral) staggered conformation (B, chiral)

As shown above, the drawing in the middle shows the molecule in its most symmetric conformation. It is planar, and has a mirror plane of symmetry. If the molecule could achieve this conformation, or even pass through it for an instant, it would not be optically active. This planar conformation is very high in energy, however, because the iodine and bromine atoms are too large to be forced so close together. The molecule can exist only in one of the two staggered conformations shown on the left and right. They are enantiomers and can be separated.

Allenes can be chiral. In allene, the central carbon atom is sp hybridized and linear, while the two outer carbon atoms are sp^2 hybridized and trigonal. The central sp hybrid carbon atom must use different p orbitals to form the π bonds with the two outer carbon atoms. The two unhybridized p orbitals on the sp hybrid carbon atom are perpendicular, so the two π bonds must also be perpendicular.

Allene itself is achiral. However, an allene having no identical substituents at both sp^2 carbons gives nonsuperimposable mirror images, and is chiral.

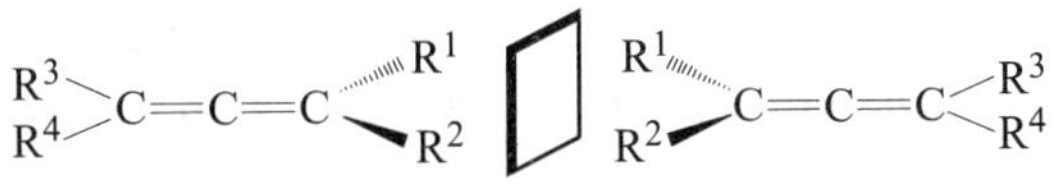

Molecules with shapes analogous to screws are also chiral, since they can be right-handed or left-handed. There are several kinds of molecules in which steric factors impose a screwlike shape. A very important case is 1,1-binaphthyl compounds. Steric interactions between the 2 and 8 hydrogens prevent the molecules from being planar, and as a result, there are two nonsuperimposable mirror image forms.

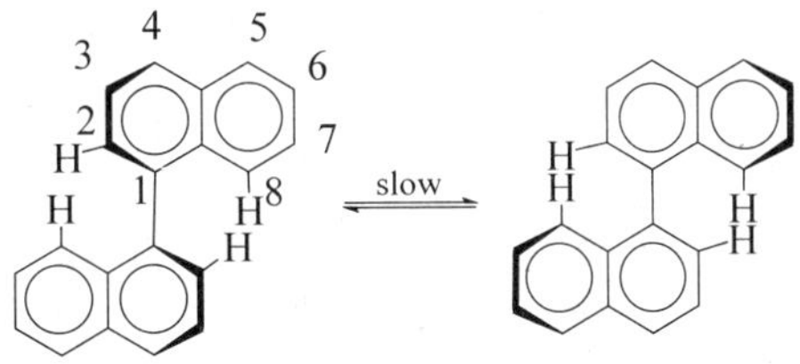

A particularly important example is the1-(2-hydroxynaphthalen-1-yl)naphthalen-2-ol, which is called BINOL. Another important type includes 1,1-binaphthyl diphosphines, which is called BINAP. BINOL and BINAP are both useful chiral ligands in organometallic compounds that serve as catalysts for hydrogenations and other reactions.

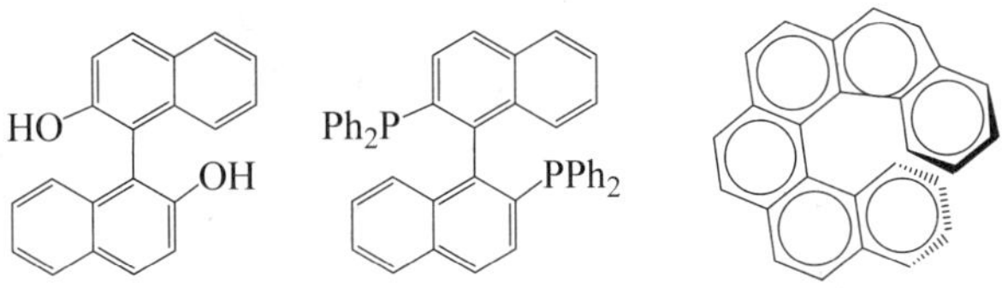

BINOL BINAP hexahelicene

A spectacular example of screw-shaped chirality is *hexahelicene*, in which the six fused benzene rings cannot be planar and is distorted to avoid bumping into each other, thus give rise to right-handed and left-handed enantiomers. The specific rotation is about 3700. Hexahelicene can be racemized by heating. The increased molecular vibration allows the two terminal rings to slip past one another. The activation energy required is 36.2 kcal/mol (1cal=4.184J).

2.1.4 *Racemate*, *Meso* compounds and *Epimer*（外消旋体，内消旋体和差向异构体）

Racemate is a mixture of equal quantities of two enantiomorphs and is optically inactive.

An achiral compound that has chirality centers is called a meso compound. Since (2*R*,3*S*) and (2*S*,3*R*)- tartaric acid are superimposable on each other, they are a single stereoisomer that we call a meso compound or meso isomer. Most meso compounds have this kind of symmetric structure, with two similar halves of the molecule having opposite configuration, and it is a stereoisomer of a compound with two or more chiral centers that is superimposable on its own mirror image. Since meso forms are stereoisomers with mirror planes of symmetry, you can identify a meso form by identifying its mirror plane.

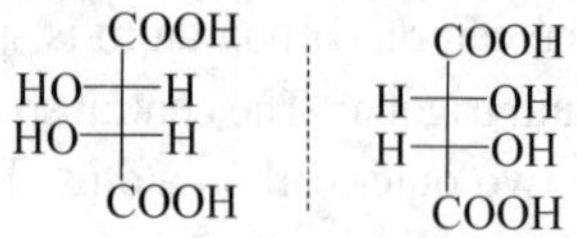

COOH | COOH
HO—H | H—OH
HO—H | H—OH
COOH | COOH
R, *S* | *S*, *R*

Epimers mean diastereomers that differ in configuration at one of two or more sterogenic units. Epimerization is generally associated with interconversion of carbohydrate diastereomers. Many *saccharide*s, such as *pentose*s and *hexose*s, exist as a mixture of an open chain form and two diasteromeric cyclic structures exhibiting an additional chiral carbon centre. The diastereomers that are formed through intramolecular hemiacetal or hemiketal formation are referred as *anomer*s.

A classical example is the interconversion between glucose and *mannose*. Epimerization generally refers to *stereomutation* of a single stereocenter in diastereomers that have at least two elements of chirality.

在偏振光通过某物质时，能使偏振光的振动平面发生旋转的性质称为旋光性，具有旋光性的物质称为旋光性物质。物质分子与其镜像不能完全重叠，它们之间相当于左手和右手的关系，这种特征称为物质的手性。具有手性的分子称为手性分子，手性分子具有旋光性，具有旋光性的分子一定是手性分子。彼此呈实物与镜像的对映关系，但又不能完全重叠的一对旋光异构体称为对映异构体，简称对映体。分子有手性，就存在对映异构体。对映异构体的物理性质和化学性质一般都相同，比旋光度相等，但旋光方向相反。对映异构体与手性物质反应速率不同，与非手性物质反应速率相同。能将分子分成互为镜像两部分的平面称为分子的对称面。从分子中任一原子或原子团向分子的中心做连线，延长此连线至等距离处，若出现相同的原子或原子团，该点称为分子的对称中心。对称面和对称中心统称对称因素。不具有对称因素的分子是手性分子，或者说手性分子不具有对称因素。

如果一化合物绕一对称轴转 360°/*n*, *n*=2, 3, 4…时会重复出现原化合物，相应地称该轴为 *n* 重对称轴。如果分子中有一条直线, 当分子以之为轴旋转 360°/*n* 后，再用一个与此直线垂直的平面进行反映，如果得到的镜像与原来的分子完全相同，这条直线就是交替对称轴。一般情况下，既没有对称面也没有对称中心，即可判断它是手性分子。手性中心可以是连接四个不同取代基的碳原子。Si、N、S、P、As 和 B 等非碳原子所形成的不对称四面体（或三角锥体）化合物如手性膦、手性氮与手性砜类化合物也具有手性。连有四个不同原子或原子团的碳原子有不对称性，称为手性碳原子或不对称碳原子，用“C*”表示，是分子的不对称中心或手性中心。手性是指整个物质的性质，手性中心是指其中某一原子。

等量对映体的混合物称为外消旋体，通常用“±”表示。外消旋体无旋光性，外消旋体与其左、右旋体的物理性质有差异，但化学性质基本相同。含有一个手性碳原子的化合物，由于不具有对称因素，一定具有旋光性。有两个旋光异构体，一个左旋体，一个右旋体，为一对对映异构体。等量混合组成一个外消旋体。如果分子有两个或两个以上手性中心（手性碳原子），并有一个内在的对称面，这个分子称为内消旋体，它不具有旋光性。同一化合物的内消旋体与其左旋体和右旋体都呈非对映体关系，它们的物理性质不同，化学性质基本相似。环状化合物的旋光异构：环烷烃分子中不同碳原子上的氢原子被其他原子或原子团取代后，同时又有手性碳原子存在时，就会产生顺反异构和旋光异构。丙二烯型化合物的旋光异构：在丙二烯分子中，如果两端碳原子上连有不同原子或原子团时，虽然分子中不含手性碳原子，但由于整个分子没有对称因素，属于手性分子，具有旋光性。一旦一端连有相同原子或原子团，就存在对称面，就不是手性分子，也就不具有旋光性。联苯型化合物的旋光异构：在联苯分子中，两个苯环直接相连，如果两个苯环的邻位上连有较大的基团时，由于空间位阻效应使两个苯环的旋转受阻，两个苯环不能处在同一平面上。如果每个苯环的邻位连有的两个基团都不相同，与丙二烯分子相似，虽然分子中不含手性碳原子，由于整个分子没有对称因素，属于手性分子，具有旋光性。一旦一个苯环的邻位上连有两个相同基团，分子就有对称面，整个分子就不是手性分子，没有旋光性。

2.2 Resolution of racemates（外消旋体的拆分）

A pair of enantiomers can be separated in several ways, of which conversion to diastereomers and separation of these by fractional crystallization are the most often used. In this method and in some of the others, both isomers can be recovered, but in some methods it is necessary to destroy one.

2.2.1 Chemical resolution of enantiomers（对映体的化学拆分）

The separation of enantiomers is called resolution. Resolution of stereoisomers can be more difficult than separation of unrelated organic compounds since most separation methods depend on differences in boiling points, or solubilities in solvents of the components of a mixture. These differences are small or non-existent for stereoisomers because they have the same mass, functional groups, and chemical structures except for configurations at chiral centers.

Individual enantiomers of a racemate have identical physical properties, so they cannot be resolved by fractional crystallization or fractional distillation. However, if we chemically convert each enantiomer into a new compound with an additional chiral atom of the same configuration by reacting with a chiral compound, the pair of enantiomers becomes a pair of diastereomers with different physical properties. Such a chiral compound is called a resolving agent.

We illustrate the process for separating a racemic mixture of (*R*)- and (*S*)-2-butanol by (+)-tartaric acid as shown below.

The diastereomers of 2-butyl tartrate have different physical properties, and they can be separated by conventional distillation, recrystallization, or *chromatography*. After separation of the diastereomers, the resolving agent is then cleaved by hydrolysis of the two diastereomers to give (*R*)-2-butanol and (*S*)-2-butanol.

(*S*)-2-butanol + (*R*, *R*)-(+)-tartaric acid $\xrightarrow{H^+}$ (*S*)-2-butyl (*R*, *R*)-tartrate $\xrightarrow{\text{separating then } H^+}$ (*S*)-2-butanol + (*R*, *R*)-(+)-tartaric acid

(*R*)-2-butanol + (*R*, *R*)-(+)-tartaric acid $\xrightarrow{H^+}$ (*R*)-2-butyl (*R*, *R*)-tartrate $\xrightarrow{\text{separating then } H^+}$ (*R*)-2-butanol + (*R*, *R*)-(+)-tartaric acid

diasteromers, not mirror images

If the racemic mixtures to be resolved are carboxylic acids, it is possible to form a salt with an optically active base. Once the two diastereomers have been separated, it is easy to convert the salts back to the free acids and the recovered base can be used again. Racemic bases can be converted to diastereomeric salts with active acids. Alcohols can be converted to diastereomeric esters (as shown above), aldehydes to diastereomeric hydrazones, and so on.

2.2.2 Chromatographic resolution of enantiomers（对映体的色谱拆分）

Chromatography is an important means of separating enantiomers on both an analytical and preparative scale. These separations are based on use of a chiral stationary phase (CSP). Chromatographic separations result from differential interactions of the enantiomers with the solid column packing material. The differential adsorption arises from the fact that the enantiomers form diastereomeric complexes with the CSP. Hydrogen bonding and aromatic-interactions often contribute to the binding. These diastereomeric complexes have different binding energies and different equilibrium constants for complexation. One of the two enantiomers will spend more time complexed with the chiral column packing.

One important type of chiral packing material is polysaccharides derivatives, which provide a chiral lattice, but separation is improved by the addition of structural features that enhance selectivity. One group of compounds includes *aroyl* esters and *carbamate*s, which are called Chiralcels; two of the most important examples are the 4-methylbenzoyl ester, called Chiralcel OJ, and the 3,5-dimethylphenyl carbamate, called Chiralcel OD. There is a related series of materials derived from amylose rather than cellulose, which have the trade name Chiralpak.

As the racemates pass through the column, the enantiomers form weak complexes, usually through hydrogen bonding, with the chiral column packing. The solvent flows continually through the column, and the dissolved enantiomers gradually move along. The more strongly adsorbed enantiomers spend more time on the stationary particles; they come off the column later than weakly adsorbed enantiomer.

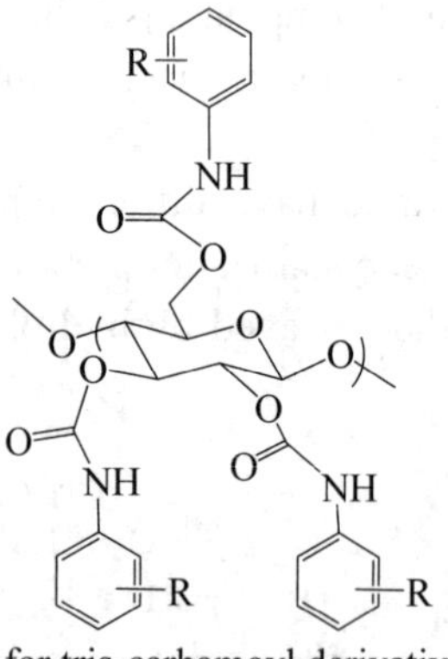

Repeating unit for tris-carbamoyl derivatives of cellulose

2.2.3 Biochemical resolution of enantiomers（对映体的生物拆分）

Biological molecules may react at different rates with the two enantiomers. For example, a certain bacterium may digest one enantiomer, but not the other. Many enzymes have been used for the selective cleavage of one enantiomeric ester.

Enzymatic resolution is based on the ability of enzymes (catalytic proteins) to distinguish between *R*- and *S*-enantiomers or between enantiotopic pro-*R* and pro-*S* groups in *prochiral* compounds. The selective conversion of pro-*R* and pro-*S* groups is often called desymmetrization or asymmetrization. Note that in contrast to enzymatic resolution, which can at best provide half the racemic product as resolved material, prochiral compounds can be completely converted to a single enantiomer, provided that the selectivity is high enough. Complete conversion of a racemic mixture to a single enantiomeric product can sometimes be accomplished by coupling an enzymatic resolution with another reaction (chemical or enzymatic) that racemizes the reactant. This is called dynamic resolution, and it has been accomplished for several arylpropanoic acids via the thioesters, using an amine to catalyze racemization. Trifluoroethyl thioesters are advantageous because of their enhanced rate of exchange and racemization.

CH_3
Ar $CSCH_2CF_3$
O

enzymatic hydrolysis selective for one enantiomer
amine racemizes unreacted thio ester

CH_3 pure *S*-enantiomer
Ar COOH
\+ CH_3
Ar $COSCH_2CF_3$

Generally speaking, the *enantioselectivity* is quite high, since enzyme-catalyzed reactions typically involve a specific fit of the reactant (substrate) into the catalytically active site. The compound to be resolved must be an acceptable substrate for the enzyme. If not, there will be no reaction with either enantiomer. The types of reactions that are suitable for enzymatic resolutions are somewhat limited. The most versatile enzymes are *esterase*s, and *lipase*s—catalyze formation or hydrolysis of esters. There are also enzymes that catalyze amide formation and hydrolysis, which can be broadly categorized as *acylase*s or *amidase*s. Others are epoxide *hydrolase*s, which open epoxide rings. Another important family is the oxido-reductases, which interconvert alcohols and carbonyl compound by oxidation and reduction.

2.2.4 Chiral recognitive resolution of enantiomers（对映体的手性识别拆分）

The use of chiral hosts to form diastereomeric inclusion compounds was mentioned above. But in some cases it is possible for a host to form an inclusion compound with one enantiomer of a racemic guest, but not the other. This is called chiral recognition. One enantiomer fits into the chiral host cavity, the other does not. More often, both diastereomers are formed, but one forms more rapidly than the other, so that if the guest is removed it is already partially resolved (this is a form of kinetic resolution). An example is use of the chiral crown ether partially to resolve the racemic amine salt. When an aqueous solution of **2** was mixed with a solution of optically active **1** in chloroform, and the layers separated, the chloroform layer contained about twice as much of the complex between **1** and (*R*)-**2** as of the diastereomeric complex. Many other chiral crown ethers and cryptands have been used, as have been cyclodextrins, cholic acid, and other kinds of hosts. Of course, enzymes are generally very good at chiral recognition, and much of the work in this area has been an attempt to mimic the action of enzymes.

Me
Ph—C—$\overset{+}{N}H_3$ PF_6^-
H

1 **2**

外消旋体的拆分方法　①转化为非对映体：基于非对映体物理性质的差别，将外消旋体制成非对映体，用常规分离手段将非对映体分开，再经一定化学方法处理使其再生成原来的对映体；②手性色谱拆分法：将外消旋体注入手性色谱柱中，由于两个对映体在手性色谱柱中的移动速度不同达到拆分的目的；③生物化学方法：利用微生物或酶在外消旋体的溶液中破坏一种对映体比另一种快的方法，分离出其中一种对映体；④用手性主体识别手性客体：

主客体之间产生手性识别作用，必须具备以下几个基本条件：主客体之间必须存在非共价键作用力的协同作用，如氢键疏水作用、π-π 堆积、范德华力作用等，而且其中有一种作用必须与手性相关。当主客体之间发生相互作用时，主体的手性基团与客体分子之间立体效应对手性识别产生极大影响，因此手性基团大小适当极为重要。主体必须具有适当的刚柔性。当主体的柔性过大，经过自身构象的调整，既能配合 *R* 构型的客体，又能配合 *S* 构型的客体,也就不能产生手性识别。主客体大小、尺寸几何形状的互补在识别过程当中也起至关重要的作用。

2.3 The stereochemistry in reaction process（反应过程中的立体化学）

2.3.1 Stereoselective and *stereospecific reaction*（立体选择性与立体专一性反应）

Synthetically useful reactions should deliver a single product or at least a preferred product and not a product mixture. If they give one product exclusively or preferentially, they are called selective reactions. One usually specifies the type of selectivity involved, depending on which of the conceivable side products do not occur at all (highly selective reaction) or occur only to a minor degree (moderately selective reaction).

A reaction that preferentially or exclusively gives one of several conceivable reaction products with different empirical formulas takes place with *chemoselectivity*. A reaction that preferentially or exclusively gives one of several conceivable constitutional isomers is regioselective. A reaction that preferentially or exclusively gives one of several conceivable stereoisomers is referred to as stereoselective. When these conceivable stereoisomers are diastereomers, we describe the reaction as a diastereoselective reaction, or *diastereoselectivity*. When the conceivable stereoisomers are enantiomers, we have an enantioselective reaction, or enantioselectivity (Fig 2.8). In a word, chemoselectivity means functional group discrimination, *regioselectivity* means product structural isomer discrimination, and stereoselectivity means product stereoisomer discrimination.

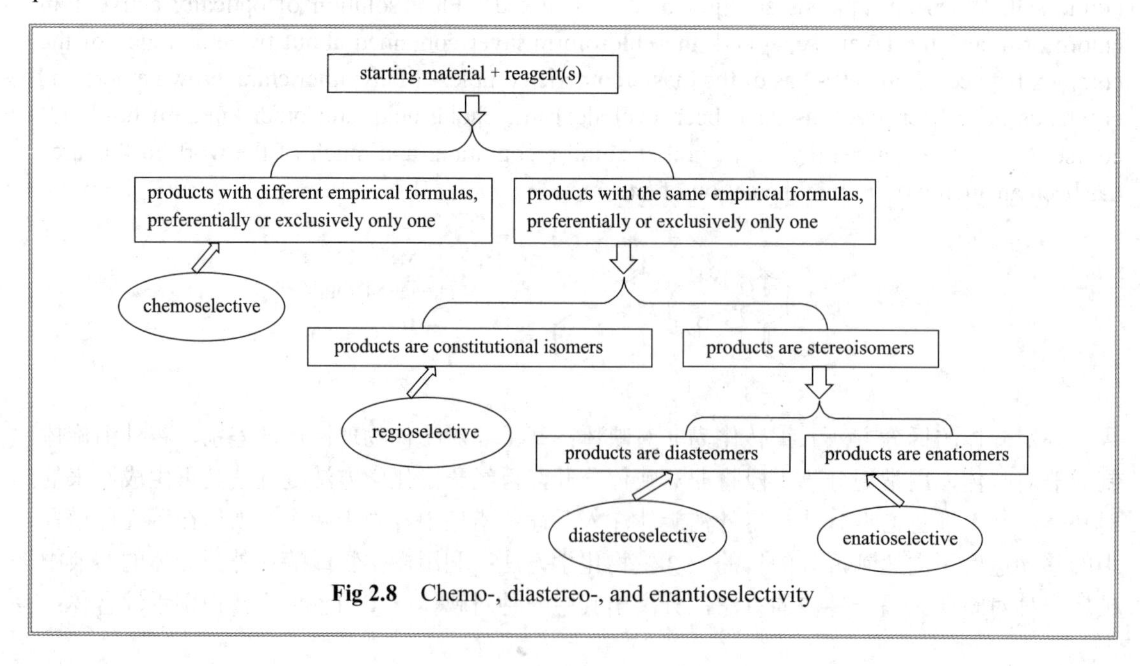

Fig 2.8 Chemo-, diastereo-, and enantioselectivity

We can fully understand the meaning of stereoselective reactions by catalytic hydrogen of alkenes and hydride reduction of cyclic ketones. Unfunctionalized alkene usually reacts by preferential *syn* delivery of hydrogen from the less hindered face of the double bond. This is determined by hydrogenation mechanism on the metal surface. Hydrogen and alkene are both adsorbed on the catalyst surface and then form addition intermediate. Adsorption to the catalyst surface normally involves the less sterically congested face of the double bond, and as a result, hydrogen is added from the less hindered face of the molecule. The degree of stereoselectivity is dependent on the reactant structure, catalysts and reaction conditions.

H_2(only the dissolved part reacts)
Cat. Pd/C, AcOEt or EtOH
(*meso*)
the same
(racemic mixture)
stereospecificity

Pt, H_2
(70~85)% (30~15)%
racemic
stereoselectivity

The stereoselectivity of nucleophilic additions to cyclic ketones has been studied extensively. The stereoselectivity in cyclohexanones is determined by the preference for approach of reactants from the axial or equatorial direction. The chair conformation of cyclohexanone places the carbonyl group in an unsymmetrical environment. The axial face has C(2,6)—H_{eq} bonds that are nearly eclipsed with the C═O bond and the C(3,5)-diaxial hydrogens point toward the trajectory for reagent approach. In contrast, the equatorial face has axial C—H bonds at an angle of roughly 120° to the carbonyl plane. There is more steric bulk, including the 3,5-axial hydrogens, on the axial face. It is observed that small nucleophiles, like $NaBH_4$ and $LiAlH_4$, prefer to approach the carbonyl group of cyclohexanone from the axial direction, to form mainly more stable equatorial alcohol. In this process, torsional effects are believed to play a major role in the preference for axial approach. In the reactant conformation, the carbonyl group is almost eclipsed by the equatorial C(2) and C(6) C—H bonds. This torsional strain is relieved by axial attack, whereas equatorial approach increases strain because the oxygen atom must move through a fully eclipsed arrangement.

$LiAlH_4$
92% 8%

eclipsed interaction increase for equatorial approach
axial approach
equatorial approach

On the other hand, the presence of axial subsituents or use of more sterically demanding reagents, such as alkylborohydrides leads to selective equatorial approach and formation of axial alcohols. This is called steric approach control and is the result of van de Waals repulsions with the 3,5-axial hydrogens. The bulkier nucleophiles encounter the 3,5-axial hydrogens on the axial approach trajectory and therefore prefer the equatorial approach.

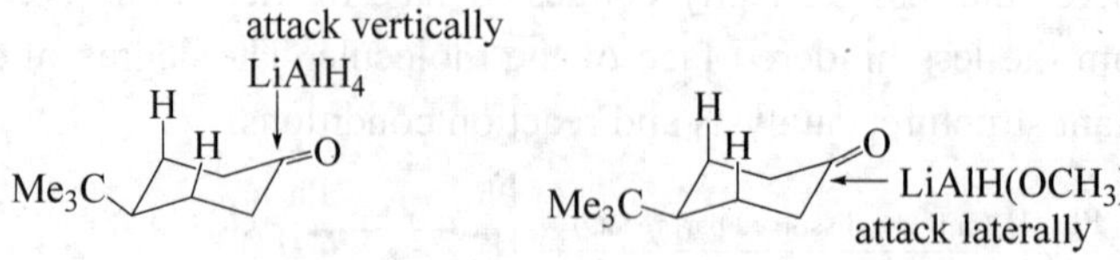

attack vertically $LiAlH_4$

Me_3C H H O Me_3C H H O ← $LiAlH(OCH_3)_3$ attack laterally

Stereoselectivity is intimately related to the mechanism of the reaction. Some reactions are stereospecific, in which stereoisomeric reactants each provide stereoisomeric products. If the starting material has two stereoisomers (such as *cis* and *trans*) and each material gives only one stereoisomer as a product of the reaction, then it is stereospecific. There are many examples of stereospecific reactions involving additions to carbon-carbon double bonds. Addition can be *anti* or *syn*, depending on the mechanism. If the mechanism specifies *syn* or *anti* addition, different products will be obtained from the *E*- and *Z*-isomers.

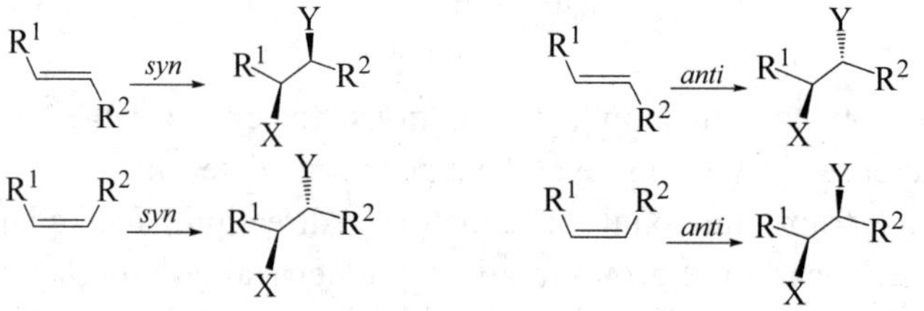

R^1 R^2 —*syn*→ R^1 R^2 X Y R^1 R^2 —*anti*→ R^1 R^2 X Y

R^1 R^2 —*syn*→ R^1 R^2 X Y R^1 R^2 —*anti*→ R^1 R^2 X Y

The bromination of substituted alkenes provides a number of examples of stereospecific reactions. These can be illustrated by considering the *Z*- and *E*-stereoisomers of disubstituted alkenes. The addition of bromine is usually stereospecifically *anti* for unconjugated disubstituted alkenes and therefore the *Z*- and *E*-alkenes lead to diastereomeric products. When both substituents on the alkene are identical, as in 2-butene, the product from the *Z*-alkene is chiral, whereas the product from the *E*-alkene is the achiral meso form.

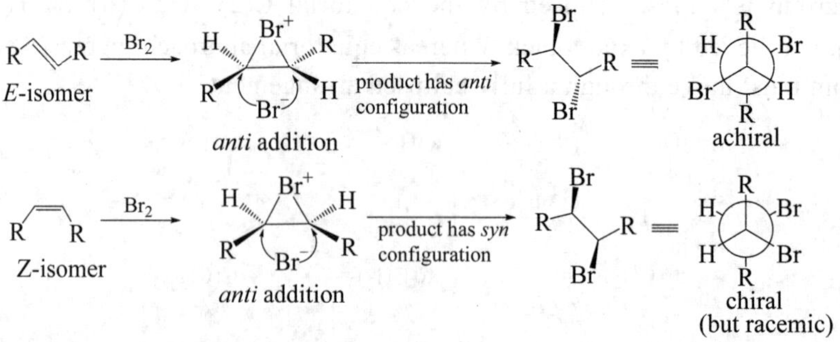

E-isomer —Br_2→ Br^+ *anti* addition → product has *anti* configuration → achiral

Z-isomer —Br_2→ Br^+ *anti* addition → product has *syn* configuration → chiral (but racemic)

Alkenes can be converted to diols with overall *anti* addition by a two-step sequence involving epoxidation and hydrolysis. Epoxidation of alkenes is a stereospecific *syn* addition that occurs as a single step. Ring opening of epoxides by hydrolysis also leads to diols. This is usually an *anti* addition with inversion of configuration at the site of nucleophilic attack, leading to overall anti dihydroxylation.

RCO_3H

anti addition

syn-isomer

A. RCO_3H

B. H_2O, H^+

racemic   racemic

Another example, the S_N2 substitution reaction results in an inversion of the configuration. It is a stereospecific reaction, also an enantiospecific reaction. The R-reactant gives the S-product and the S-reactant gives the R-product. So if we begin with an enantiopure reactant, we end up with enantiopure reactant.

$Bu_4N^+ AcO^-$

DMF

2.3.2 Asymmetric synthesis（不对称合成）

The objective of stereoselective synthesis is to produce compounds as pure diastereomers (diastereomerically pure) and/or pure enantiomers (enantiomerically pure). Enantioselective synthesis, also called chiral synthesis, asymmetric synthesis, is organic synthesis which introduces one or more new elements of chirality in a substrate molecule and which produces the stereoisomeric (enantiomeric or diastereoisomeric) products in unequal amounts. It is the synthesis of a compound by a method that favors the formation of a specific enantiomer or diastereomer. It is a particular case of stereoselective reactions.

Enantioselective synthesis is a key process in modern chemistry and is particularly important in the field of pharmaceuticals, as the different enantiomers or diastereomers of a molecule often have different biological activities.

The quality of enantioselective reactions is numerically expressed as the so-called *enantiomeric excess* (*ee*). It is equal to the yield of the major enantiomer minus the yield of the minor enantiomer in the product whose total yield is normalized to 100%.

$$ee = \frac{\text{major} - \text{minor}}{\text{major} + \text{minor}} \times 100\%$$

For example, in the Sharpless epoxidation of allyl alcohol, *S*- and *R*- glycidol are formed in a ratio of 19:1. For a total glycidol yield standardized to 100%, the *S*-glycidol fraction (95% yield) thus exceeds the *R*-glycidol fraction (5% yield) by 90%. Consequently, *S*-glycidol is produced with an *ee* of 90%.

tert-BuOOH

$Ti(OiPr)_4$

L-(+)-DET

95 : 5

i.e. *ee*= 90%

A racemic mixture has equal amounts of enantiomers, so the *ee* is zero. A pure enantiomer has *ee* of 100%.

In a reaction in which two diastereomeric products are possible, the percent *diastereomeric excess* is given by diastereomeric *de*:

$$de = \frac{|D_1 - D_2|}{|D_1 + D_2|} \times 100\% = |D_1\% - D_2\%|$$

where D_1 and D_2 are the mole fractions of the two diastereomeric products. If a reaction can produce more than two diastereomers, the ratio should itself be reported.

No reaction can produce an excess of one enantiomer unless there is at least one chiral component involved. That means no optically active material can be generated if all the starting material, reagents and conditionsare either achiral or racemic i.e. optically inactive. Stated in another way, if a chiral compound is synthesized from achiral or racemic reactants, reagents, and catalysts, then it will be formed as a racemate. Ultimately to generate non-racemic material (material which is optically active), it is necessary to utilize molecules from the chiral pool i.e. from the vast array of enantiopure and enantioenriched molecules which occur in nature. This may mean that one directly uses a substance from the chiral pool as a reagent, or maybe the substance is used as a catalyst to make a non-racemic reagent which is subsequently used for a different transformation. However, somewhere along the way a molecule from the chiral pool will have been utilized.

In principle, asymmetric synthesis involves the formation of a new stereogenic unit in the substrate under the influence of a chiral group. These methods can be divided into four major classes, depending on how this influence is exerted: ①chiral substrate-controlled methods; ② chiral auxiliary-controlled methods; ③ chiral reagent-controlled methods, and ④ chiral catalyst-controlled methods.

(1) Substrate-controlled asymmetric methods（底物控制的不对称合成方法）

The substrate-controlled reaction is often called the first generation of asymmetric synthesis. It is based on intramolecular contact with a stereogenic unit that already exists in the chiral substrate. Formation of the new stereogenic centre most occurs by reaction of the substrate with an achiral reagent at a diastereotopic site controlled by a nearby stereogenic unit.

$$S^* \longrightarrow P^*$$

In the following illustration, two hydride reductions of chiral methyl ketones are shown. Since each ketone has an existing stereogenic site, and since the reduction creates a new chiral center, diastereomeric products are possible. When the existing stereogenic center is located next to the carbonyl group, it may influence the proportion of product diastereomers to a significant degree. Because the new stereogenic centers are vicinal, this is termed as 1,2-diastereoselectivity, and is the same for both an enantiomerically pure or a racemic reactant. However, if the stereogenic center is far away from the carbonyl group, it has a negligible influence on the reduction, and a nearly 50:50 mixture of diastereomers is produced.

1-phenyl-2-butanone $\xrightarrow[\text{ether}]{LiAlH_4}$ 3-phenyl-2-butanol isomers: 30% *syn* {(*S*,*S*) and (*R*,*R*) diastereomer}; 70% *anti* {(*R*,*S*) and (*S*,*R*) diastereomer}

$\xrightarrow[\text{ether}]{LiAlH_4}$ 7-phenyl-2-octanol: 50:50 mixture of diastereomers

A number of models have been proposed to explain the *diastereoselectivity* of the first reduction. Three models will be considered here. For general reference, the substituents on the chiral center adjacent to the carbonyl group are labeled L (large), M (medium) and S (small), reflecting their approximate size. The original structure of these models assumed an orthogonal (90°) approach of the nucleophile to the plane of the carbonyl group, and a reactant-like transition state. In the following diagram this has been changed to the preferred Bürgi-Dunitz approach. Each model predicts the correct configuration of the favored diastereomer from $LiAlH_4$ reduction of 3-phenyl-2-butanone (upper equation above with L=C_6H_5, M=CH_3 and S=H). This diastereoselectivity is commonly termed as Cram or Felkin selectivity.

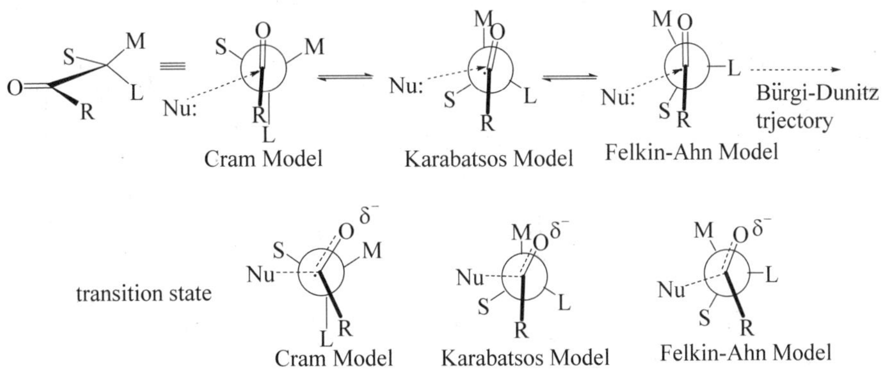

S M O L R Nu: Cram Model Karabatsos Model Felkin-Ahn Model Bürgi-Dunitz trjectory

transition state Cram Model Karabatsos Model Felkin-Ahn Model

D. J. Cram proposed the model shown above in 1952. It suffers from steric hindrance of substituent L with the R substituent on the carbonyl group as well as torsional strain. A more favored conformation was chosen by G. Karabatsos, as shown by the center model. Here the torsional strain is reduced, and the selectivity in reactions of phenylacetaldehydes bearing different α-substituents was rationalized more correctly than by Cram's model. The most recent model (on the right) is that proposed by H. Felkin and elaborated by N. T. Ahn. In this case, overlap of the carbonyl π*-orbital with the C—L σ*-orbital provides electronic stabilization. These models all require classification of S, M and L substituents, occasionally a tricky process, and assume a reactant-like transition state for the reduction.

(2) Chiral auxiliary-controlled asymmetric synthesis（手性辅助-控制的不对称合成方法）

The auxiliary-controlled reaction is referred to as the second generation of asymmetric synthesis. The same point to the first generation is that the asymmetric control is achieved intramolecularly by a chiral group in the substrate. The difference is that the directing group, the "chiral auxiliary", is deliberately attached to the original achiral substrate in order to direct the enantioselective reaction. If this group exerts a controlling influence, and if the diastereoselectivity of the reaction is excellent, the product should be obtained as a single enantiomer. A substituent serving this purpose is commonly called a chiral auxiliary. The chiral auxiliary group may then be removed, yielding the final enantiomerically pure product.

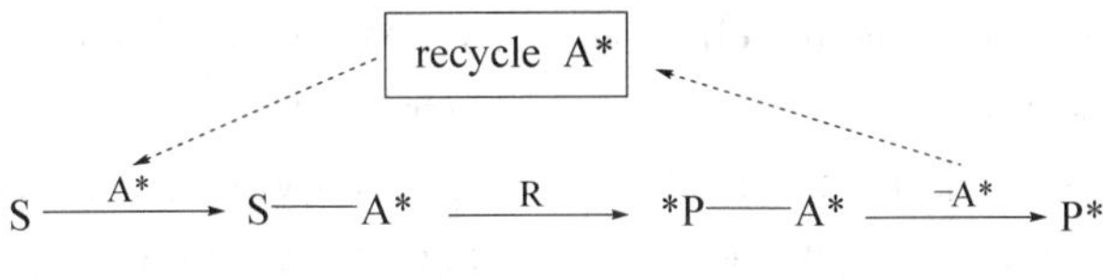

recycle A*

$$S \xrightarrow{A^*} S—A^* \xrightarrow{R} {}^*P—A^* \xrightarrow{-A^*} P^*$$

The third-generation method is an achiral substrate that is directly converted to the chiral product using a chiral reagent. The typical feature is that the sterocontrol happens intermolecularly.

Take famous Evans auxiliaries as examples. The oxazolidinone auxiliary is usually incorporated as an imide derivative, shown for propionic acid in the equation below. A *Z*-enolborinate is prepared in the usual way, and this reacts with a number of achiral aldehydes with very high enantioselectivity. Although the enolborinate by itself might be expected to exist in a chelated form, with two B—O bonds, the aldol reaction requires a reorganization of this chelation in order to activate the aldehyde carbonyl group for nucleophilic addition. As shown by the formula in brackets, the free oxazolidinone ring has rotated 180° from its chelated position in order to minimize dipole repulsion. Steric hindrance by the pendent isopropyl group directs the reaction to the 2*S*,3*R* product.

i-PrCHO

Z-enolate　　bonding to si-face of enolate　　*syn*(2*R*,3*S*), >99.5% *ee*

PhCHO

Z-enolate　　bonding to re-face of enolate　　*syn*(2*S*,3*R*), >99.5% *ee*

CH_3ONa, CH_3OH

Another example is the successful synthesis of frontalin, an aggregation pheromone of the Southern Pine Beetle—the most destructive beetle to pine forests in southeastern United States, utilizing the chiral auxiliary 8-phenylmenthol.

RMgBr, −78℃　　$LiAlH_4$　　O_3, −78℃

(−)-frontalin

(3) Chiral reagent-controlled asymmetric synthesis（手性试剂控制的不对称合成）

Above we have looked at the use of chiral substrates that either include all the necessary stereocenters or are used to control the introduction of new stereocenters. In the above-mentioned asymmetric synthesis, there is a drawback that stoichiometric amounts of enantiomerically pure compounds are required. Can we prepare chiral compounds if there is no chirality in the substrates? An important development in recent years has been the third method, and it involves the reaction of a substrate with a chiral reagent. Ideally we would like to control the introduction of chirality regardless of any already possessed by the substrate.

For example, the CBS catalyst or Corey-Bakshi-Shibata catalyst is an important asymmetric catalyst derived from proline. It finds many uses in organic reactions such as the CBS reduction, Diels-Alder reactions and [3+2] cycloadditions. Proline, a naturally occurring chiral compound, is readily and cheaply available. It transfers its stereocenter to the catalyst which in turn is able to drive an organic reaction enantioselectively to one of two possible enantiomers. This selectivity is due to steric strain in the transition state that develops for one enantiomer but not for the other. In the following reduction of ketone to alcohol, the amine lone pair of the CBS reagent fills the p-orbital of borane, and the resulting borane adduct is activated enough to make the reduction of ketones possible. Compared with the above two types of asymmetric methods, only catalytic amounts of CBS are required in these reactions.

Accordingly, this reagent has been extensively applied in natural product synthesis, for instance in the recently published total syntheses of (–)-laulimalide, brevetoxin-B, (+)-tanikolide, and bistramide A.

L-proline

S-(−)-CBS reagent

Borane adduct

S-(−)-CBS (0.1 eq)BH_3(1 eq)

99% yield
97%*ee*

R-(+)-CBS (0.05 eq)BH_3 (1 eq)

95% yield
94%*ee*

SmithKline Beecham blood pressure drug

(4) Chiral catalyst-controlled asymmetric synthesis（手性催化剂控制的不对称合成）

The recent most significant advance in asymmetric synthesis has been the application of chiral catalysts to induce the conversion of achiral substrates to chiral products. This catalytic asymmetric synthesis is that only small amounts of chiral catalysts are needed to generate large quantities of chiral products. The enormous economic potential of asymmetric catalysis has made it one of the most extensively explored areas of research in recent years.

Metal catalyzed enantioselective synthesis was pioneered by William S.Knowles, Ryōji Noyori and K. Barry Sharpless, for which they received the 2001 Nobel Prize in Chemistry. Knowles and Noyori began with the development of asymmetric hydrogenation, which they developed independently in 1968. Knowles replaced the achiral triphenylphosphine ligands in Wilkinson's catalyst with chiral phosphine ligands. This experimental catalyst was employed in an asymmetric hydrogenation with a modest 15% enantiomeric excess. Knowles was also the first to apply enantioselective metal catalysis to industrial-scale synthesis; while working for the Monsanto

Company he developed an enantioselective hydrogenation step for the production of L-DOPA, utilizing the DIPAMP ligand.

Ph, CH$_2$, O, OH; $RhCl_2L_2$(0.15mol%), H_2; Ph, *, O, OH, Me, H; L=:P Ph, Me, i-Pr

Noyori devised a copper complex using a chiral Schiff base ligand, which he used for the metal-carbenoid cyclopropanation of styrene. In common with Knowles' findings, Noyori's results for the enantiomeric excess for this first-generation ligand were disappointingly low: 6%. However continued research eventually led to the development of the Noyori asymmetric hydrogenation reaction.

1 eq N_2 COOEt, 60℃; COOMe + Ph, N, O, Cu, O, N, Ph 0.01 eq; COOMe

一种反应物能够生成两种以上的立体异构体时，其中一种立体异构体生成较多的反应为立体选择性反应。一定立体异构的原料在某种条件下只取得一定构型产物的反应称立体专一性反应。由于作用物、试剂、催化剂、偏光等因素产生的不对称反应环境，得到不等量的对映体或非对映体产物的反应为不对称合成。一个成功的不对称合成的标准是：高的 *ee* 值（对映体过量百分率）；手性试剂易得，最好可以循环使用；*R* 与 *S* 异构体都可以分别制得；最好是催化性的合成。

对映体过量百分率 (*ee*)或光学纯度百分率 (O.P.)

$$ee = \frac{[R]-[S]}{[R]+[S]} \times 100\% = R\% - S\%$$

$$\text{O.P.}=([\alpha]_{测}/[\alpha]_{纯})\times 100\%=ee$$

非对映体过量百分率 (*de*):一个非对映体对另一个非对映体的过量。

$$de = \frac{[S,S]-[S,R]}{[S,S]+[S,R]} \times 100\%$$

进行不对称合成常见的方法有：①手性底物控制的不对称合成；②手性源反应物经不对称反应得到新的不对称产物；③手性辅助试剂的控制的不对称合成，借助手性辅助试剂与反应底物作用生成手性中间体，经不对称反应得到新的不对称产物；④手性试剂的不对称反应。当底物与手性试剂反应时，得到了新的不对称产物。当一对对映体与手性试剂反应时，反应速度不相等。一对对映体与非手性试剂反应时，反应速率完全相等；⑤利用不对称催化的不对称合成。利用底物经不对称反应时加入少量手性催化剂，使之与底物或试剂形成高反应活性的中间体催化剂作为模板控制反应物的对映面，不对称反应得到新的不对称产物。手性催化剂可循环使用，具有手性放大效应。不对称催化反应是产生大量手性化合物最经济和实用的技术。

Problems

1. Give the R or S for the chiral atoms of the following compounds.

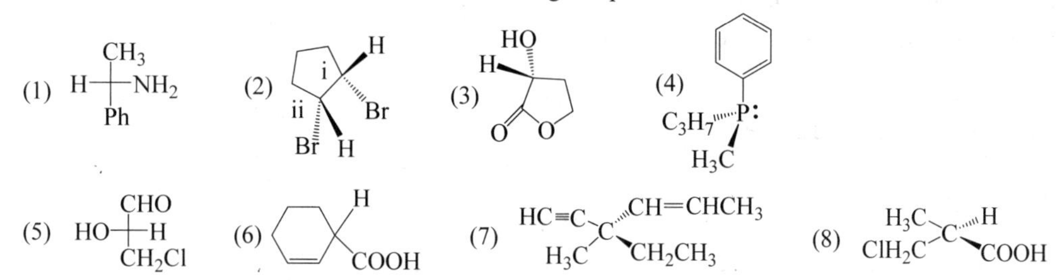

2. Judge if those compounds are chiral or not.

(1) 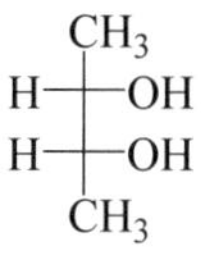 CH_3; H—OH; H—OH; CH_3

(2) CH_3; H—OH; HO—H; CH_3

(3) COOH; H—OH; H—OH; CH_2OH

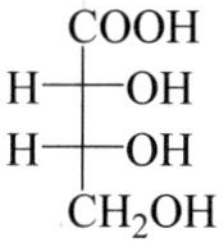

(4) 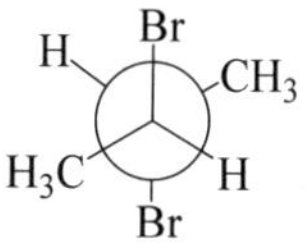 Br, H, CH_3, H_3C, H, Br

(5) 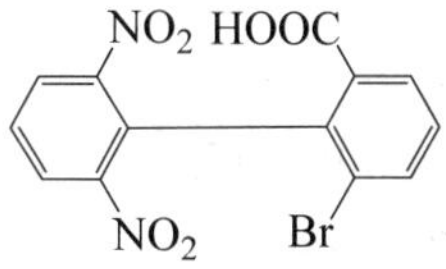 NO_2 HOOC; NO_2 Br

(6) 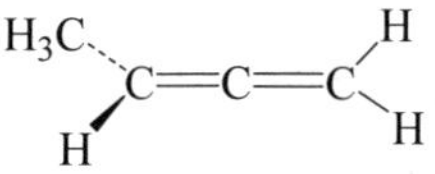 H_3C, H, C=C=C, H, H

(7) 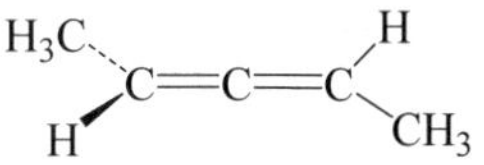 H_3C, H, C=C=C, H, CH_3

(8) 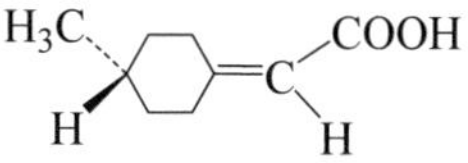 H_3C, H, =C, COOH, H

(9) 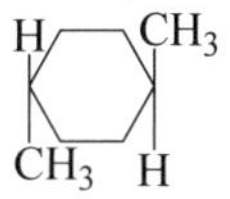 H, CH_3, CH_3, H

(10) 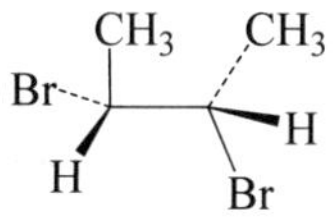 CH_3 CH_3, Br, H, H, Br

3. Write down the products of the following reactions.

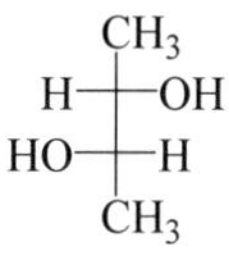

(1) 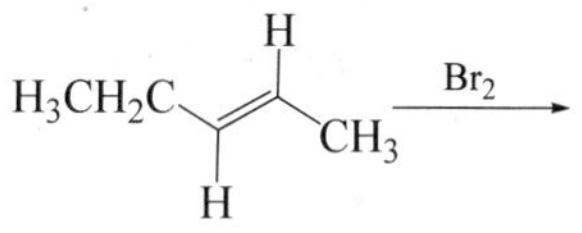 H_3CH_2C, H, H, CH_3 $\xrightarrow{Br_2}$

(2) O $\xrightarrow{LiAlH_4}$

(3) O $\xrightarrow{LiBH(s\text{-}Bu)_3}$

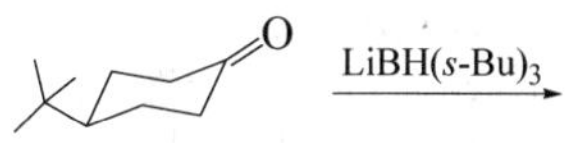

(4) 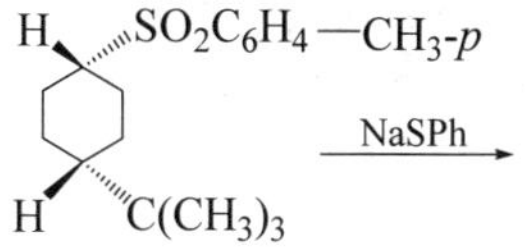 H, $SO_2C_6H_4—CH_3$-*p*; H, $C(CH_3)_3$ $\xrightarrow{NaSPh}$

(5) 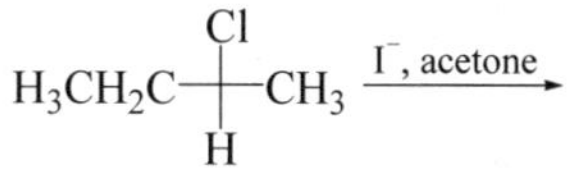 H_3CH_2C—C(Cl)(H)—CH_3 $\xrightarrow{I^-,\ acetone}$

(6) 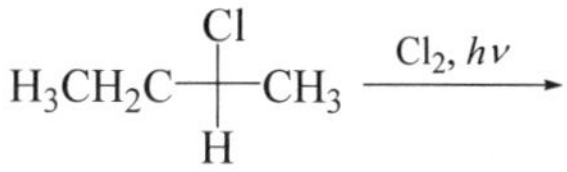 H_3CH_2C—C(Cl)(H)—CH_3 $\xrightarrow{Cl_2,\ h\nu}$

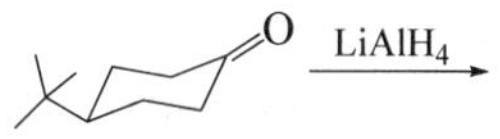

4. Write formulas for all of the isomers of each of the following. Designate pairs of enantiomers and achiral compounds where they exist.

(1) 1-Bromo-2-chlorocyclohexane

(2) 1-Bromo-4-chlorocyclohexane

5. Explain the following stereoselective reactions.

(1) 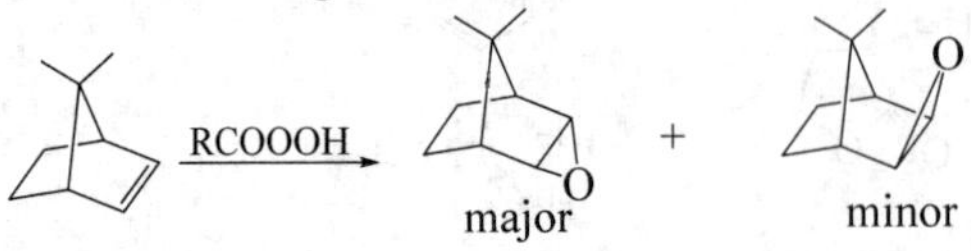 RCOOOH → O major + O minor

(2) 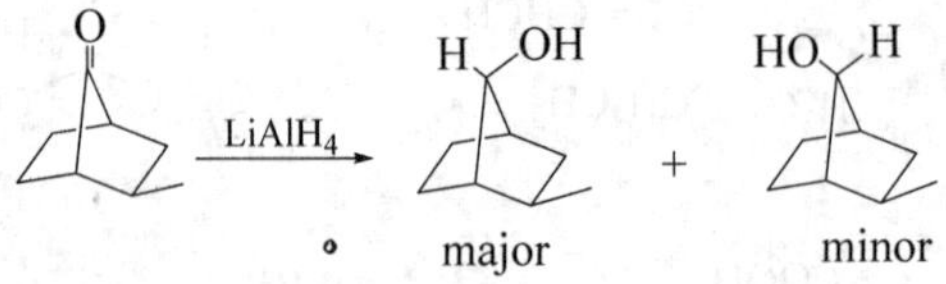 O LiAlH$_4$ → H OH major + HO H minor

Vocabulary

isomer ['aisəmər] 同分异构体
constitutional [ˌkɒnstə'tjuːʃənəl] 构造
stereoisomer [ˌsterɪoʊ'aɪsəmə] 立体异构体
conformational [ˌkɒnfɔː'meɪʃənəl] 构象
enantiomer [ɪ'næntiːəmə] 对映体
diastereomer [daɪə'stɪərɪoʊmə] 非对映体
stereogenic center 不对称中心
alternating axis of symmetry 交替对称轴
chirality [kaɪ'rælətɪ]手性
superimpose [ˌsuːpərɪm'poʊz] 重叠
achiral [ə'kaɪərəl] 非手性的
epinephrine [ˌepə'nefrɪn] 肾上腺素
meso ['mesoʊ] 中间，中等
allene ['æliːn] 丙二烯
hexahelicene [heksəhelɪ'siːn] 六螺烯
racemate ['ræsəmeɪt] 外消旋体
epimer['epɪmə] 差向异构体
saccharide ['sækəˌraɪd] 糖类；糖化物
pentose ['pentoʊs] 戊糖
hexose ['heksoʊs] 己糖
anomer [æ'nəʊmər] 异头物
glucose ['gluːkoʊs] 葡萄糖
mannose ['mænˌoʊs] 甘露糖
stereomutation [stɪərɪəmjʊ'teɪʃən] 立体改变
chromatography [ˌkroʊmə'tɑːgrəfi] 色谱法
carbamate ['kɑːbəˌmeɪt] 氨基甲酸酯
aroyl [æ'rɔɪl] 芳香烃酰基
esterase ['estəˌreɪs] 酯酶
lipase ['laɪpeɪz] 脂肪分解酵素
acylase ['æsɪleɪs] 酰基转移酶
amidase [æ'maɪdeɪz] 酰胺酶
hydrolase ['haɪdrəleɪs] 水解酶
prochiral [prəʊt'ʃaɪərəl] 前手性的
enantioselectivity [enəntiːəʊsɪlek'tɪvɪtɪ] 对映选择性
diastereoselectivity [diəsteriːəʊsɪlek'tɪvɪtɪ] 非对映立体选择性
regioselectivity [riːdʒiːəʊsɪlek'tɪvɪtɪ] 区域选择性
stereoselectivity[stɪərɪəsɪlek'tɪvɪtɪ]立体选择性
chemoselectivity [ʃemɒsɪlek'tɪvɪtɪ] 化学选择性
stereospecific reaction 立体专一性的反应
asymmetric synthesis 不对称合成
enantiomeric excess 对映体过量
diastereomeric excess 非对映体过量

Chapter 3

Mechanisms of Organic Reactions
（有机反应机理）

Mechanisms of organic reactions are the integration of many elementary reaction in the process of the reaction. It is a specific pathway through which reactants are converted into products. Sometimes we also call it reaction path. The analytical process of organic reaction mechanism is one by one analysis of elementary reactions. Every elementary reaction must conform to the objective law of electronic motion. The conclusion of mechanism analysis must be consistent with the experimental results of organic reaction.

3.1 Types of reaction mechanisms（反应机理的类型）

3.1.1 Classification by structural change（基于结构改变的分类）

According to structural changes between reactants and products, there are four types of organic reactions: additions, eliminations, substitutions, and rearrangements.

When two reactants add together to form a single new product with no atoms "left over", *addition reactions* occur. These reactions can take place by three mechanistic possibilities.

Electrophilic addition (heterolytic) $A{=}B + Y{-}W \longrightarrow W{-}A{-}B{-}Y$

$$HCl + H_2C{=}CH_2 \longrightarrow CH_3CH_2Cl$$

Nucleophilic addition (heterolytic) $A{=}B + Y{-}W \longrightarrow W{-}A{-}B{-}Y$

$$CH_3{-}C({=}O){-}H + HCN \xrightarrow{OH^-} CH_3{-}C(OH)(CN){-}H$$

Free-radical addition (homolytic) $A{=}B + Y{-}W \xrightarrow{-W\cdot} \dot{A}{-}B{-}Y \longrightarrow W{-}A{-}B{-}Y$

$$H_3C{-}CH{=}CH_2 + HBr \xrightarrow[\text{peroxide}]{} H_3C{-}\dot{C}H{-}CH_2Br \xrightarrow{HBr} H_3C{-}CH_2{-}CH_2Br + Br\cdot$$

The reactions in which a small molecule is removed from adjacent carbon atoms resulting in the formation of additional (multiple) bond between them are called *elimination* reactions. These reactions can take place by either heterolytic or *pericyclic* mechanisms. Free-radical β eliminations are extremely rare. In heterolytic eliminations W and Y may or may not leave simultaneously and may or may not combine.

$$W-A-B-Y \longrightarrow A=B + WY$$

$$H_3C-CH_2Cl \xrightarrow{NaOH} H_2C=CH_2 + HCl$$

The reactions in which an atom or group of atoms in a molecule is replaced or substituted by different atoms or group of atoms are called substitution reaction. *Substitution reactions* can be classified as nucleophilic, electrophilic or free-radical substitution.

Nucleophilic substitution $A-X + Y^- \longrightarrow A-Y + X^-$

$$CH_3CH_2-Cl + OH^- \longrightarrow CH_3CH_2-OH + Cl^-$$

Electrophilic substitution $A-X + Y^+ \longrightarrow A-Y + X^+$

$$C_6H_6 + NO_2^+ \longrightarrow C_6H_5-NO_2 + H^+$$

Free-radical substitution $A-X + Y\cdot \longrightarrow A-Y + X\cdot$

$$C_6H_5CH_3 + Cl\cdot \longrightarrow C_6H_5CH_2Cl + H\cdot$$

When a single reactant undergoes a reorganization of bonds and atoms to yield a single *isomeric* product, *rearrangement reactions* occur. There are three types, depending on how many electrons the *migrating* atom or group carries with it.

Migration with electron pair (nucleophilic) $W-A-B \longrightarrow A-B-W$

$$H_3C-C(CH_3)_2-\overset{+}{C}H_2 \longrightarrow H_3C-\overset{+}{C}(CH_3)-CH_2-CH_3$$

Migration with one electron (free-radical) $W-A-\dot{B} \longrightarrow \dot{A}-B-W$

$$H_3C-CH(CH_3)-\dot{C}H_2 \longrightarrow H_3C-\dot{C}(CH_3)-CH_2-H$$

Migration without electrons (electrophilic; rare) $W-A-\ddot{B} \longrightarrow \ddot{A}-B-W$

$$H_3C-CH(CH_2)-CH_3 \text{ (cyclic)} \longrightarrow H_3C-\bar{C}H_2-CH_3 \text{ (cyclic)}$$

Some common reactions may actually be a combination of reaction types. The reaction of an ester with ammonia to give an amide, as shown below, appears to be a substitution reaction. However, it is actually two reactions, an addition followed by an elimination.

$$R-\overset{\displaystyle O}{\overset{\|}{C}}-O-CH_3 \xrightarrow{NH_3} \left[R-\overset{\displaystyle OH}{\underset{\displaystyle NH_2}{\overset{|}{\underset{|}{C}}}}-O-CH_3 \right] \xrightarrow{-CH_3OH} R-\overset{\displaystyle O}{\overset{\|}{C}}-NH_2$$

The addition of water to a *nitrile* does not seem to fit any of the above reaction types, but it is a slow addition reaction followed by a rapid rearrangement, which is called *tautomerization*.

$$R-C\equiv N \xrightarrow{H_2O} \left[R-C(=N-H)(O-H) \right] \longrightarrow R-\overset{\displaystyle O}{\overset{\|}{C}}-NH_2$$

3.1.2 Classification by reaction type（基于反应类型的分类）

(1) Lewis acid-base reactions（路易斯酸碱反应）

According to the Lewis theory, an acid is an electron pair acceptor, and a base is an electron pair donor. The product of a Lewis acid-base reaction is a neutral, *dipolar* or charged complex, which may be a stable covalent molecule. As shown at the following drawing, coordinate covalent bonding of a phosphorous Lewis base to a boron Lewis acid creates a complex in which the formal charge of boron is negative and that of phosphorous is positive. In this complex, boron acquires a neon valence shell *configuration* and phosphorous an argon configuration. If the substituents (R) on these atoms are not large, the complex will be favored at equilibrium. However, steric hindrance of bulky substituents may prohibit complex formation.

$$BR_3 + :PR_3 \rightleftharpoons R_3B{-}PR_3$$

Lewis Acid　Lewis Base　Complex Formation

$$H_3C-\underset{\displaystyle CH_3}{\overset{\displaystyle CH_3}{\underset{|}{\overset{|}{C}}}}-Cl + AlCl_3 \rightleftharpoons H_3C-\underset{\displaystyle CH_3}{\overset{\displaystyle CH_3}{\underset{|}{\overset{|}{C^+}}}} + AlCl_4^-$$

Lewis Base　Lewis Acid

Many *carbocations* may also function as Brønsted acids. The following equation illustrates this dual behavior. In its Brønsted acid role the carbocation donates a proton to the base (hydroxide anion), and is converted to a stable neutral molecule having a C=C double bond.

$$H_3C-\overset{+}{C}(CH_3)-CH_3 + OH^- \begin{cases} \longrightarrow H_3C-C(=CH_2)-CH_3 + H_2O \\ \longrightarrow H_3C-C(CH_3)_2-OH \end{cases}$$

The term *electrophile* corresponds to a Lewis acid, and *nucleophile* corresponds to a Lewis base.

Electrophile: An electron deficient atom, ion or molecule that has an affinity for an electron pair, and will bond to a base or nucleophile, such as H^+, H_3O^+, NO_2^+, R_3C^+, ArN_2^+, HX (X=Cl, Br, I), X_2 (X=Cl, Br, I), CH_3X, $(CH_3)_2SO_4$, RCHO, RCOR, RCN, BH_3, BF_3, $AlCl_3$, FeF_3, etc.

Nucleophile: An atom, ion or molecule that has an electron pair that may be donated in bonding to an electrophile (or Lewis acid), such as HO^-, RO^-, Cl^-, Br^-, I^-, CN^-, H^-, $RCOO^-$, RS^-, RSO_2^-, $S_2O_3^{2-}$, NH_2^-, NO_2^-, NaN_3, NCO^-, NCS^-, RLi, RMgX, NaC≡CH, RCuLi, NH_3, RNH_2, R_2NH, R_3N, H_2O, ROH, etc.

(2) Oxidation and reduction reactions（氧化还原反应）

To determine whether a carbon atom has undergone a *redox* change during a reaction, we can simply note any changes in the number of bonds to hydrogen and the number of bonds to more *electronegative* atoms such as O, N, F, Cl, Br, I and S that have occurred. Bonds to other carbon atoms are ignored. This count should be conducted for each carbon atom undergoing any change during a reaction.

If the number of hydrogen atoms bonded to a carbon increases, and/or if the number of bonds to more electronegative atoms decreases, the carbon atom has been reduced (i.e. it is in a lower oxidation state).

If the number of hydrogen atoms bonded to a carbon decreases, and/or if the number of bonds to more electronegative atoms increases, the carbon atom has been oxidized (i.e. it is in a higher oxidation state).

If there has been no change in the number of such bonds, then the carbon has not changed its oxidation state.

These rules are illustrated by the following four addition reactions involving the same starting material, cyclohexene. In the addition of hydrogen both carbon atoms are reduced, and the overall reaction is a reduction reaction. *Peracid* epoxidation and addition of bromine oxidize both carbon atoms, so these are oxidation reactions. Addition of HBr reduces one of the double bond carbon atoms and oxidizes the other; consequently, there is no overall redox change in the substrate molecule.

Since metals such as lithium and magnesium are less electronegative than hydrogen, their covalent bonds to carbon are polarized so that the carbon is negative (reduced) and the metal is positive (oxidized). Thus, Grignard reagent formation from an alkyl halide reduces the substituted carbon atom. In the following equation and half-reactions, the carbon atom is reduced and the magnesium is oxidized.

$$H_3C{-}Br + 2e^- \longrightarrow H_3C{:}^- + Br^-$$
$$Mg - 2e^- \longrightarrow Mg^{2+}$$
$$\overline{H_3C{-}Br + Mg \longrightarrow H_3C{-}Mg{-}Br}$$

3.1.3 Classification by *functional group*（基于官能团的分类）

Functional groups are atoms or small groups of atoms (usually two to four) that exhibit a characteristic reactivity when treated with certain reagents. A particular functional group will almost always display its characteristic chemical behavior when it is present in a compound. Because of this, the discussion of organic reactions is often organized according to functional groups. Table 3.1 summarizes the general chemical behaviors of the common functional groups.

Table 3.1 The characteristic reactions by functional groups

Functional Class	Formula	Characteristic Reactions
Alkanes	C—C,C—H	Substitution (of H, commonly by Cl or Br) Combustion (conversion to CO_2 & H_2O)
Alkenes	C═C—C—H	Addition Substitution (of α-H)
Alkynes	C≡C—H	Addition Substitution (of H)
Alkyl Halides	H—C—C—X	Substitution (of X) Elimination (of HX)
Alcohols	H—C—C—O—H	Substitution (of H); Substitution (of OH) Elimination (of H_2O); Oxidation (elimination of α-H)
Ethers	(α)C—O—R	Substitution (of OR); Substitution (of α-H)
Amines	C—NRH	Substitution (of H); Addition (to N); Oxidation (of N)
Benzene Ring	C_6H_6	Substitution (of H)
Aldehydes	(α)C—CH═O	Addition; Oxidation Substitution (of α-H)
Ketones	(α)C—CR═O	Addition Substitution (of α-H)
Carboxylic Acids	(α)C—CO_2H	Substitution (of H); Substitution (of OH) Substitution (of α-H); Addition (to C═O)
Carboxylic Derivatives	(α)C—CZ═O (Z=OR, Cl, NHR, etc.)	Substitution (of Z); Substitution (of α-H) Addition (to C═O)

按反应物与产物之间的结构改变可把有机反应分为四大类：加成反应、消除反应、取代反应、重排反应。其中加成反应包括亲电型、亲核型、自由基型加成反应，反应物不饱和度减少。取代反应包括亲电型、亲核型、自由基型取代反应，反应物的不饱和度不发生变化。消除反应中反应物不饱和度减小，分为离子消除、协同消除或α-消除、β-消除。重排反应中碳骨架发生变化，分子的不饱和度不变，有离子重排、自由基重排和协同重排。

按反应类型可把有机反应分为路易斯酸碱反应、氧化还原反应。

按化学键断裂和形成方式可把有机反应分为离子反应（异裂历程）、自由基反应（均裂反应）、分子反应（协同反应）。共价键发生异裂形成了正负离子，有离子参与的反应叫离子反应。共价键发生均裂形成两个自由基，如烯的反马氏加成即过氧化反应是自由基反应（均裂反应）。共价键的断裂与形成是同时（协同）进行的，反应一步完成，叫协同反应。经过一个环状过渡态，一步形成产物，过程无任何中间体的反应叫周环反应。S_N2、E2、Diels-Alder均叫协同反应，但只有 Diel-Alder 反应叫周环反应。

3.2 The properties and characteristics of organic reactions（有机反应的性质和特点）

It is important to consider the various properties and characteristics of a reaction that may be observed and/or measured as the reaction proceeds.

3.2.1 Reactants and reagents（反应物与试剂）

(1) Reactant structure（反应物结构）

Variations in the structures of reactants may have a marked influence on the course of a reaction, even though the functional group is unchanged. For example, reaction of 1-bromopropane with *sodium cyanide* proceeds smoothly to yield *butanenitrile*, whereas 1-bromo-2,2-dimethylpropane fails to give any product and the starting material is recovered.

$$CH_3CH_2CH_2—Br + CN^- \xrightarrow{\text{alcohol}} CH_3CH_2CH_2—CN + Br^-$$

$$(CH_3)_3CCH_2—Br + CN^- \xrightarrow{\text{alcohol}} \text{No Reaction}$$

(2) Reagent characteristics（试剂特点）

Apparently minor changes in a reagent may lead to a significant change in the course of a reaction. For example, 2-bromopropane gives a substitution reaction with *sodium methylthiolate* but undergoes predominant elimination on treatment with sodium methoxide.

$$(H_3C)(H)C{=}CH_2 + CH_3OH + Br^- \xleftarrow{CH_3—O^-} (H_3C)_2CH—Br \xrightarrow{CH_3—S^-} (H_3C)_2CH—SCH_3 + Br^-$$

3.2.2 Product selectivity（产物的选择性）

(1) *Regioselectivity*（区域选择性）

It is often the case that addition and elimination reactions may result in more than one product. Thus 1-butene might add HBr to give either 1-bromobutane or 2-bromobutane, depending on which carbon of the double bond receives the hydrogen and which carbon of the double bond receives the bromine. If one possible product out of two or more is formed preferentially, the reaction is said to be regioselective.

$$CH_2{=}CH—CH_2—CH_3 + HBr \longrightarrow \underset{\text{minor}}{CH_2Br—CH_2—CH_2—CH_3} + \underset{\text{major}}{CH_3—CHBr—CH_2—CH_3}$$

(2) *Stereoselectivity*（立体选择性）

If the reaction products are such that *stereoisomers* may be formed, a reaction that yields one stereoisomer preferentially is said to be stereoselective.

$$CH_3CHBrCH_2CH_3 \xrightarrow[\triangle]{KOH\text{-}C_2H_5OH} \underset{\text{major}}{(H_3C)(H)C{=}C(H)(CH_3)} + \underset{\text{minor}}{(H_3C)(H)C{=}C(H)(CH_3)} + \underset{\text{minor}}{H_3CH_2CHC{=}CH_2}$$

The major product for the elimination reaction of 2-bromobutane under KOH/EtOH is

(*E*)-but-2-ene. (*Z*)-but-2-ene and 1-butene are minor products of this reaction.

(3) *Stereospecificity*（立体专一性）

This term is applied to cases in which stereoisomeric reactants behave differently in a given reaction. For example, in the addition of bromine to cyclohexene, *cis* and *trans*-1,2-dibromocyclohexanes are both possible products of the addition. Since the *trans*-isomer is the only isolated product, this reaction is stereospecific.

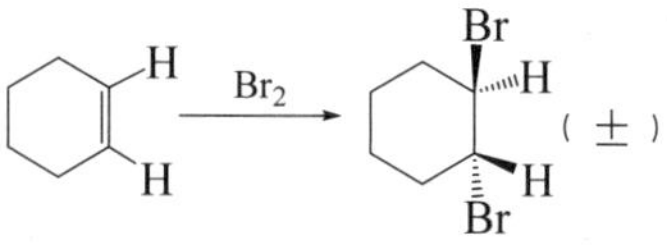

1) Formation of different stereoisomeric products is shown in the reaction of enantiomeric 2-bromobutane isomers with sodium methylthiolate.

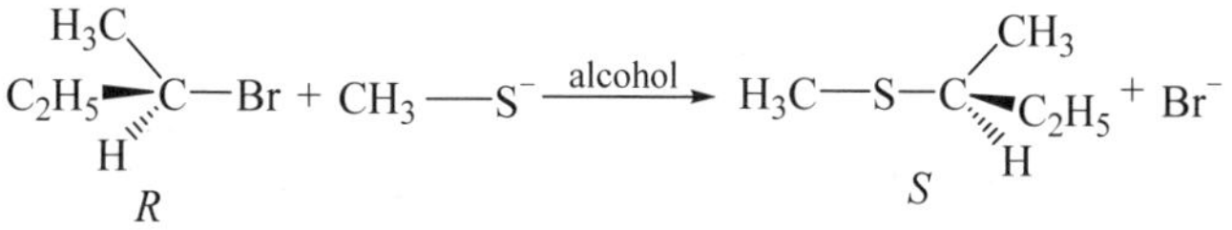

Here, the (*R*)-reactant gives the configurationally inverted (*S*)-product, and (*S*)-reactant produces (*R*)-product.

2) Different rates of reactions are shown in the base-induced elimination of *cis* and *trans*-4-tert-butyl cyclohexyl bromide.

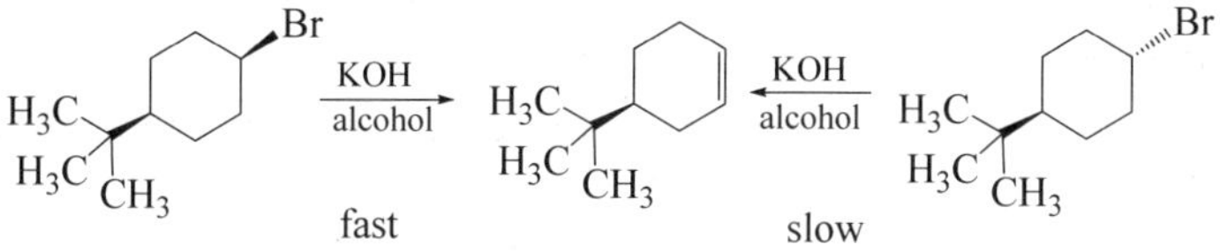

3) Different reaction paths lead to different products, as in the base-induced elimination of *cis* and *trans*-2-methylcyclohexyl bromide.

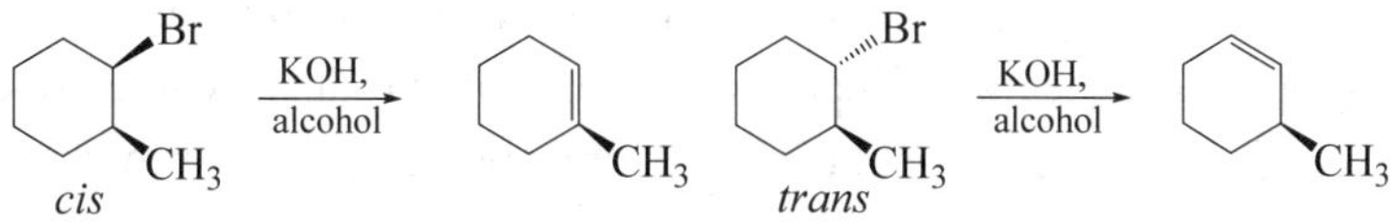

3.2.3 Reaction rates and *thermodynamic stability*（反应速率与热力学稳定性）

Product composition at the end of the reaction may be governed by thermodynamic control and the stability differences among the competing products. Alternatively, product composition may be governed by competing rates of formation of products, which is called kinetic control.

Let us consider cases (a) to (c) in Fig 3.1. In case (a), the $\Delta Gs^{\neq}$ for formation of the competing transition states $A^{\neq}$ and $B^{\neq}$ from the reactant R are substantially less than the $\Delta Gs^{\neq}$ for formation of $A^{\neq}$ and $B^{\neq}$ from A and B, respectively. If the latter two $\Delta Gs^{\neq}$ are sufficiently large that the competitively formed products B and A do not return to R, the ratio of the products A to B will not depend on their relative stabilities, but on the relative rate of formation. This is a case of *kinetic control*. The relative amounts of products A and B depend on the *relative activation barriers* $\Delta G^{\neq}_{A}$ and $\Delta G^{\neq}_{B}$ and not on the relative stability of products A and B. The diagram shows $\Delta G^{\neq}_{B} < \Delta G^{\neq}_{A}$, so the major product will be B, even though it is less stable than A.

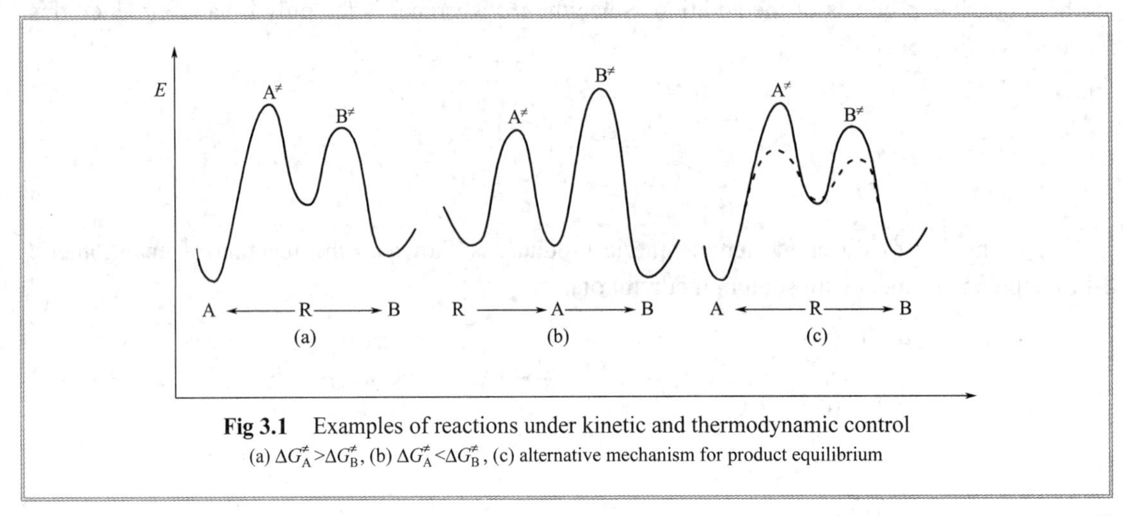

Fig 3.1 Examples of reactions under kinetic and thermodynamic control
(a) $\Delta G^{\neq}_{A} > \Delta G^{\neq}_{B}$, (b) $\Delta G^{\neq}_{A} < \Delta G^{\neq}_{B}$, (c) alternative mechanism for product equilibrium

Case (b) represents a situation of two successive reactions. The $\Delta G^{\neq}$ for formation of the $B^{\neq}$ from A is slightly larger than that for formation of $A^{\neq}$ from R. The reaction might be governed by either kinetic or thermodynamic factors. Conversion of R to A will be slightly faster than conversion of A to B. If the reaction conditions are carefully adjusted it will be possible for A to accumulate and not proceed to the more stable product B. Under such conditions, A will be the dominant product and the reaction will be under kinetic control. If under more energetic conditions, for example, at a higher temperature, A will be transformed into B. Then the reaction will be under thermodynamic control. A and B will equilibrate and the product ratio will depend on the *equilibrium constant* determined by ΔG for the reaction $A \rightleftharpoons B$.

In case (c), the solid energy profile represents the same situation of kinetic control as shown in (a), with product B being formed because $\Delta G^{\neq}_{A} > \Delta G^{\neq}_{B}$. The dashed energy profile represents a different of conditions for the same transformation, such as addition of catalyst or change of solvent that substantially reduces the energy of $A^{\neq}$ and $B^{\neq}$ such that interconversion of A and B is fast. This will result in formation of the more stable product A, even though the barrier to formation of B remains lower. Under these circumstances, the reaction is under thermodynamic control.

Therefore, whenever competing or successive reaction products come to equilibrium, the product compositions reflect relative stability and are subjected to thermodynamic control. If product composition is governed by competing rates, the reaction is under kinetic control. A given reaction may be subject to either thermodynamic or kinetic control, depending on the conditions.

Kinetic versus thermodynamic control can be illustrated by the formation of *enolate anions* from unsymmetrical ketones. Most ketones can give rise to more than one enolate. The ratio among the possible enolates that are formed depends on the reaction conditions. This can be illustrated for the case of 2-hexanone (**1**). If the base is a strong, sterically hindered one, such as *lithium diisopropylamide* (LDA), and the solvent is *aprotic*, the major enolate formed is **3** in the diagram below. If a protic solvent or a weaker base is used, the dominant enolate is **2**. Under the latter conditions, equilibration can occur by reversible formation of the enol. Enolate **3** is the kinetic enolate, but enolate **2** is thermodynamically favored.

$$CH_3-\overset{O}{\overset{\|}{C}}-(CH_2)_3CH_3 \; (\mathbf{1})$$

$$\underset{\text{thermodynamic enolate } \mathbf{2}}{CH_3-\overset{O^-}{\overset{|}{C}}=CH(CH_2)_2CH_3} \xleftarrow{\text{weak base}} \mathbf{1} \xrightarrow{\text{LDA}} \underset{\text{kinetic enolate } \mathbf{3}}{H_2C=\overset{O^-}{\overset{|}{C}}-(CH_2)_3CH_3}$$

The α-hydrogen of the methyl group are less sterically hindered than the α-hydrogen of the butyl group in compound **1**. As a result, removal of one methyl hydrogen as a proton is faster than removal of one butyl hydrogen. This effect is magnified when the base is sterically bulky and is particularly strong, the enolate will not be reconverted to the ketone because the enolate is too weak a base to regain the proton. These conditions correspond to Figure 3.1 (a) and represent a case of kinetic control. If a weaker base is used or if the solvent is protic, protons can be transferred reversibly between the isomeric enolates and the base (because the base strengths of the enolate and the base are comparable). Under these conditions the more stable enolate will be dominant because the enolates are in equilibrium. The more substituted enolate **2** is the more stable one, just as more substituted alkenes are more stable than terminal alkenes. This corresponds to case (c) in Figure 3.1, and product (enolate) equilibration occurs through rapid proton exchange. In protic solvents this exchange can occur through the enols.

反应物与试剂：官能团与进攻试剂相同，但反应物结构不同，反应机理就可能不同。不同的进攻试剂，相同的反应物结构，反应机理也可能不同。

产物的选择性：包括区域选择性、立体选择性与立体专一性。

反应速率与热力学稳定性：反应的产物可能受热力学控制或者动力学控制，反应物 A 在一定条件下按相反方向转变成两种产物 B 和 C，速率常数 $k_B<k_C$（形成 C 快于 B）；平衡常数 $K_B>K_C$（B 比 C 更稳定）；产物的比例由反应速率控制的过程叫动力学控制或速率控制反应。产物的比例由其相对热力学稳定性来控制的反应叫热力学控制或平衡控制的反应。

3.3 How do organic reactions occur: mechanisms（有机反应如何发生：机理）

An overall description of how a reaction occurs is called a reaction mechanism. A mechanism describes in detail exactly what takes place at each stage of a chemical transformation—which bonds are broken and in what order, which bonds are formed and in what order, and what the relative rates of the steps are.

We can divide organic mechanisms into three basic types, depending on how the bonds break. If a bond breaks in such a way that both electrons remain with one fragment, the mechanism is called *heterolytic*. If a bond breaks in such a way that each fragment gets one electron, free radicals are formed and such reactions are said to take place by *homolytic* or free-radical mechanisms. There is a third type of mechanism in which electrons (usually six, but sometimes some other number) move in a closed ring. There are no intermediates, ions or free radicals, and it is impossible to say whether the electrons are paired or unpaired. Reactions with this type of mechanism are called pericyclic.

Homolytic bond breaking: A: B ⟶ A· + B· (radical)

Heterolytic bond breaking: A : B ⟶ A^+ + :B^- (polar)

Pericyclic bond breaking:

Conversely, a covalent two-electron bond can form in two ways: A bond can form in an electronically symmetrical, homogenic way when one electron is donated to the new bond by each reactant, or a bond can form in an electronically unsymmetrical, heterogenic way when both bonding electrons are donated to the new bond by one reactant.

Homogenic bond making: A· + ·B ⟶ A:B

Heterogenic bond making: A^+ + :B^- ⟶ A:B

Pericyclic bond making:

Processes that involve symmetrical bond breaking and making are called *radical reactions*. Processes that involve unsymmetrical bond breaking and making are called polar reactions. Polar processes are the most common reaction type in organic chemistry. Certain bonds in a molecule, particularly the bonds in functional groups are often polar bonds. When carbon bonds to a more electronegative atom such as chlorine or oxygen, the bond is polarized so that the carbon bears a partial positive charge (δ^+), and the electronegative atom bears a partial negative charge (δ^-). When carbon bonds to a less electronegative atom such as a metal, the opposite polarity results in.

δ^- Cl, δ^+ C, H, H, H δ^+ Li, δ^- C, H, H, H

What effect does bond polarity have on chemical reaction? The fundamental characteristic of all *polar reactions* is that electron-rich sites in one molecule react with electron-poor sites in another molecule. Bonds are made when an electron-rich atom donates a pair of electrons to an electron-poor atom, and bonds are broken when one atom or group leaves with both electrons from the former bond.

Chemists normally indicate the electron movement that occurs during a polar reaction by using curved arrows. A curved arrow means that an electron pair moves from the atom (or bond) at the tail of the arrow to the atom or bond at the head of the arrow.

A^+ + :B^- ⟶ A—B

Electrophile Nucleophile

3.3.1 Radical reactions（自由基反应）

A free-radical process consists of at least two steps. The first step involves the formation of free radicals, usually by homolytic cleavage of bond:

A—B ⟶ A • + B •

This is called *an initiation step*. It may happen spontaneously or may be induced by heat or light, depending on the type of bond. Peroxides, including hydrogen peroxide, dialkyl, diacyl, alkyl acyl peroxides, and *peroxyacids* are the most common source of free radicals. Organic compounds with low-energy bonds, such as *azo* compounds, are also used. Molecules that are cleaved by light are most often chlorine, bromine, and various ketones.

The second step involves the destruction of free radicals. This usually happens by a process opposite to the first, that is, a combination of two like or unlike radicals to form a new bond:

$$A\cdot + B\cdot \longrightarrow A{-}B$$

This type of step is called *termination*, and it ends the reaction as far as these particular radicals are concerned. However, it is not often that termination follows directly upon initiation. Most radicals are very reactive and will react with the first available species with which they come in contact. The product may be one particle, as in the addition of a radical to a π-bond, which is another free radical. Or abstraction of an atom such as hydrogen gives two particles, RH and the new radical.

$$R\cdot + >C{=}C< \longrightarrow >C(R){-}\dot{C}<$$

$$R\cdot + R'H \longrightarrow RH + R'\cdot$$

In these examples, new radicals are generated. This type of step is called *propagation*, since the newly formed radical can now react with another molecule and produce another radical, and so on, until two radicals meet each other and terminate the sequence. The process just described above is called a chain reaction, and there may be hundreds or thousands of propagation steps between an initiation and a termination.

3.3.2 Polar reactions （极性反应）

The reaction of ethylene with HCl is a typical polar reaction.

$$\underset{\text{Electrophile}}{HCl} + \underset{\text{Nucleophile}}{CH_2{=}CH_2} \longrightarrow CH_3CH_2Cl$$

C═C double bonds have greater electron density than single bonds-four electrons in a double bond versus only two electrons in a single bond. In addition, the electrons in the π bond are accessible to external reagents because they are located above and below the plane of the double bond rather than between the nuclei such as the σ bond. Both electron richness and electron accessibility lead to the prediction that C═C bonds should behave as nucleophiles. That is, the chemistry of alkenes should involve reaction of the electron-rich double bond with electron-poor reagents (electrophiles).

As a strong acid, HCl is a powerful proton (H^+) donor. Because a proton is positively charged and electron-poor, it is good electrophile. Thus, the reaction of H^+ with ethylene is a typical electrophile-nucleophile combination, characteristic of all polar reactions.

As shown in Fig 3.2 The electrophilic addition reaction takes place in two steps, beginning when the alkene reacts with H^+. Two electrons from the ethylene π bond move to form a new σ bond between the H^+ and one of the ethylene carbon atoms. The other ethylene carbon atom, having lost its share of the π electrons, is now left with a vacant p orbital, has only six valence electrons, and carries a positive charge. In the second step, this positively charged species carbocation is itself an electrophile that accepts an electron pair from the nucleophilic Cl^- anion to form a C—Cl bond and give the neutral addition product.

Carbocation

The electrophile H^+ is attacked by the π electron of the double bond, and a new C—H σ bond is formed. This leaves the other carbon atom with a positive charge and a vacant p orbital

Cl^- donates an electron pair to the of the positively charged carbon atom, forming a C—Cl σ bond and yielding the neutral addition product

Fig 3.2 The mechanism of the electrophilic addition of ethylene with HCl.

反应物变成产物所经历的全过程，包括试剂的进攻、反应物的断键（异裂过程、均裂过程）、反应中间体的形成、产物的生成（新键形成的异裂过程、均裂过程）等整个过程。自由基反应历程包括引发、链增长、终止等阶段。

3.4 Describing a reaction: intermediates（反应的描述：中间体）

There are many types of reaction intermediates that are involved in organic reactions such as carbocations, *carbanion*s, radicals, *carbenes*, *nitrenes*, benzynes etc. They are usually very short-lived, and most exist only as intermediates that are quickly converted to more stable molecules.

carbocation　carbanion　radical　carbene　nitrene　benzyne

3.4.1 Carbocations（碳阳离子）

Carbocations have a vacant orbital that bears a positive charge. For trivalent carbon, the preferred hybridization is sp^2 for the positive charge to be located in an unhybridized p orbital.

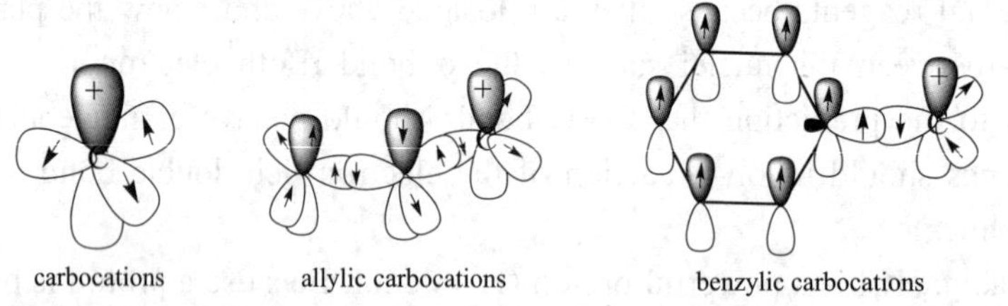

Among simple alkyl carbocations the order of stability is tertiary > secondary > primary > CH_3^+ due to the electron-donating effect of alkyl substituents. Carbocations that are adjacent to C═C bonds, as in allylic and *benzylic* carbocations, are strongly stabilized by delocalization, where the positive charge is spread over several atoms instead of being concentrated on one. Substituents such as oxygen and nitrogen that have nonbonding electrons are very strongly stabilizing toward carbocations. This is true both for ether-type oxygen and carbonyl-type oxygen (in *acylium* ions). Even fluorine and the other halogens are stabilizing by delocalization.

The stabilizing effect of delocalization can be seen even with certain functional groups that are normally considered to be electron withdrawing. For example, computations indicate that *cyano* and *carbonyl* groups have a stabilizing resonance effect. This is opposed by an inductive effect, so the net effect is destabilizing, but the resonance component is stabilizing.

$$\gt\overset{+}{C}-C(=O)- \longleftrightarrow \gt C=C(-\overset{+}{O})- \qquad \gt\overset{+}{C}-C\equiv N \longleftrightarrow \gt C=C=\overset{+}{N}$$

A number of methods can be used to generate carbocations.

(1) A direct ionization, in which a leaving group attached to a carbon atom leaves with its pair of electrons, such as in *solvolysis* reactions of alkyl halides or *sulfonate esters*:

$$R-X \longrightarrow R^+ + X^-$$

(2) Ionization after an initial reaction that converts one functional group into a leaving group, as in protonation of an alcohol to give an *oxonium ion* or conversion of a primary amine to a *diazonium salt*. They ionize to the corresponding carbocation:

$$R-OH \xrightarrow{H^+} R-\overset{+}{O}H_2 \longrightarrow R^+ + H_2O$$

$$R-NH_2 \xrightarrow{HONO} R-N_2^+ \longrightarrow R^+ + N_2$$

(3) A proton or other positive species adds to one atom of an alkene or alkyne, leaving the adjacent carbon atom with a positive charge.

$$\gt=CR_2 \xrightarrow{H^+} \gt\overset{+}{}-CHR_2$$

$$-\equiv CR \xrightarrow{H^+} -\overset{+}{=}C(R)H$$

(4) A proton or other positive species adds to one atom of a C=X bond, where X=O, S, N in most cases, leaving the adjacent carbon atom with a positive charge. When X=O, S, this ion is resonance stabilized, as shown. When X=NR, protonation leads to an *iminium ion*, with the charge localized on the nitrogen.

$$\gt C=X \xrightarrow{H^+} \gt C=\overset{+}{X}-H \longleftrightarrow \gt\overset{+}{C}-X-H \quad (X = O, S)$$

Carbocations are most often short-lived transient species and react further without being isolated.

3.4.2 Carbanions（碳负离子）

By definition, every carbanion possesses an unshared pair of electrons and is therefore a base. When a carbanion accepts a proton, it is converted to its conjugate acid.

The structure of simple unsubstituted carbanions, such as methyl carbanions, is that the central carbon is sp^3 hybridized, with the unshared pair occupying one apex of the tetrahedron.

If the carbanion conjugate with adjacent unsaturated atom, the structure of the carbanion is planar and the central carbon is sp^2 hybridized, such as in the allylic carbanions.

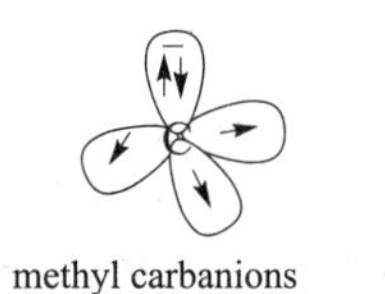
methyl carbanions

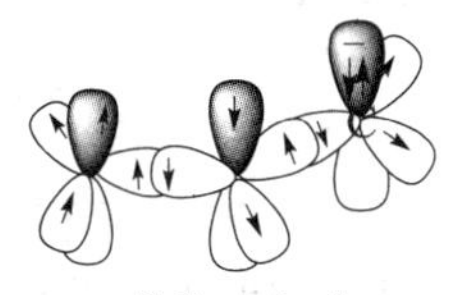
allylic carbanions

The stability of the carbanion is directly related to the strength of the conjugate acid. The weaker the acid, the greater the base strength and the lower the stability of the carbanion is. The lower the stability, the more willing the carbanion is to accept a proton from any available source, and hence to end its existence as a carbanion. Stability of simple carbanions decreases in the order methyl > primary > secondary > tertiary.

Many carbanions are far more stable than the simple kind mentioned above. The increased stability is due to certain structural features:

(1) Conjugation of the unshared pair with an unsaturated bond:

$$R(Y{=})C{-}\bar{C}R_2 \longleftrightarrow R({}^{-}Y)C{=}CR_2$$

In cases where a double or triple bond is located adjacent to the carbanionic carbon, the ion is stabilized by resonance in which the unshared pair overlaps with the π electrons of the double bond.

(2) Carbanions increase in stability with an increase in the amount of s character at the carbanionic carbon. Increased s character means that the electrons are closer to the nucleus and hence of lower energy. Thus the order of stability is

$$RC{\equiv}C^{-} > R_2C{=}CH^{-} \sim Ar^{-} > R_3C{-}CH_2^{-}$$

(3) Stabilization by sulfur or phosphorus.

Attachment to the carbanionic carbon of a sulfur or phosphorus atom causes an increase in carbanion stability, although the reasons for this are in dispute.

$$O{=}S(O)(R){-}C^{-}R_2 \longleftrightarrow {}^{-}O{-}S(O)(R){=}CR_2 \longleftrightarrow \text{etc}$$

(4) Field effects

Most of the groups that stabilize carbanions by resonance effects have electron-withdrawing field effects and thereby stabilize the carbanion further by spreading the negative charge, although it is difficult to separate the field effect from the resonance effect. However, in a *nitrogen ylide* $R_3N^{+}{-}C^{-}R_2$, where positively charged nitrogen is adjacent to the negatively charged carbon, only the field effect operates. Ylides are more stable than the corresponding simple carbanions. Carbanions are stabilized by a field effect if there is any heteroatom (O, N, or S) connected to the carbanionic carbon, provided that the heteroatom bears a positive charge in at least one important canonical form, for example,

$$Ar{-}C({=}O){-}N(Me){-}CH_2^{-} \longleftrightarrow Ar{-}C(O^{-}){=}N^{+}(Me){-}CH_2^{-}$$

Carbanions are generated in two principal ways.

(1) A group attached to a carbon leaves without its electron pair:

$$R{-}H \longrightarrow R^{-} + H^{+}$$

The leaving group is most often a proton. This is a simple acid–base reaction, and a base is required to remove the proton. However, other leaving groups are known:

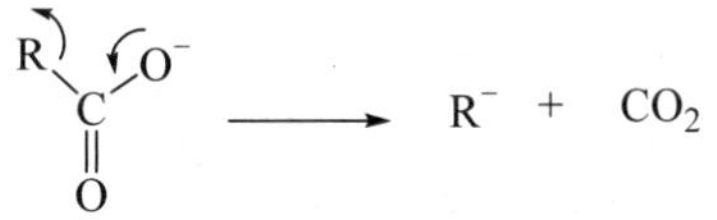

$$R-C(=O)-O^- \longrightarrow R^- + CO_2$$

(2) A negative ion adds to a C═C or C≡C bond.

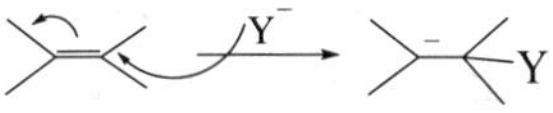

3.4.3 Radicals（自由基）

Carbon radicals have an unpaired electron in a nonbonding orbital. The simple alkyl radicals have sp^2 bonding and the structure is planar, with the odd electron in a p orbital. Radical stabilization is often defined by comparing C—H *bond dissociation energies* (BDE).

As with carbocations, the stability order of free radicals is tertiary > secondary > primary > CH_3 • .

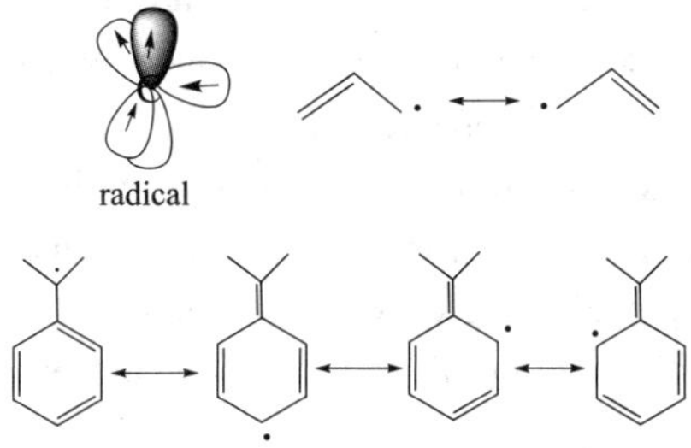

radical

The stabilizing effects of vinyl groups (in *allylic* radicals) and phenyl groups (in benzyl radicals) are large due to the resonance.

For radicals, nearly all functional groups have stabilizing effect. For example, carbonyl and cyano have a stabilizing effect on a radical intermediate at an adjacent carbon.

Free radicals are formed from molecules by breaking a bond so that each fragment keeps one electron. The energy necessary to break the bond is supplied in one of two ways.

(1) Thermal *cleavage*

Subjection of any organic molecule to a temperature high enough in the gas phase results in the formation of free radicals. Two common examples are cleavage of *diacyl peroxides* to acyl radicals that decompose to alkyl radicals and cleavage of azo compounds to alkyl radicals.

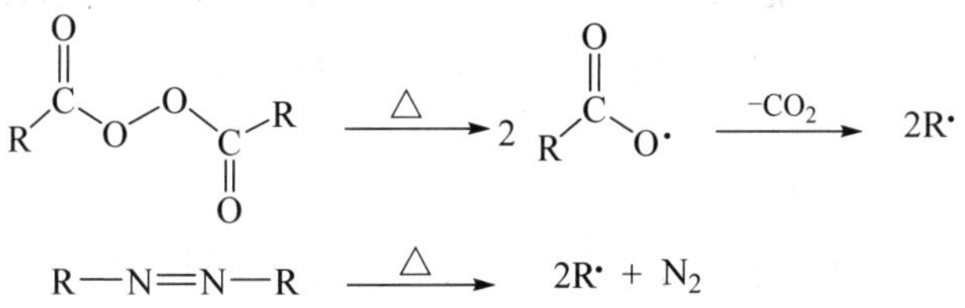

$$RC(=O)O-OC(=O)R \xrightarrow{\triangle} 2\,RC(=O)O\cdot \xrightarrow{-CO_2} 2R\cdot$$

$$R-N=N-R \xrightarrow{\triangle} 2R\cdot + N_2$$

(2) Photochemical cleavage

The energy of light of 600～300 nm is 200～400 kJ/ mol, which is of the order of magnitude of covalent-bond energies. Typical examples are photochemical cleavage of alkyl halides in the presence of triethylamine, alcohols in the presence of mercuric oxide and iodine, alkyl *4-nitrobenzenesulfonates*, chlorine, and of ketones:

$$Cl_2 \xrightarrow{h\nu} 2Cl\cdot$$

$$R_2C{=}O \xrightarrow[\text{vapor phase}]{h\nu} RC^{\cdot}{=}O + R\cdot$$

Radicals are also formed from other radicals, either by the reaction between a radical and a molecule (which must give another radical, since the total number of electrons is odd) or by cleavage of a radical to give another radical, for example,

$$Ph{-}C(=O){-}O\cdot \longrightarrow Ph\cdot + CO_2$$

3.4.4 Carbenes（卡宾）

Carbenes are highly reactive species, practically all having lifetimes considerably under 1 s. The parent species $:CH_2$ is usually called *methylene*. Its derivative $:CCl_2$ is generally known as dichlorocarbene, it can also be called dichloromethylene.

The two nonbonded electrons of a carbene can be either paired or unpaired. If they are paired, the species is spectrally a *singlet*. Two unpaired electrons appear as a triplet. The singlet species has a lower energy than the triplet species. Triplet $:CH_2$ is a bent molecule, with an angle of ~136° , and singlet $:CH_2$ is also bent, with an angle of ~103° . The singlet species can convert to the triplet species under the conditions of inert gas.

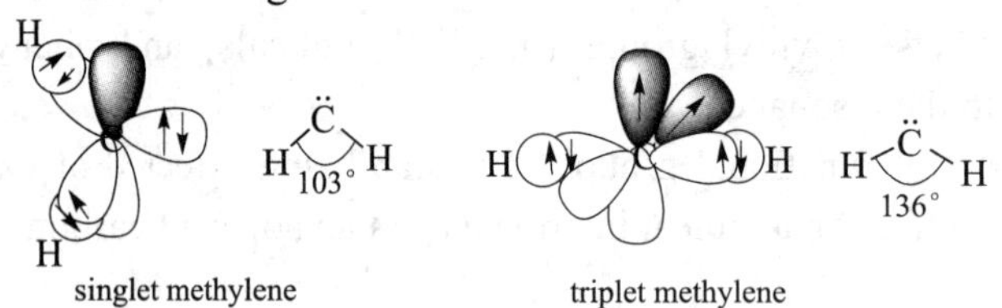

singlet methylene　　triplet methylene

Carbenes are mainly formed in two ways.

(1) In an elimination, a carbon loses a group without its electron pair, usually a proton, and then a group with its electron pair, usually a halide ion:

$$R_2CH{-}Cl \xrightarrow{-H^+} R_2\bar{C}{-}Cl \xrightarrow{-Cl^-} R_2C:$$

The most common example is the formation of dichlorocarbene by treatment of chloroform with a base and *geminal alkyl dihalides* with Me_3Sn^-, but many other examples are known, such as

$$CCl_3{-}COO^- \xrightarrow{\triangle} :CCl_2 + CO_2 + Cl^-$$

$$\text{(phenanthrene-fused cyclopropane with } t\text{-Bu)} \xrightarrow{h\nu} :CH{-}C(CH_3)_3 + \text{phenanthrene}$$

(2) *Disintegration* of compounds containing certain types of double bonds:

$$R_2C{=}Z \longrightarrow R_2C: + Z$$

The two most important ways of $:CH_2$ formations are the *photolysis* of *ketene* and the isoelectronic decomposition of *diazomethane*.

$$CH_2{=}C{=}O \xrightarrow[\text{pyrolysis}]{h\nu} :CH_2 + C{\equiv}O$$

$$CH_2{=}\overset{+}{N}{=}N^- \xrightarrow[\text{pyrolysis}]{h\nu} :CH_2 + N{\equiv}N$$

3.4.5 Nitrenes（乃春）

Nitrenes are the nitrogen analogs of carbenes, and most of what we have said about carbenes also applies to them. Nitrenes are too reactive for isolation under ordinary conditions, although nitrenes are more stable than carbenes. Nitrenes can be generated in both triplet and singlet states. The ground state of most nitrenes is a triplet.

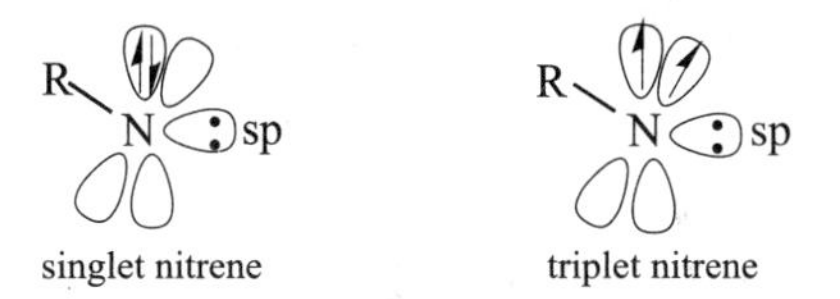

The two principal means of generating nitrenes are analogous to those to form carbenes.

(1) Elimination. An example is

$$R(H)N{-}OSO_2Ar \xrightarrow{\text{base}} R{-}N: + BH + ArSO_3^-$$

(2) Breakdown of certain double bond compounds. The most common method of forming nitrenes is photolytic or thermal decomposition of *azides*.

$$R{-}N{=}\overset{+}{N}{=}\overset{-}{N} \xrightarrow{\triangle \text{or } h\nu} R{-}N: + N_2$$

3.4.6 Benzynes（苯炔）

The species represented by a Lewis structure with two double bonds and one triple bond in a six-membered ring is called benzyne. The triple bond of benzyne is quite different from a normal triple bond because of the angle constraints (a triple bond prefers 180° bond angles) imposed by the six-membered ring. It is an extremely reactive species and has only a fleeting existance. Although normal triple bonds do not react with nucleophiles, the high reactivity of benzyne allows a nucleophile to attack to form a carbanion.

A summary of common reactive intermediates and their properties are listed in Table 3.2.

Table 3.2 Common reactive intermediates and their properties

Intermediates	Stability	Properties
carbocations	3° >2° >1° > $^+CH_3$	electrophilic strong acids
radicals	3° >2° >1° > $\cdot CH_3$	electron deficient
carbanions	$^-CH_3$ >1° >2° >3°	nucleophilic strong bases
carbenes	$:CH_2 > :CR_2 > :CAr_2 > :CX_2$ (X=Cl, Br)	both electrophilic and nucleophilic

反应中间体（或中间体）是化学反应的中间产物，一般不稳定，难以分离。常见的中间体包括碳正离子、碳负离子、自由基、卡宾、乃春、苯炔等。

碳正离子（碳阳离子）是含有正电的碳的活性中间体，通常碳为 sp^2 杂化，与三个基团结合，留下一对垂直于平面的 p 轨道。一般有能稳定正电荷的基团的碳正离子具有较高的稳定性。一般而言，三级碳正离子的稳定性大于二级碳正离子；二级碳正离子的稳定性大于一级碳正离子。

碳负离子指含有一个连有三个基团，并且带有一对孤对电子的碳的活性中间体。碳负离子带有一个单位负电荷，通常是四面体构型，其中孤对电子占一个 sp^3 杂化轨道。通过比较相应酸的酸性大小，可以大致判断碳负离子的稳定性大小。一般地，具有能稳定负电荷的基团的碳负离子具有较高的稳定性。这些基团可以是苯基、电负性较强的杂原子 [如 O、N，基团如—NO_2、—C(═O)—、—CO_2R、—SO_2—、—CN 和—$CONR_2$ 等] 或末端炔烃。

自由基 （游离基）是含有一个不成对电子的原子团，通常碳为 sp^2 杂化。碳自由基无空轨道，带一个未成对电子。自由基的活性最强，通常不能稳定存在。三级碳自由基的稳定性大于二级碳自由基；二级碳自由基的稳定性大于一级碳自由基。由于共振效应的存在，几乎所有官能团对碳自由基都有稳定作用。

卡宾（碳烯、碳宾）是含二价碳的电中性化合物。卡宾是由一个碳和其他两个基团以共价键结合形成的，碳上还有两个自由电子。卡宾有两种结构，在光谱学上分别称为单线态和三线态。单线态卡宾中，中心碳原子是 sp^2 杂化，有一对未成键电子；三线态卡宾有两个自由电子，可以是 sp^2 或直线形的 sp 杂化。卡宾的稳定性顺序排列如下：

$$H_2C: < ROOCCH: < PhCH: < BrCH: < ClCH: < Br_2C: < Cl_2C:$$

氮烯（乃春、氮宾）是卡宾（碳烯）的氮类似物，通式为 R—N:。其中氮周围有 6 个电子，具亲电性，也分为单线态结构和三线态结构。

苯炔是从芳环去除两个邻位取代基后（剩下含两个电子的两个轨道）得到的电中性活性中间体。在苯炔中多出的 π 键是定域的，与芳环的其他 π 键呈正交。

3.5 Methods of determining mechanisms（确定反应机理的方法）

There are a number of commonly used methods for determining mechanisms. In most cases, one method is not sufficient, and the problem is generally approached from several directions.

3.5.1 Identification of products（通过产物确定）

Any mechanism proposed for a reaction must account for all the products obtained and for their relative proportions, including products formed by side reactions.

$$(H_3C)_3C—Br + H_2O \longrightarrow (H_3C)_3C—OH + HBr$$

The product is an alcohol, so this is a nucleophilic substitution reaction.

$$(H_3C)_3C—Br + NaOH \xrightarrow{C_2H_5OH} (H_3C)_2C{=}CH_2 + NaBr + H_2O$$

This is an elimination reaction, as the product is an alkene.

3.5.2 Determination of the presence of an intermediate（通过测定中间体确定）

(1) Isolation of an intermediate（中间体的分离）

It is sometimes possible to isolate an intermediate from a reaction mixture by stopping the reaction after a short time or by the use of very mild conditions.

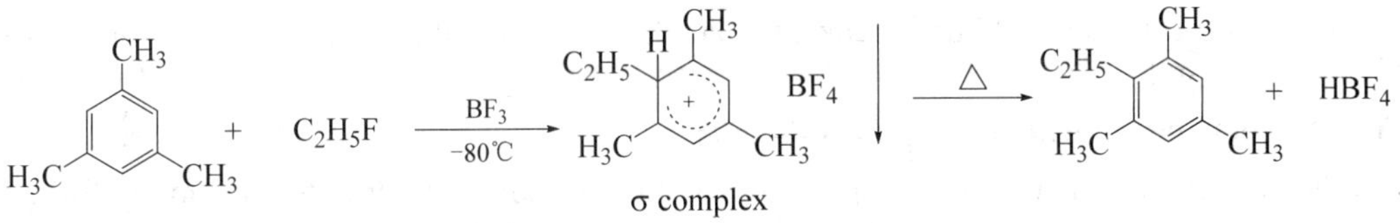

The σ complex (the intermediate) in the above electrophilic substitution reaction has been isolated as a yellow solid with the m. p. of −15℃.

(2) Detection of an intermediate（中间体的测定）

In many cases, an intermediate cannot be isolated, but can be detected by IR, NMR, or other spectra. For example, free radical and triplet intermediates can often be detected by ESR and by CIDNP.

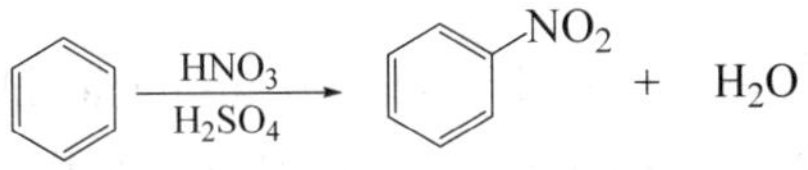

The detection by Raman spectra of NO_2^+ was regarded as strong evidence that this is an intermediate in the nitration of benzene.

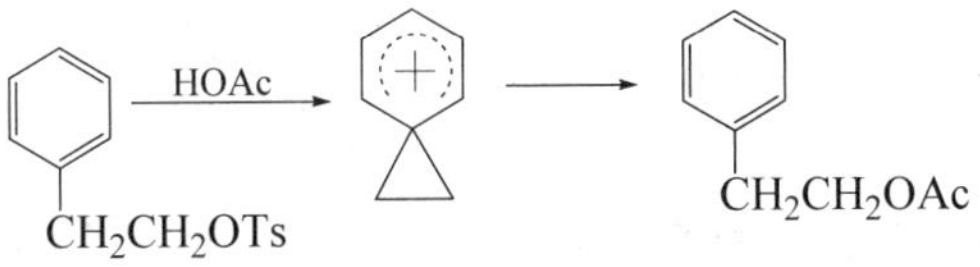

The *phenonium ion* intermediate has been detected by ^1H NMR, which proved this is a neighboring group participation of phenyl group in this nucleophilic substitution reaction.

(3) *Trapping* of an intermediate（中间体的捕获）

In some cases, the suspected intermediate is known to be one that reacts in a given way with a certain compound. The intermediate can then be trapped by running the reaction in the presence of that compound. For example, *benzynes* react with dienes in the Diels–Alder reaction. In any reaction where a benzyne is a suspected intermediate, the addition of a diene and the detection of the Diels–Alder adduct indicate that the benzyne was probably present.

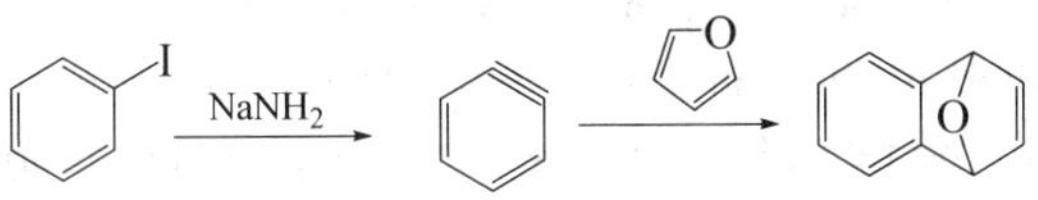

3.5.3 The study of catalysis（催化反应研究）

The mechanism of a reaction can be obtained from a knowledge of which substance catalyze the reaction, which inhibit it, and which do neither. Of course, just as a mechanism must be compatible with the products, so must it be compatible with its catalysts. If the reaction can proceed via light or peroxide, it may be proposed as a radical reaction. If it is an acid-catalyzed reaction, it may be a reaction involving in a positive intermediate. If it is a base-catalyzed reaction, it may be a reaction involving in a negative intermediate.

3.5.4 *Isotopic labeling*（同位素标记）

Much useful information has been obtained by using molecules that have been isotopically labeled and tracing the path of the reaction in that way. An example is the hydrolysis of esters.

$$R-C(=O)-OR' + H_2O \longrightarrow R-C(=O)-OH + ROH$$

Which bond of the ester is broken, the acyl–O or the alkyl–O bond? The answer is found by the use of $H_2{}^{18}O$. If the acyl–O bond breaks, the labeled oxygen will appear in the acid; otherwise it will be in the alcohol.

3.5.5 Stereochemical evidence（立体化学证据）

If the products of a reaction are capable of existing in more than one stereoisomeric form, the form that is obtained may give information about the mechanism. For example, the fact that *cis*-2-butene treated with $KMnO_4$ gives *meso*-2,3-butanediol and not the *racemic* mixture is evidence that the two OH groups attack the double bond from the same side.

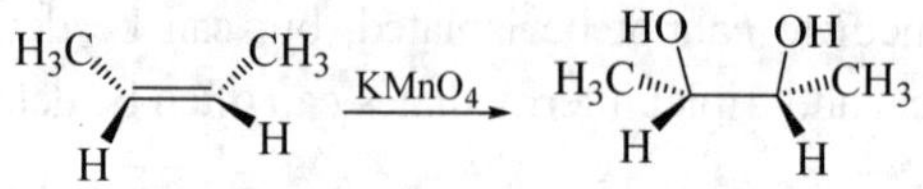

H_3C, CH_3, H, H (cis-2-butene) $\xrightarrow{KMnO_4}$ HO, OH, H_3C, CH_3, H, H (meso-2,3-butanediol)

Much useful information has been obtained about nucleophilic substitution, elimination, rearrangement, and addition reactions from this type of experiment.

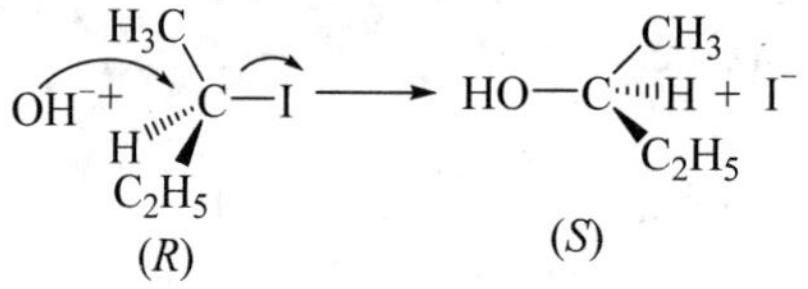

$OH^- + H_3C(H)(C_2H_5)C{-}I$ (R) $\longrightarrow$ $HO{-}C(CH_3)(H)(C_2H_5)$ (S) $+ I^-$

The Walden inversion of this nucleophilic substitution implied it's a S_N2 mechanism, not a S_N1 mechanism.

3.5.6 Kinetic study（动力学研究）

The rate of a homogeneous reaction is the rate of disappearance of a reactant or appearance of a product. The rate nearly always changes with time, since it is usually proportional to concentration and the concentration of reactants decreases with time. However, the rate is not always proportional to the concentration of all reactants. In some cases, a change in the concentration of a reactant produces no change at all in the rate, while in other cases the rate may be proportional to the concentration of a substance (a catalyst) that does not even appear in the *stoichiometric equation*. A study of which reactants affect the rate often tells a good deal about the mechanism.

If the rate is proportional to the change in concentration of only one reactant, the reaction is called a first-order reaction. The rate of a second-order reaction is proportional to the concentration of two reactants, or to the square of the concentration of one. For reactions that take place in more than one step, the order for each step is the same as the *molecularity* for that step. If any one step of a mechanism is considerably much slower than all the others (this is usually the case), the rate of the overall reaction is essentially the same as that of the slow step, which is consequently called the rate determining step.

Problems

1. Propose a reaction mechanism for the following reactions.

(1) $H_3C{-}CH(C_6H_5){-}CH(OTs){-}CH_2CH_3 \xrightarrow{HOAc} H_3C{-}CH(C_6H_5){-}CH(OAc){-}CH_2CH_3 + C_2H_5{-}CH(C_6H_5){-}CH(OAc){-}CH_3$

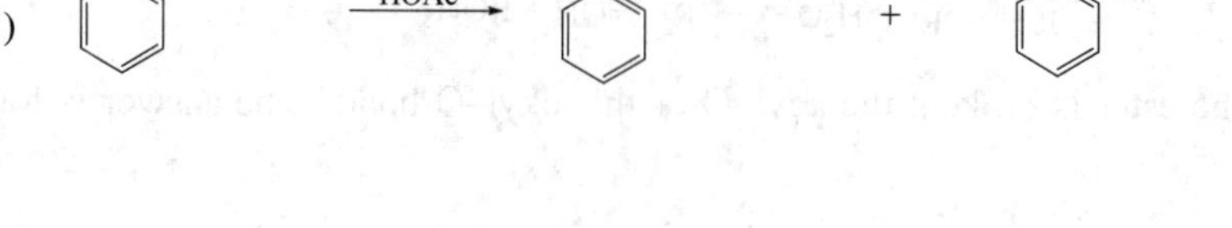

(2) 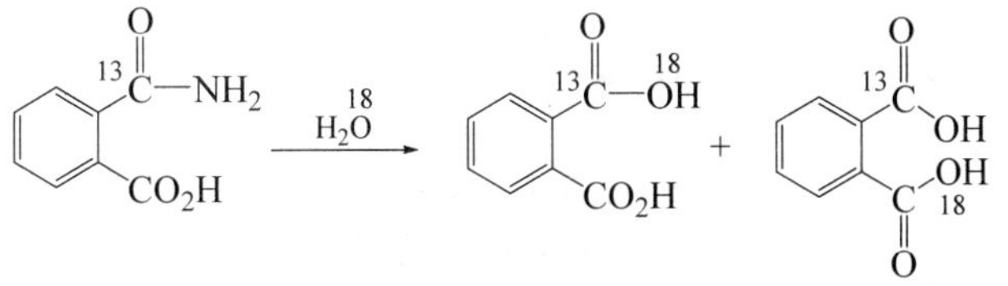 [chemical equation: 2-carbamoyl(13C)benzoic acid + $H_2{}^{18}O$ → 2-carboxybenzoic acid with 13C-18OH + 2-carboxybenzoic acid with 13C-OH and C-18OH]

2. Treatment of *t*-butyl alcohol with concentrated HCl gas gives t-butyl chloride.

$(H_3C)_3C—OH + H^+ + Cl^- \longrightarrow (H_3C)_3C—Cl + H_2O$

When the concentration of H^+ is doubled, the reaction rate doubles. When the concentration of *t*-butyl alcohol is tripled, the reaction rate triples. When the chloride ion concentration is quadrupled, however, the reaction rate is unchanged. Write the rate equation for this reaction.

3. In the presence of a small amount of bromine, the following light-promoted reaction has been observed.

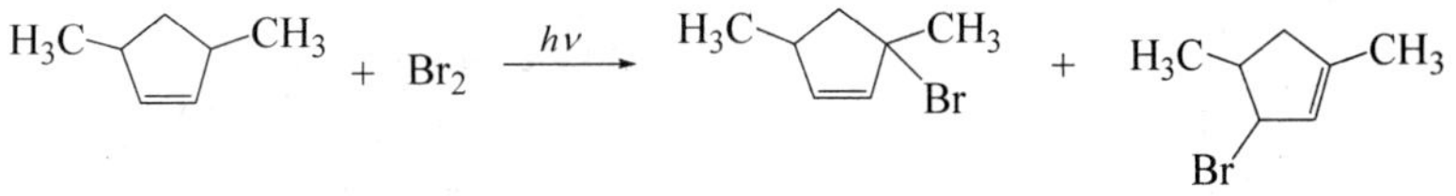

[chemical equation: 3,5-dimethylcyclopentene + Br_2 —$h\nu$→ 3-bromo-3,5-dimethylcyclopentene + 3-bromo-1,4-dimethylcyclopentene]

Write a mechanism for this reaction. Your mechanism should explain how both products are formed.

4. Explain the basicity and acidity of the following compounds.

(1) A.  [quinuclidine] B. 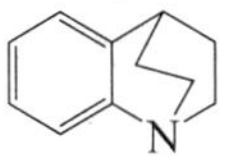 [benzo-fused quinuclidine] C. 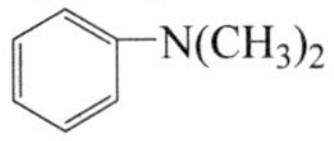 [$C_6H_5N(CH_3)_2$]

pK_a 10.58 7.79 5.06

(2) pK_a [o-hydroxybenzoic acid] < [p-hydroxybenzoic acid, HO–C_6H_4–CO_2H]

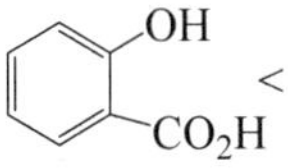

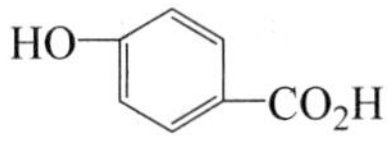

确定反应机理的方法如下。

（1）产物的鉴定

研究任何反应的中间过程之前，对产物（包括副产物）的确证是首要的；

（2）中间体的确证

方法包括：①中间体的分离　中间体活性高，寿命短，难以分离，但某些活性中间体可以在特殊条件下分离出来；②中间体的检测　多数中间体不能分离，但可利用 IR、NMR、MS、EPR、Raman、XPS 等波谱跟踪反应以检测中间体的存在；③中间体的捕获　如果推测到一个反应可能存在某一种中间体时，可在反应中加入另一种物质作为捕获剂，当它与不稳定中间体作用后再分离出预测化合物来证明；

（3）催化剂的研究

根据反应所需催化剂类型，往往可大致推测反应的历程，光或过氧化物的催化反应一般为自由基历程，能被酸催化的反应可能有正离子中间体形成，能被碱催化的反应可能有负离子中间体形成；

（4）同位素标记

用同位素标记的化合物作反应物，反应后测定产物中同位素的分布，往往可以为反应历程的确定提供有用的信息；

（5）立体化学

根据化合物构型的变化来推断反应物变化的方式，键的形成和断裂的方向等；

（6）热力学方法

通过热力学计算可得反应的 ΔH、ΔS 和 ΔG，从而推断反应的机理；

（7）动力学研究

大多数有机反应是经过多步完成的，整个反应的决速步骤是速率最慢的一步反应，通过动力学研究可知哪些分子和有多少分子参与了决速步骤的信息。从决速步骤的速率方程可知道反应级数和参与反应的分子数等信息。

Vocabulary

addition reactions 加成反应
elimination [ɪˌlɪmɪ'neɪʃn] 消除
pericyclic 周环的
substitution reactions 取代反应
isomeric [ˌaɪsə'merɪk] 异构的
rearrangement reactions 重排反应
migrate [maɪ'greɪt] 迁移
nitrile ['naɪtrɪl] 腈
tautomerization [tɔːtəməraɪ'zeɪʃən] 互变
dipolar 偶极的
configuration [kənˌfɪgə'reɪʃn] 构造
carbocation [kɑːbə'keɪʃən] 碳正离子
electrophile [ɪ'lektrəfaɪl] 亲电试剂
nucleophile ['njuːklɪəfaɪl] 亲核试剂
redox 氧化还原
electronegative 电负性的
peracid ['pɜːæsɪd] 过酸
functional group 官能团
sodium cyanide 氰化钠
butanenitrile 丁腈
sodium methylthiolate 甲硫醇钠
regioselectivity [riːdʒiːəʊsɪlek'tɪvɪtɪ] 区域选择性
stereoselectivity [stɪərɪəsɪlek'tɪvɪtɪ] 立体选择性
stereoisomer [ˌsterɪəʊ'aɪsəmə] 立体异构体
stereospecificity [stɪərɪəʊspesɪ'fɪsɪtɪ] 立体专一性
thermodynamic stability 热力学稳定性
kinetic control 动力学控制
relative activation barrier 相对活化能垒
equilibrium constant 平衡常数
enolate anion 烯醇负离子
lithium diisopropylamide 二异丙胺基锂
aprotic [ˌeɪ'prəʊtɪk] 非质子的
heterolytic [hetərə'lɪtɪk] 异裂的
homolytic [həʊmə'lɪtɪk] 均裂的
radical reaction 自由基反应
polar reaction 极性反应
an initiation step 引发阶段
peroxyacid 过酸
azo ['æzəʊ] 含氮的
termination [ˌtɜːmɪ'neɪʃn] 终止
propagation [ˌprɒpə'geɪʃn] 增长
intermediate 中间体
carbanion ['kɑːbənaɪən] 碳负离子
carbene ['kɑːbiːn] 碳烯
nitrene ['nɪtriːn] 氮烯
benzylic ['benzɪlɪk] 苄基的
acylium 酰基的
cyano [saɪ'ænəʊ] 氰基的
carbonyl ['kɑːbənɪl]羰基的
solvolysis [sɒl'vɒlɪsɪs] 溶剂分解反应
sulfonate ester 磺酸酯
oxonium ion [ɔk'səunjəm 'aɪən] 氧鎓离子
diazonium salt [ˌdaiə 'zəuniəm sɔːlt] 重氮盐
iminium ion 亚胺离子
nitrogen ylide 氮叶立德
bond dissociation energies (BDE) 键离解能
allylic ['æləˌlɪk] 烯丙基的
diacyl peroxides 二酰基过氧化物
cleavage ['kliːvɪdʒ] 分裂
4-nitrobenzenesulfonates 4-硝基苯磺酸盐
methylene ['meθɪliːn] 亚甲基
singlet ['sɪŋglət] 单线态
geminal alkyl dihalide 二烷基二卤化物
disintegration [dɪsˌɪntɪ'greɪʃn] 裂解
photolysis [fəʊ'tɒlɪsɪs] 光解作用
ketene ['kiːtiːn] 乙烯酮
diazomethane 重氮甲烷
azide ['æzaɪd] 叠氮化物
phenonium ion 苯鎓离子
trap [træp] 捕获
benzyne ['benzaɪn] 苯炔
isotopic labeling 同位素示踪
racemic [rə'siːmɪk] 外消旋的
stoichiometric equation 化学计算方程式
molecularity [məlekjʊ'lærɪtɪ] 分子性

Chapter 4

Nucleophilic Substitution（亲核取代反应）

In *nucleophilic* aliphatic substitutions, the leaving group (the *nucleofuge*) comes away with an electron pair, and the attacking reagent (the *nucleophile*) brings an electron pair to the substrate, using this pair to form the new bond:

General formula:

$$\underset{}{Nu\!:} + \underset{\text{Substrate}}{R{-}L} \longrightarrow R{-}Nu + \underset{\text{Leaving group}}{L^-}$$

Nucleophile (Nu) may be neutral or negatively charged; RL may be neutral or positively charged. Nucleophilic substitution reactions may involve several different combinations of charged and uncharged species as reactants. The equations in Scheme 4.1 illustrate the four most common types.

A. Neutral reactant + anionic nucleophile $R{-}X + Y^- \longrightarrow R{-}Y + X^-$

$$CH_3CHBrCN + NaI \longrightarrow CH_3CHICN + NaBr$$

$$RI + NaOH \longrightarrow ROH + NaI$$

B. Neutral reactant + neutral nucleophile $R{-}X + Y \longrightarrow R{-}Y^+ + X^-$

$$R{-}X + Y{-}H \longrightarrow R{-}Y + H{-}X$$

$$CH_3CH_2I + (C_2H_5)_3N \longrightarrow (C_2H_5)_3N^+C_2H_5I^-$$

$$C_6H_5C(CH_3)_2Cl \xrightarrow{C_2H_5OH} C_6H_5C(CH_3)_2OC_2H_5 + HCl$$

C. Cationic reactant + neutral nucleophile $R{-}X^+ + Y \longrightarrow R{-}Y^+ + X$

$$R{-}\overset{+}{N}Me_3 + H_2S \longrightarrow R{-}\overset{+}{S}H_2 + NMe_3$$

D. Cationic reactant + anionic nucleophile $R{-}X^+ + Y^- \longrightarrow R{-}Y + X$

$$R{-}\overset{+}{N}Me_3 + OH^- \longrightarrow ROH + NMe_3$$

Scheme 4.1 Representative nucleophilic substitution reactions

In all cases, nucleophile must have an unshared pair of electrons, so that all nucleophiles are Lewis bases. When nucleophile is the solvent, the reaction is called *solvolysis*. The reactions in Scheme 4.1 show the relationship of reactants and products in nucleophilic substitution reactions, but say nothing about mechanisms. In order to understand the mechanisms of such reactions, let us review the limiting cases as defined by Hughes and Ingold: the S_N1 (substitution-nucleophilic-unimolecular) and the S_N2 (substitution-nucleophilic-bimolecular) mechanisms.

亲核取代反应的类型

类型 A：中性底物+阴离子亲核试剂　$Nu^- + R—L \longrightarrow R—Nu + L^-$

类型 B：中性底物+中性亲核试剂　$Nu: + R—L \longrightarrow R—Nu^+ + L^-$

类型 C：中性亲核试剂+阳离子型底物　$Nu: + R—L^+ \longrightarrow R—Nu^+ + L:$

类型 D：阴离子亲核试剂+阳离子型底物　$Nu^- + R—L^+ \longrightarrow R—Nu + L:$

4.1 Mechanisms of nucleophilic substitutions（亲核取代反应的机理）

4.1.1 The S_N1 mechanism（单分子亲核取代反应的机理）

S_N1 consists of two steps. The first step, a unimolecular dissociation of the substrate to form a carbocation as the key intermediate, is rate-determining. The second is a rapid reaction between the intermediate carbocation and the nucleophile. The reactive nature of the carbocation can be expressed by its *electrophilic* character, or electrophlicity.

$$\text{Step 1}\quad R—X \xrightleftharpoons{\text{slow}} R^+ + X^-$$
$$\text{Step 2}\quad R^+ + Y^- \xrightarrow{\text{fast}} R—Y$$

The rate of the S_N1 reaction is *proportional* to the concentration of the substrate but not the concentration of the nucleophile. It follows a *first-order rate equation*. S_N1 rate = k_r[substrate].

The S_N1 mechanism is an ionization mechanism. For many carbocations there are two competing processes that lead to other products: *rearrangement* and elimination. There is carbocation rearrangement in the following S_N1 reaction.

$$CH_3CH(Br)CH(CH_3)CH_3 \xrightarrow{H_2O} CH_3C(CH_3)(OH)CH_2CH_3$$

A reasonable mechanism for this observation assumes rate-determining ionization of the substrate as the first step followed by a hydride shift that converts the secondary carbocation to a more stable tertiary one. The tertiary carbocation then reacts with water to yield the major product.

$$CH_3CH(CH_3)CH(Br)CH_3 \xrightarrow{\text{slow}} CH_3C(CH_3)(H)—\overset{+}{C}HCH_3 \xrightarrow{\text{fast}} CH_3\overset{+}{C}(CH_3)—CH(H)CH_3 \xrightarrow{H_2O} CH_3C(CH_3)(OH)CH_2CH_3$$

4.1.2 The S_N2 mechanism（双分子亲核取代反应的机理）

S_N2 mechanism is a single-step process in which both the substrate and the nucleophile are involved in the *transition state*. Cleavage of the bond between carbon and the leaving group is assisted by formation of a bond between carbon and the nucleophile. The new bond forms at the same time when the old bond is broken. In fact, the nucleophile "pushes off" the leaving group from its point of attachment to carbon. For this reason, the S_N2 mechanism is sometimes referred to as a direct *displacement* process.

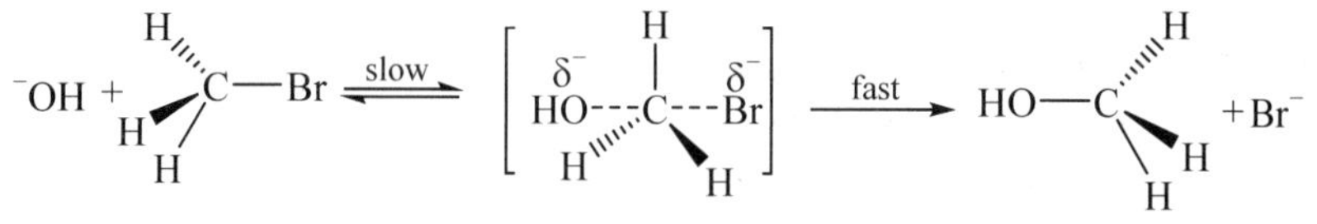

The rate of the S_N2 reaction is proportional to the concentrations of the substrate and the nucleophile. It follows a second-order rate equation. S_N2 rate = k_r[substrate][nucleophile].

The S_N2 reaction takes place in one step with no intermediates. No rearrangement is possible in the S_N2 reaction.

4.1.3 The S_Ni mechanism（分子内亲核取代反应的机理）

S_Ni stands for nucleophilic substitution *internal*. In S_N1 mechanism, nucleophilic substitution proceeds with *retention of configuration* and part of the leaving group must be able to attack the substrate, detaching itself from the rest of the leaving group in the process. The first step is the same as the first step of the S_N1 mechanism which dissociate into an intimate ion pair. But in the second step part of the leaving group attacks the carbon center, necessarily from the front since it is unable to get to the rear, which results in retention of configuration.

The S_Ni mechanism is relatively rare. An example is as follows:

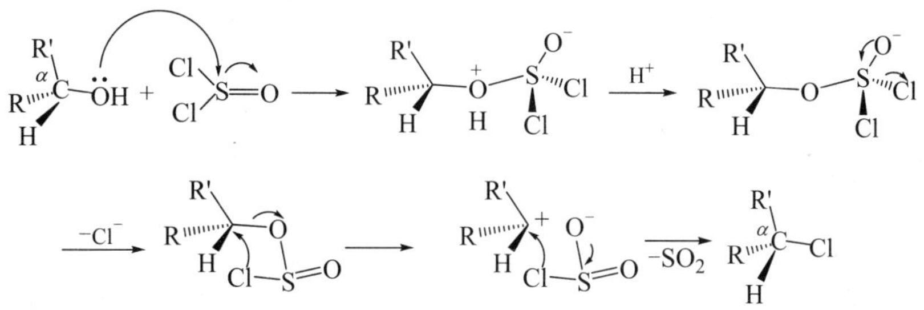

In this mechanism, the addition of pyridine to the mixture of alcohol and thionyl chloride results in the formation of alkyl halide with inverted configuration. Inversion of configuration is resulted in because the pyridine reacts with ROSOCl first. The free Cl^- in this process attacks the carbon center from the rear of the intermediates.

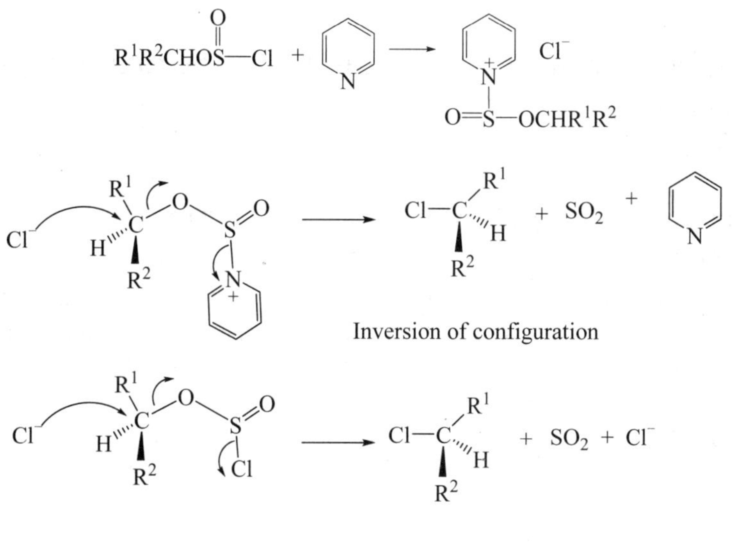

Inversion of configuration

亲核取代反应机理

（1）单分子亲核取代反应——S_N1 历程

$$C-X \xrightarrow[\text{慢}]{-X^-} C^+ + X^-$$

$$C^+ \xrightarrow[\text{快}]{Nu^-} Nu-C + C-Nu$$

动力学特征：$v=k[RX]$。动力学上为一级反应，因而反应速率只与反应底物有关而与亲核试剂的性质和浓度无关。

立体化学特征：光学活性的 RX 经 S_N1 反应后得到外消旋产物。碳正离子中间体中心碳为 sp^2 杂化的平面结构和空 p 轨道，理论上产物是外消旋化的。这取决于底物结构和试剂性质，通常情况是外消旋化与构型翻转同时存在。

反应特征：反应分两步进行，生成碳正离子是决速步骤，可能有重排产物生成。

（2）双分子亲核取代反应——S_N2 历程

$$:Nu^- + C-X \longrightarrow [Nu\cdots C\cdots X]^{\neq} \longrightarrow Nu-C + X^-$$

动力学特征：$v=k[RX][Nu^-]$。

立体化学特征：光学活性的 RX 经 S_N2 反应，生成构型完全转化的产物（即 Walden 转化）。

反应特征：反应连续、缓慢、一步完成。亲核试剂从离去基团背面进攻。过渡态为五配位的三角双锥构型。

亲核取代反应与 β-消除反应是一组竞争反应，一般情况下：

① 叔卤代烷较易发生消除反应，而伯卤代烷较易发生取代反应，仲卤代烷处于二者之间。

② 亲核试剂的亲核性强有利于取代反应；亲核试剂的碱性强有利于消除反应。

（3）分子内亲核取代反应——S_Ni 历程

分子的一部分作为亲核试剂发生了亲核取代反应，形成了紧密离子对，亲核试剂从失去的一侧进攻碳正离子（溶剂分子还未介入），导致手性中心底物产物构型保持。

4.2 Stereochemistry of nucleophilic substitutions（亲核取代反应的立体化学）

Stereochemistry of the nucleophilic reaction is related to its mechanism.

4.2.1 Stereochemistry of S_N1 reactions（单分子亲核取代反应的立体化学）

A carbocation intermediate forms in the *rate-determining step*. Nucleophile attacks planar carbocation from either side to yield a pair of enantiomers when the center atom is a chiral center.

$$R^1R^2R^3C-L \longrightarrow R^1R^2R^3C^+ \xrightarrow{Nu^-} R^1R^2R^3C-Nu\ (50\ \%) + Nu-CR^1R^2R^3\ (50\ \%)$$

(Racemization)

In fact, the stereochemical evidence for the S_N1 mechanism is less clear-cut than it is for the S_N2 mechanism. In a few cases, a small amount of retention of configuration has been found. These results have led to the conclusion that in many S_N1 reactions at least some of the products are not formed from free carbocations but rather from ion pairs. Because of the *shielding effect* of leaving group, nucleophile has more opportunity to attack centre carbon from the back of leaving group, this increases the ratio of inversion configuration product.

49 %　　51 %

4.2.2 Stereochemistry of S_N2 reactions（双分子亲核取代反应的立体化学）

The nucleophile attacks the substrate from the side opposite to the leaving group. This is called substitution with inversion of configuration. The configuration of product is inverted—*Walden inversion*.

$$\text{H}-\overset{C_2H_5}{\underset{CH_3}{|}}-\text{Br}\ (R)\xrightarrow[S_N2]{NaOH,\ H_2O}\text{HO}-\overset{C_2H_5}{\underset{CH_3}{|}}-\text{H}\ (S)$$

Walden inversion

As the nucleophile attacks and the leaving group leaves, the other three groups move from one side to the other, much like an umbrella blowing inside out in strong winds.

4.3 Nucleophiles and nucleophilicity（亲核试剂和亲核性）

4.3.1 Nucleophiles（亲核试剂）

(1) Oxygen-bearing nucleophiles: H_2O, ROH, OH^-, RO^-, ArO^-, ^-OOCR, $^-ONO_2$, SO_3.

(2) Sulfur-bearing nucleophiles: H_2S, S^{2-}, RS^-, RSO_2^-, SO_3^{2-}, $S_2O_3^{2-}$, HS^-, RSH, ArS^-, SCN^-, etc.

(3) Nitrogen-bearing nucleophiles: HN_3, RNH_2, R_2NH, R_3N, $NH_2—NH_2$, $ClSO_2NCO$, Li_3N, NO_2^-, NaN_3, NCO^-, NCS^-, etc.

(4) Halogen-bearing nucleophiles: HX, X^-, LiI, $SOCl_2$, PCl_3, PBr_3, PI_3, PCl_5, PBr_5, PI_5, etc.

(5) Hydrides as nucleophiles: NaH, $LiAlH_4$, AlH_3, $NaBH_4$, $LiAlH(O—t\text{-}Bu)_3$ etc.

(6) Carbon-bearing nucleophiles: RMgX, R_2CuLi, RNa, RK, RLi, ylides, R_3B, R_3Al, ArCu, $RC{\equiv}C^-$, CN^- and various carbanions which are *transform*ed from reactive secondary carbon.

4.3.2 Trends in nucleophilicity（亲核性的影响因素）

The term nucleophilicity refers to the capacity of Lewis base to participate in a nucleophile substitution reaction and is contrasted with basicity, which is defined by the position of an equilibrium reaction with a proton donor. Nucleophilicity is used to describe trends in the rates of substitution reactions that are attributable to properties of nucleophile. Several properties can influence nucleophilicity. Those considered to be most significant are: the solvation energy of the nucleophile; the electronegativity of the attacking atom; the *polarizability* of the attacking atom and the steric bulk of nucleophile. Let us consider how each of these factors affects nucleophilicity.

(1) Solvent effects on nucleophilicity

The nucleophilicity of anions is very dependent on the degree of solvation, particularly in *protic* solvents. The protons in the form of O—H or N—H groups can form hydrogen bonds to negatively charged nucleophiles. Small anions are solvated more strongly than large anions in a protic solvent because the solvent approaches a small anion more closely and forms stronger hydrogen bonds. The enhanced solvation of smaller anions in protic solvents, requiring more energy to *strip off* their solvent molecules, reduces their nucleophilicity. This trend *reinforce*s the trend in polarizability: The polarizability increases with increasing atomic number, and the solvation energy in a protic solvent decreases with increasing atomic number. Therefore, nucleophilicity (in a protic solvent) generally increases down a column in the periodic table, as long as we compare similar species with similar charges. In methanol, the relative reactivity order is $CH_3CO_2^- < Cl^- < Br^- < N_3^- < CH_3O^- < CN^- \approx SCN^- < I^- < CH_3S^-$.

In contrast with protic solvent, *aprotic* solvents enhance the nucleophilicity of anions. An anion is more reactive in an aprotic solvent because it is not so strongly solvated. However, most polar, ionic reagents are not soluble in simple aprotic solvents such as alkanes. Polar aprotic solvent such as CH_3CN, DMF, CH_3COCH_3 etc has strong dipole moments to enhance solubility without forming hydrogen bonds with anions. These solvent effects are more pronounced for small basic anions than for large weakly basic anions. In DMSO, the relative reactivity order is $CH_3O^- > CH_3S^- \approx CN^- > CH_3COO^- > N_3^- > Cl^- > Br^- > SCN^- > I^-$. This order is roughly the order of basicity.

(2) More electronegative elements hold their electrons more tightly and are not as good nucleophiles as less electronegative atoms.

A. As long as the nucleophilic atom is the same, the more basic the nucleophile, the more reactive it is:

$$RO^- > HO^- >> RCO_2^- > ROH > H_2O$$

$$CH_3O^- > PhO^- > CH_3CO_2^- > NO_2^-$$

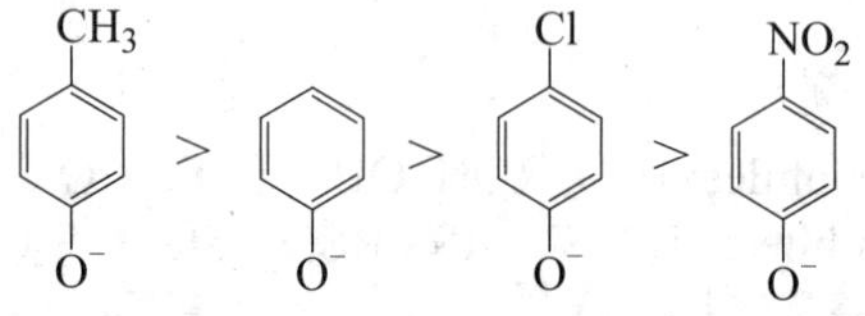

$$p\text{-}CH_3C_6H_4O^- > C_6H_5O^- > p\text{-}ClC_6H_4O^- > p\text{-}NO_2C_6H_4O^-$$

Species with a negative charge is a stronger nucleophile than a similar neutral species. In particular, a base is a stronger nucleophile than its conjugate acid.

$$HO^- > H_2O \quad CH_3O^- > CH_3OH \quad RS^- > RSH \quad NH_2^- > NH_3$$

B. When the nucleophilic atoms in nucleophiles are different, the strength of nucleophile is sometimes different from basicity:

Basicity: $C_2H_5O^- > I^-$ Nucleophilicity: $I^- > C_2H_5O^-$

C. Nucleophilicity decreases from left to right in the periodic table, following the electronegativity from left to right. The more electronegative elements have more tightly held nonbonding electrons that are less reactive toward forming new bonds.

$$H_3C^- > H_2N^- > OH^- > F^- \quad NH_3 > H_2O \quad (CH_3CH_2)_3P > (CH_3CH_2)_2S$$

(3) Nucleophilicity increases down the periodic table, following the increase in size and polarizability, whereas basicity decreases.

$$F^- < Cl^- < Br^- < I^- \quad R{-}O^- < R{-}S^- < R{-}Se^- \quad (CH_3CH_2)_3N < (CH_3CH_2)_3P$$

Polarizability describes the ease of distortion of the electron density of the nucleophile. Because the S_N2 process requires bond formation by an electron pair from the nucleophile, the more easily distorted the attacking atom, the better its nucleophilicity.

(4) When the *steric hindrance* of nucleophile is large, it will decrease the strength of nucleophilicity because bulky groups on the nucleophile hinder the close approach to the attacking center. The larger the steric hindrance is, the weaker the nucleophilicity the reagent is.

$$CH_3O^- > C_2H_5O^- > (CH_3)_2CHO^- > (CH_3)_3CO^-$$

$$C_2H_5NH_2 > (C_2H_5)_2NH > (C_2H_5)_3N$$

试剂的亲核性

具有亲核性的试剂叫亲核试剂，亲核性是指亲核试剂与亲电子原子的成键能力。

亲核试剂可以是中性分子或负离子，主要指含有进攻原子如氧、硫、氮、磷、羰负离子、卤素、氢负离子等的试剂。

亲核试剂均为路易斯碱，而碱性(质子理论)是指试剂与质子的结合能力，多些情况下二者的强度一致。

（1）亲核原子相同时，亲核性大小与碱性一致。取代基的供电性越强，亲核性和碱性越强。

$RO^- > HO^- \gg RCO_2^- > ROH > H_2O$

$CH_3O^- > PhO^- > CH_3CO_2^- > NO_2^-$

CH_3（O^-） > （O^-） > Cl（O^-） > NO_2（O^-）

亲核原子不同时，亲核性大小与碱性有时不一致。

碱性：$C_2H_5O^- > I^-$　　亲核性：$I^- > C_2H_5O^-$

（2）亲核性原子为同周期元素的原子时，其亲核性与碱性基本一致。

$$H_3C^- > H_2N^- > OH^- > F^- \quad NH_3 > H_2O \quad (CH_3CH_2)_3P > (CH_3CH_2)_2S$$

（3）同族元素原子的亲核性大小与碱性相反，即原子半径较小者，碱性较大，而亲核性则较小。试剂的可极化度越大，亲核性越强。

亲核性：$PhSe^- > PhS^- > PhO^-$

亲核性(在质子性溶剂中)：$I^- > Br^- > Cl^- > F^-$

（4）空间效应

位阻越大，亲核性越小。

（5）溶剂对亲核性的影响

极性质子性溶剂如甲醇、水等，使亲核试剂溶剂化增加亲核试剂的稳定性，降低其亲核活性。在甲醇中，亲核性 $CH_3CO_2^- < Cl^- < Br^- < N_3^- < CH_3O^- < CN^- \approx SCN^- < I^- < CH_3S^-$。

极性非质子溶剂如 DMF、DMSO 和 HMPT 等，亲核试剂完全暴露，未溶剂化，其亲核性没有降低。在 DMSO 中，亲核性 $CH_3O^- > CH_3S^- \approx CN^- > CH_3COO^- > N_3^- > Cl^- > Br^- > SCN^- > I^-$。

4.4 The factors that can influence the rates of S_N1, S_N2 reactions（影响单分子、双分子亲核取代反应速率的因素）

4.4.1 Effect of nucleophile（亲核试剂的影响）

The nucleophile takes part in the slow step (the only step) of S_N2 reaction but not in the slow step of S_N1. The nature of the nucleophile strongly affects the rate of the S_N2 reactions. A strong nucleophile is much more effective than a weak one in attacking an electrophilic center. A strong nucleophile promotes the S_N2 but not S_N1. Weak nucleophiles often go by the S_N1 mechanism if the substrate is secondary or tertiary.

4.4.2 Effect of solvent（溶剂的影响）

The effect of solvent polarity on the rate of S_N1 reactions depends on whether the substrate is neutral or positively charged (Table 4.1). According to the S_N1 mechanism, a molecule of a neutral substrate ionizes to a positively charged carbocation and a negatively charged ion in the rate-determining step. As the substrate approaches the transition state for this step, a partial positive charge is developed on carbon and a partial negative charge on the leaving group. Polar and nonpolar solvents are similar in their interaction with the starting materials, but differ markedly in how they affect the transition state. A solvent with a low *dielectric constant* has little effect on the energy of the transition state, whereas one with a high dielectric constant stabilizes the charge-separated transition state, lowers the activation energy, and increases the rate of reaction.

Table 4.1 Transition states for S_N1 reactions of charged and uncharged substrates, and for S_N2 reactions of four charge types

	Reactants and transition states	Charge in transition state relative to starting materials	How an increase in solvent polarity affects the rate
S_N2	A. $R—X + Y^- \longrightarrow \overset{\delta^-}{Y}\cdots R\cdots X^{\delta^-}$	dispersed	small decrease
	B. $R—X + Y \longrightarrow \overset{\delta^+}{Y}\cdots R\cdots X^{\delta^-}$	increased	large increase
	C. $R—X^+ + Y \longrightarrow \overset{\delta^+}{Y}\cdots R\cdots X^{\delta^+}$	dispersed	small decrease
	D. $R—X^+ + Y^- \longrightarrow \overset{\delta^+}{Y}\cdots R\cdots X^{\delta^-}$	decreased	large decrease
S_N1	$R—X \longrightarrow \overset{\delta^+}{R}\cdots X^{\delta^-}$	increased	large increase
	$R—X^+ \longrightarrow \overset{\delta^+}{R}\cdots X^{\delta^+}$	dispersed	small decrease

For S_N2 reactions, the effect of the solvent depends on which of the four charge types the reaction belongs to. In types A and C, an initial charge is dispersed in the transition state, so the reaction is hindered by polar solvents. In type D, initial charges are decreased in the transition state, so that the reaction is even more hindered by polar solvents. Only type B, where the reactants are uncharged but the transition state has built up a charge, is aided by polar solvents.

4.4.3 Effect of leaving-group（离去基团的影响）

The nature of the leaving group influences the rate of the nucleophilic substitution proceeding by either the direct displacement or ionization mechanism. Since the leaving group departs with the pair of electrons from the covalent bond to the reacting carbon atom, a correlation with both bond strength and anion stability is expected. Table 4.2 lists estimated relative rates of solvolysis of 1-phenylethyl esters and halides in 80% aqueous ethanol at 75℃. The reactivity of the leaving groups generally parallels their electron-accepting capacity.

Table 4.2 Relative solvolysis of 1-phenylethyl esters and halides

Leaving group	k_{rel}	Leaving group	k_{rel}
$CF_3SO_3^-$ (triflate)	1.4×10^8	$CF_3CO_2^-$	2.1
p-nitrobenzen*sulfonate* (nosylate)	4.4×10^5	Cl^-	1.0
p-toluenesulfonate (*tosylate*)	3.7×10^4	F^-	9×10^{-6}
$CH_3SO_3^-$ (mesylate)	3.0×10^4	*p*-nitrobenzoate	5.5×10^{-6}
I^-	91	$CH_3CO_2^-$	1.4×10^{-6}
Br^-	14		

From Table 4.2 we can see that sulfonate esters are especially useful reactants in nucleophilic substitutions. They can be prepared from alcohols by reactions that do not directly involve the carbon atom at which substitution is to be affected. The latter feature is particularly important in cases where the stereochemical and structural integrity of the reactant must be maintained. Sulfonate esters are usually prepared by reaction of an alcohol with *sulfonyl* halide in the presence of pyridine. Tertiary alcohols are more difficult to convert to sulfonate esters and their high reactivity often makes them difficult to isolate.

$$ROH + R'SO_2Cl \xrightarrow{\text{pyridine}} ROSO_2R'$$

$$(H_3C)(H)(PhCH_2)C{-}OH + TsCl \xrightarrow[25℃]{\text{pyridine}} (H_3C)(H)(PhCH_2)C{-}OTs + HCl$$

The S_N1 and S_N2 mechanisms differ in their sensitivity to the nature of the leaving group. S_N1 mechanism should exhibit a greater dependence on leaving group ability because it requires cleavage of the bond to the leaving group without assistance of the nucleophile. Table 4.3 lists k_{tos}/k_{Br} for solvolysis of RX in 80% ethanol.

Table 4.3 k_{tos}/k_{Br} for solvolysis of RX in 80% ethanol

R	k_{tos}/k_{Br}
methyl	11
ethyl	10
isopropyl	40
t-butyl	4000
1-adamantyl	9750

There are smaller differences in reactivity between tosylate and bromide for the systems of S_N2 mechanism, because the nucleophile assists in bond breaking. The largest differences can be seen for tertiary systems (S_N1 mechanism), where nucleophilic participation is minimal.

A poor leaving group can be made more reactive by coordination to an electrophile. Hydroxyl group is a very poor leaving group. *Protonation* of the hydroxyl group provides the much better leaving group—water. Primary alcohols can be converted to alkyl bromides by heating sodium bromide and sulfuric acid or with concentrated hydrobromic acid.

4.4.4 Effect of substrate（底物的影响）

The structure of substrate is an important factor in determining which of these substitution mechanisms might operate.

In S_N1 reactions, the rate-determining step is carbocation formation, if the substrate can easily ionize and undergo S_N1 substitution because tertiary carbocations are low in energy. We know that a branching increase the stability of alkyl cations and the reactivity of substrate in S_N1 reactions is: tertiary > secondary > primary > methyl.

The steric crowding that influences reaction rates in S_N2 processes plays less role in S_N1 reactions. In S_N2 reactions, bulk groups hinder the nucleophile approach the substrate from the opposite side of the leaving group, so that they decrease the reaction rates. So the reactivity of substrate in S_N2 reactions is: methyl > primary > secondary > tertiary.

4.4.5 Effect of conjugation（共轭效应的影响）

As we know, the *benzylic* and *allylic* cations are stabilized by electron *delocalization*. So the nucleophilic substitution reactions of the ionization type proceed more rapidly in these systems than in alkyl systems. Direct displacement reactions also take place particularly rapidly in benzylic and allylic systems; for example, allyl chloride is 33 times more reactive than ethyl chloride toward iodide ion in acetone. The π systems of benzylic and allylic groups provide extended conjugation. This conjugation can stabilize the transition.

Substitution by the S_N1 mechanism proceeds slowly on α-halo derivatives of ketones, aldehydes, acids, esters, nitriles and related compounds. Such substituents destabilize a carbocation intermediate, but substitution by the S_N2 mechanism proceeds especially readily in these systems. Table 4.4 indicates some representative relative reaction rates.

Table 4.4 Effect of α-EWG on relative rates of S_N2 reactions

$$Z—CH_2Cl + I^- \longrightarrow Z—CH_2I + Cl^-$$

Z	relative reaction rate
$CH_3CH_2CH_2$	1
$PhSO_2$	0.25
$CH_3C{=}O$	3.5×10^4
$PhC{=}O$	3.2×10^4
$C{\equiv}N$	3×10^3
$CH_3CH_2OC{=}O$	1.7×10^3

Note：EWG means electron withdrawing group.

Steric effect may be responsible for part of the observed acceleration, since a sp^2 carbon, such as in a carbonyl group, offers less steric resistance to the approaching nucleophile than an alkyl. The major effect is electronic. The adjacent π LUMO of the carbonyl group can interact with the electron density that builds up at the pentacoordinate carbon in the transition state. The sulfonyl and trifluoro groups at adjacent carbon which cannot take part in this type of π conjugation retard the rate of S_N2 reactions.

亲核取代反应的影响因素

（1）亲核试剂的亲核性对亲核取代反应的影响

亲核试剂的亲核性对 S_N1 几乎无影响（但影响产物的分配），对 S_N2 影响较大。

亲核试剂的亲核性越大，越有利于 S_N2；亲核试剂的体积越小，越有利于 S_N2。

（2）溶剂效应对亲核取代反应的影响

增加溶剂极性和溶剂的离子-溶剂化能力，导致 S_N1 反应速率显著增大。

对于反应物为中性底物+阴离子亲核试剂或者阳离子性底物+中性亲核试剂的 S_N2 反应，溶剂极性增大，不利于反应；极性减小，有利于 S_N2 反应。

对于反应物为中性底物+中性亲核试剂的 S_N2 反应，溶剂极性增大，有利于反应。

反应物为阳离子型底物+阴离子型亲核试剂的 S_N2 反应，溶剂极性降低，有利于反应。

（3）离去基对亲核取代反应的影响

S_N1 和 S_N2 的决速步骤中均存在 C—X 断裂，故 C—X 愈易断裂，S_N1 和 S_N2 反应的反应愈快。离去基碱性越弱，离去能力越强，离去能力：I^- > Br^- > Cl^- > F^-，强碱 OH^-为不好的离去基，H_2O 碱性弱，为好的离去基。

（4）底物的结构对亲核取代反应的影响

RX 发生 S_N1 反应的活性主要取决于生成的碳正离子的稳定性。凡是能稳定碳正离子的各种因素均有利于 S_N1 的进攻，中心碳原子上连接的供电基团，能分散碳正离子上的正电荷，提高碳正离子的稳定性。烯丙型 RX、苄基型 RX＞3° RX＞2° RX＞1° RX＞乙烯型 RX。底物的空间位阻越大，反应越快。当杂原子如 O、N、S 等原子直接与中心碳相连时，按 S_N1 进行，且反应很快。卤原子位于桥头碳上时，只能按 S_N1 进行，但反应极慢。桥头碳难以形成平面碳正离子，刚性越大（桥原子数减小），反应速率越小。

底物对 S_N2 反应的主要影响因素为空间效应。发生 S_N2 反应的底物中 α-碳原子所连烃基的体积越小，α-碳原子周围的空间位阻越小，越有利于亲核试剂进攻 α-碳原子。RX 发生 S_N2 反应的活性顺序：CH_3X＞1° RX＞2° RX＞3° RX。

（5）共轭效应对亲核取代反应的影响

发生 S_N1 反应的底物中，如果能通过电子共振（离域）使碳正离子稳定，反应活性更高。离去基所连碳的 α 位为酮基、羧酸、酯、氰等时，碳正离子不稳定，S_N2 反应活性增大。

4.5 Neighboring-group participation effect（邻基参与效应）

When a molecule that reacts by nucleophilic substitution also contains a nucleophilic substituent, it is often observed that the rate and stereochemistry of the nucleophilic substitution are strongly affected. There is usually a group with an unshared pair of electrons β to the leaving group (or sometimes farther away) in the substrate. The mechanism operating in such cases is called the neighboring-group mechanism which consists essentially of two S_N2 substitutions. In the first step of this reaction, the neighboring group acts as a nucleophile, pushing out the leaving group but still retaining attachment to the molecule. In the second step, the external nucleophile displaces the neighboring group by a backside attack:

Step 1

Step 2

The reason attack by Z is faster than that by Y is that the group Z is more available. In order for Y to react, it must collide with the substrate, but Z is immediately available by virtue of its position. The common neighboring groups are: COO^- (but not COOH), COOR, COAr, OCOR, OR, OH, O^-, NH_2, NHR, NR_2, NHCOR, SH, SR, S^-, SO_2Ph, I, Br and Cl. The effectiveness of halogens as neighboring groups decreases in the order I > Br > Cl.

It was shown that the *threo* pair of 3-bromo-2-butanol treated with HBr gave the racemic mixture of 2,3-dibromobutane, while the *erythro* pair gave the *meso* isomer.

CH3
H—Br　　HBr　　H—Br　Br—H
HO—H　——→　Br—H + H—Br
CH3　　　　　CH3　　CH3
threo　　　(racemic mixture)

CH3　　　　　CH3
H—Br　HBr　H—Br
H—OH ——→ H—Br
CH3　　　　　CH3
erythro　　　(*meso*)

CH3 / H—Br / HO—H / CH3 ≡ Br, H, CH3, H, CH3, OH —H^+→ Br, H, CH3, H, CH3, OH_2^+ —$-H_2O$→ H, Br^+, H, H_3C, a, b, CH_3, Br^- —a→ Br, H, CH3, H, CH3, Br ≡ CH3 / H—Br / Br—H / CH3; —b→ Br, H_3C, H, H_3C, H, Br ≡ CH3 / Br—H / H—Br / CH3

Either of the two *threo* isomers alone gave not just one of the enantiomeric dibromides, but the *dl* pair. The reason for this is that the intermediate present after the attack by the neighboring group is symmetrical, so the external nucleophile Br^- can attack both carbon atoms well equally.

CH3 / H—Br / H—OH / CH3 ≡ Br, H, CH3, OH, H, CH3 ≡ H_3C, H, Br, OH, H, CH3 ——→ H_3C, H, Br, OH_2^+, H, CH3

CH3 / Br—H / Br—H / CH3 ≡ Br, H_3C, H, Br, H_3C, H ≡ Br, H_3C, H, H, CH3, Br ←a— H_3C, H, a, Br^-, b, Br^+, H, CH3

CH3 / H—Br / H—Br / CH3 (*meso*) ≡ Br, H, CH3, Br, H, CH3 ≡ H_3C, H, Br, Br, H, CH3 ←b—

The rates of solvolysis of the *cis* and *trans* isomers of 2-acetoxy cyclohexyl *p*-toluenesulfonate differ by a factor of about 670, the *trans* isomer being more reactive. Besides the difference in rate, there is an obvious difference in stereochemistry. The diacetate obtained from the *cis* isomer is the *trans* compound (inversion), whereas retention of configuration is observed for the *trans* isomer. The assistance provided by acetoxy group facilitates the ionization of the tosylate group, resulting in the rate enhancement. The back-side participation by the adjacent acetoxy group is both sterically and energetically favorable. The cation formed by participation is stabilized by both acetoxy oxygen atoms and is far more stable than a secondary carbocation. The resulting acetoxonium intermediate is subsequently opened by nucleophilic attacking with inversion at either of the two equivalent carbons. The product is racemic *trans*-diacetate.

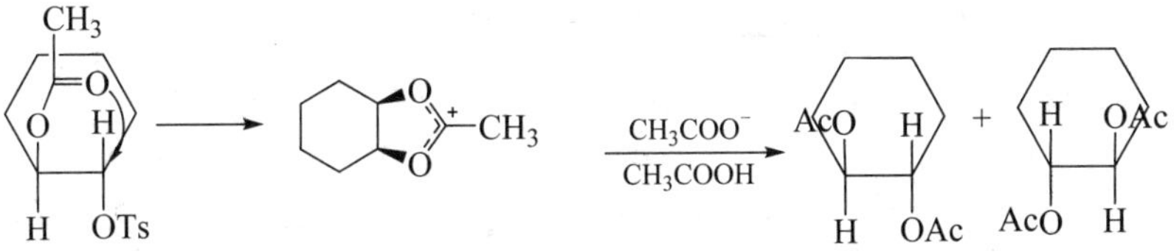

2-acetoxycyclohexyl *p*-toluenesulfonate

The alkoxide ions formed by deprotonation in basic solution are more effective nucleophiles than hydroxyl groups. Neighboring group participation is involved in the following reaction to give a product with the retention of configuration.

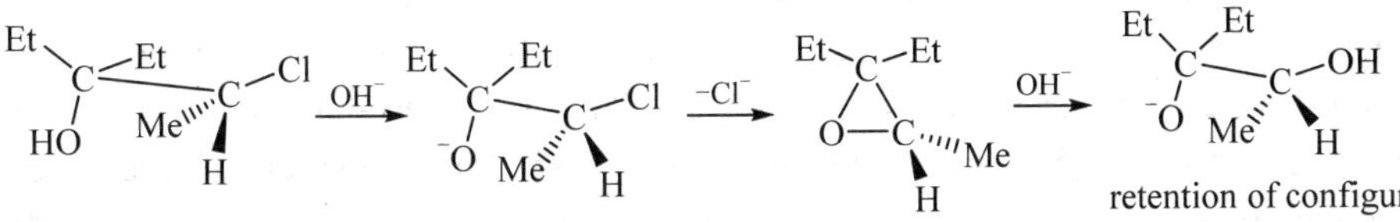

retention of configuration

The most striking evidence that the π electrons of C═C bonds can act as a neighboring group is that *acetolysis* of *anti* isomer of 7-norbornenyl tosylates is 10^{11} times faster than that of the saturated analog and proceeds with retention of configuration. These results can be explained by participation of the π electrons of C═C bonds to give the 7-norbornenyl ion, which is stabilized by delocalization of positive charge. In contrast, the *syn* isomer, where the double bond is not in a position to participate in the ionization step, reacts 10^{11} times slower than the *anti* isomer. The reaction product in this case is derived from a rearranged carbocation ion that is stabilized by virtue of being allylic.

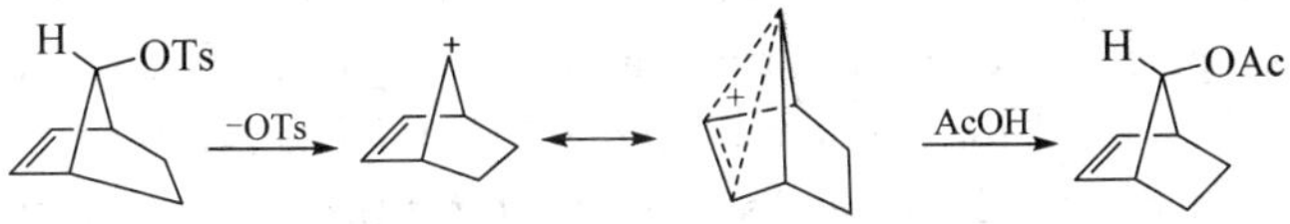

anti-7-norbornenyl tosylate 7-norbornenyl cation

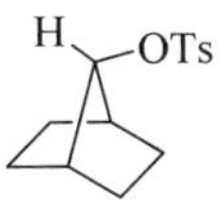

bicyclo[2.2.1]heptan-7-yl-4-methylbenzenesulfonate

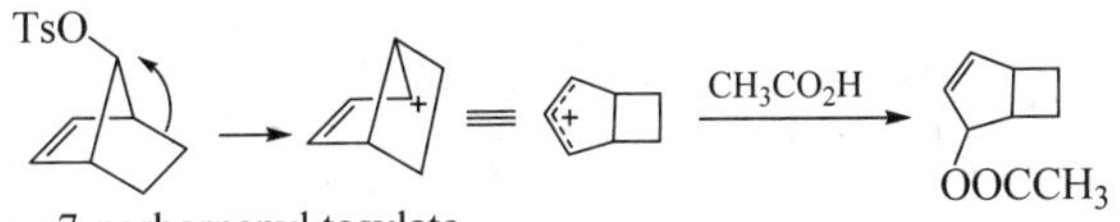

syn-7-norbornenyl tosylate

In classical carbocations the positive charge is localized on one carbon atom or delocalized by resonance involving an unshared pair of electrons or a double or triple bond in the allylic position. Carbocations is an important intermediate in organic reaction.

Intermediates formed by neighboring-group participation by C═C π bonds and C—C and C—H σ bonds, were called nonclassical (or bridged) carbocations. In a *nonclassical carbocation*, the positive charge is delocalized by a double or triple bond that is not in the allylic position or by a single bond. Examples are the 7-norbornenyl cation, the norbornyl cation etc.

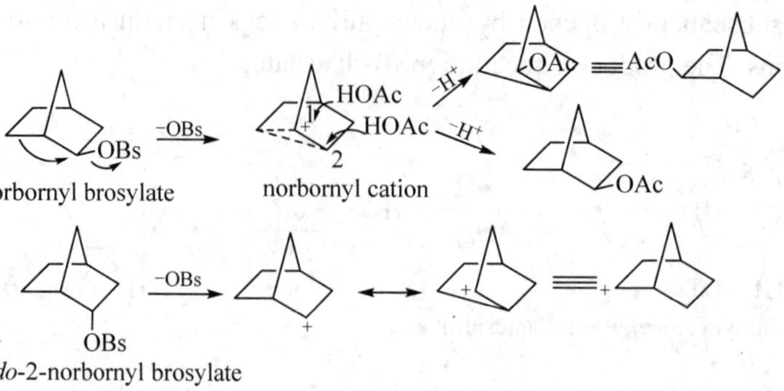

exo-2-norbornyl brosylate —OBs→ norbornyl cation

endo-2-norbornyl brosylate

Solvolysis in acetic acid of optically active *exo*-2-norbornyl *brosylate* gave a racemic mixture of the two *exo* acetates; no *endo* isomers were formed. Furthermore, *exo*-2-norbornyl brosylate solvolyzed~350 times faster than its *endo* isomer. These results were interpreted by Winstein and Trifan as indicating that the 1,6 bond assists in the departure of the leaving group and that a nonclassical intermediate norbornyl cation is involved. They reasoned that solvolysis of the *endo* isomer is not assisted by the 1,6 bond because it is not in a favorable position for backside attack, and that consequently solvolysis of the *endo* isomer takes place at a "normal" rate. Therefore the much faster rate for the solvolysis of the *exo* isomer must be caused by *anchimeric assistance*. The stereochemistry of the product is also explained by the intermediacy of norbornyl cation, since in norbornyl cation the 1 and 2 positions are equivalent and would be attacked by the nucleophile with equal facility, but only from the *exo* direction in either case. Incidentally, acetolysis of *endo*-2-norbornyl brosylate also leads exclusively to the *exo* acetates, so that in this case Winstein and Trifan postulated that a classical ion.

As the properties of a cyclopropane ring are in some ways similar to those of a double bond. Therefore a suitably placed cyclopropyl ring can also be a neighboring group. Thus *endo*-anti-tricyclo[3.2.1.$0^{2,4}$]octan-8-yl *p*-nitrobenzoate **1** solvolyzed ~10^{14} times faster than **2**. Obviously, a suitably placed cyclopropyl ring can be even more effective as a neighboring group than a double bond. The need for suitable placement is emphasized by the fact that **3** solvolyzed only about five times faster than **2**, while **4** solvolyzed three times slower than **2**.

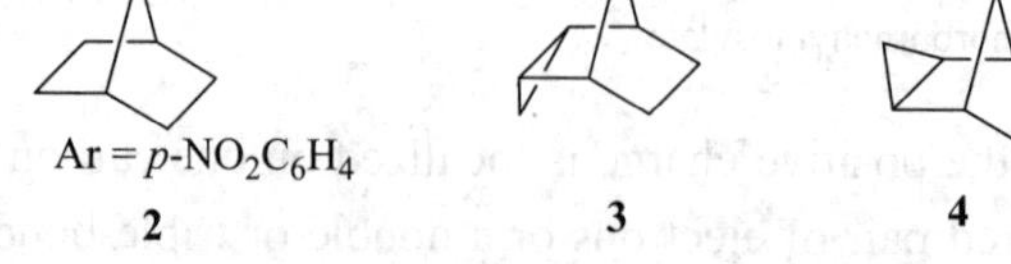

Ar = *p*-$NO_2C_6H_4$

1

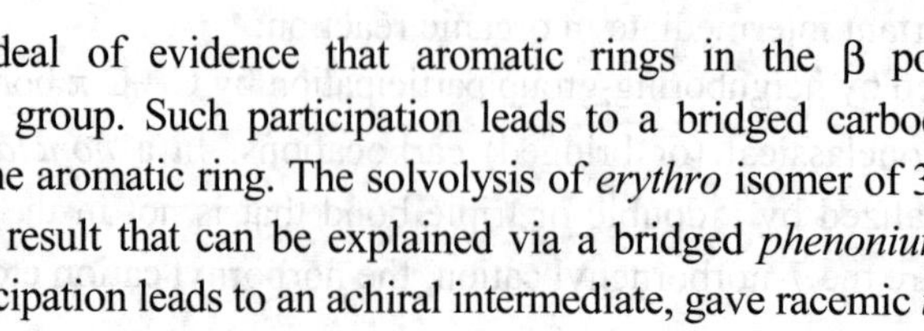

Ar = *p*-$NO_2C_6H_4$

2 **3** **4**

There is a great deal of evidence that aromatic rings in the β position can function as neighboring-participation group. Such participation leads to a bridged carbocation with the positive charge delocalized into the aromatic ring. The solvolysis of *erythro* isomer of 3-phenyl-2-butyl tosylate gave largely retention, a result that can be explained via a bridged *phenonium* ion intermediate. The *threo* isomer, where participation leads to an achiral intermediate, gave racemic *threo* product.

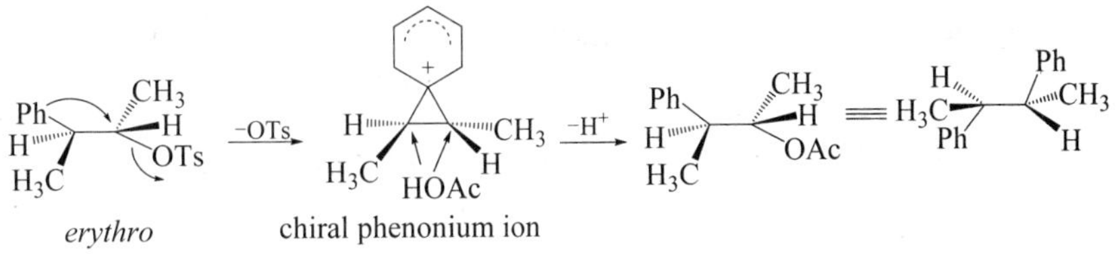

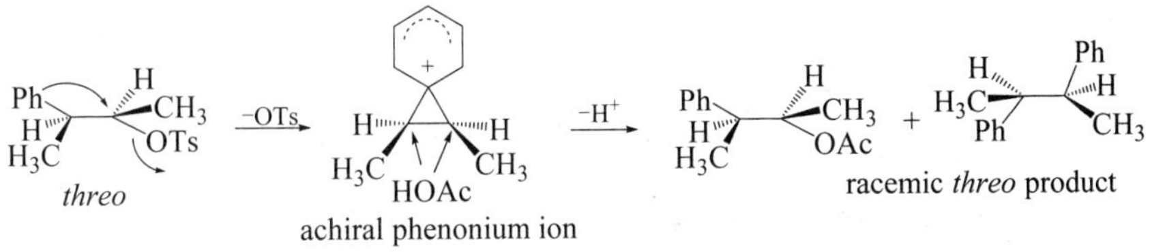

erythro　chiral phenonium ion

threo　achiral phenonium ion　racemic *threo* product

邻基参与对亲核取代反应的影响

如果亲核性取代基与离去基共处于同一分子中，亲核性取代基参与了另一部位上的取代反应，称为邻基参与作用或邻基协助。邻基参与作用是一种分子内 S_N2 过程，参与基团与离去基团在分子内处于反式共平面的构象才能发生参与作用，一般生成三元环或五元环状化合物或环状中间体，参与基团带有π电子或未共用电子对，参与基团有 COO^-(COOH 除外)、COOR、COAr、OCOR、OR、OH、O^-、NH_2、NHR、NR_2、NHCOR、SH、SR、S^-、SO_2Ph、I、Br、Cl、Ph、C═C。邻基参与使反应速率异常增大。只有一个手性碳且反应中心为手性碳时，邻基参与的结果将使构型保持。

Problems

1. In each of the following indicate which reaction will occur faster. Explain your reason.

(1) $CH_3CH_2CH_2CH_2Br$ or $CH_3CH_2CH_2CH_2I$ with sodium cyanide in dimethyl sulfoxide;

(2) 1-Chloro-2-methylbutane or 1-chloropentane with sodium iodide in acetone;

(3) Hexyl chloride or cyclohexyl chloride with sodium azide in aqueous ethanol;

(4) Solvolysis of 1-bromo-2,2-dimethylpropane or tert-butyl bromide in ethanol;

(5) Solvolysis of isobutyl bromide or sec-butyl bromide in aqueous formic acid;

(6) Reaction of 1-chlorobutane with sodium acetate in acetic acid or with sodium methoxide in methanol;

(7) Reaction of 1-chlorobutane with sodium azide or sodium *p*-toluenesulfonate in aqueous ethanol.

2. Predict whether the following reactions would occur by S_N2 or S_N1 mechanisms:

(1) 1-Bromo-3-methylbutane and $NaOCH_3$;

(2) 3-Bromo-3-methylbutane and CH_3OH.

Explain your prediction.

3. Under conditions of photochemical chlorination, $(CH_3)_3CCH_2C(CH_3)_3$ gave a mixture of two monochlorides in a 4:1 ratio. The structures of these two products were assigned on the basis of their S_N1 hydrolysis rates in aqueous ethanol. The major product (compound A) underwent hydrolysis much more slowly than the minor one (compound B). Deduce the structures of compounds A and B.

4. Write an equation, clearly showing the stereochemistry of the starting material and the product, for the reaction of (*S*)-1-bromo-2-methylbutane with sodium iodide in acetone. What is the configuration (*R* or *S*) of the product?

5. If the temperature is not kept below 25℃ during the reaction of primary alcohols with *p*-toluenesulfonyl chloride in pyridine, it is sometimes observed that the isolated product is not the desired alkyl *p*-toluenesulfonate but is instead the corresponding alkyl chloride. Suggest a mechanistic explanation for this observation.

6. Give a mechanistic to explain for the following equations.

(1) $(CH_3)_2CHCHBrCH_3 + Ag^+ \xrightarrow[H_2O]{C_2H_5OH} (CH_3)_2CHCH(CH_3)OC_2H_5 + (CH_3)_2C(OH)CH_2CH_3$

(2) $(CH_3)_3CCH_2Br + Ag^+ \xrightarrow[H_2O]{C_2H_5OH} (CH_3)_2C(OH)C_2H_5 + (CH_3)_2C(OC_2H_5)CH_2CH_3 + (CH_3)_2C{=}CHCH_3$

(3) 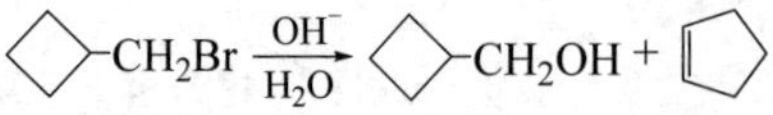 cyclobutyl-$CH_2Br \xrightarrow[H_2O]{OH^-}$ cyclobutyl-CH_2OH + cyclopentene

(4) 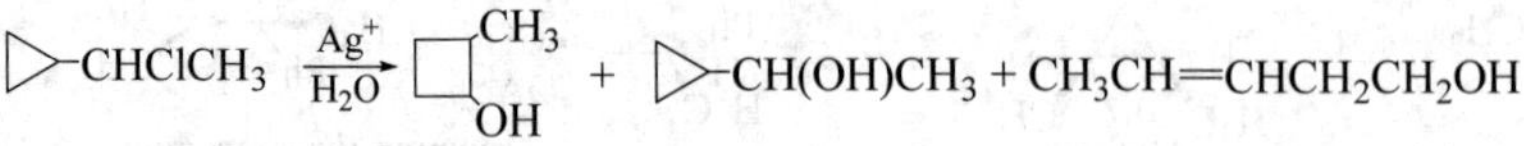 cyclopropyl-$CHClCH_3 \xrightarrow[H_2O]{Ag^+}$ 2-methylcyclobutanol + cyclopropyl-$CH(OH)CH_3 + CH_3CH{=}CHCH_2CH_2OH$

7. Complete the following reaction equations.

(1) $(CH_3)_2CHCH_2CH_2Cl \xrightarrow{NaCN} (\quad) \xrightarrow[Ni]{H_2} (\quad)$

(2) $(CH_3)_3N + CH_3CHBrCH_2CH_3 \longrightarrow (\quad) \xrightarrow[\triangle]{NaOH} (\quad)$

(3) 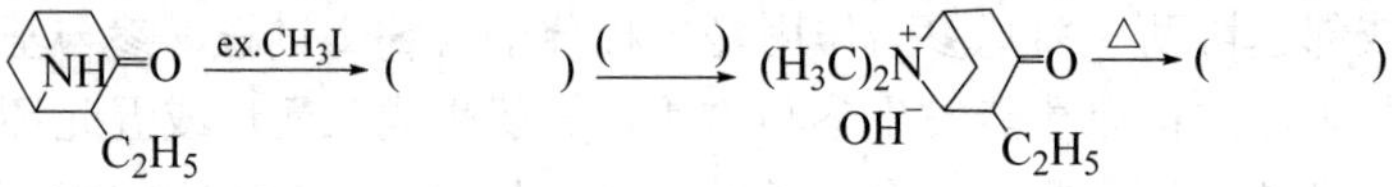 bicyclic amine (NH, C=O, C_2H_5) $\xrightarrow{ex.CH_3I} (\quad) \xrightarrow{(\quad)} (H_3C)_2\overset{+}{N}$ bicyclic ketone (C_2H_5) $OH^- \xrightarrow{\triangle} (\quad)$

(4) $2CH_3I + Na_2S \longrightarrow (\quad)$

(5) $ClCH{=}CHCH_2Cl + CH_3COONa \longrightarrow (\quad)$

(6) $2CH_3I + NaOOCCOONa \longrightarrow (\quad)$

(7) $CH_3CSSNa + CH_3I \longrightarrow (\quad)$

(8) 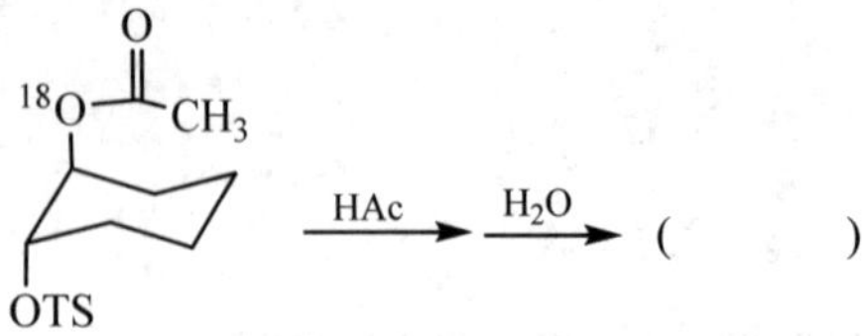 cyclohexane with ^{18}O-C(=O)CH_3 and OTS substituents $\xrightarrow{HAc} \xrightarrow{H_2O} (\quad)$

(9) 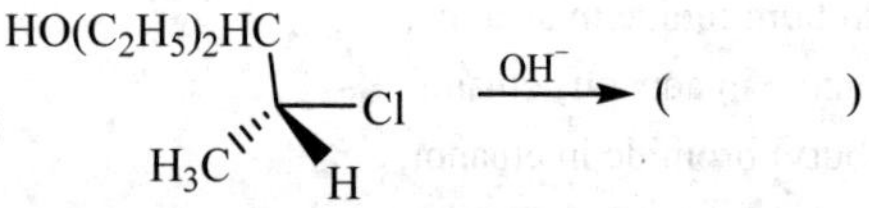 $HO(C_2H_5)_2HC$–C(Cl)(H)(H_3C) $\xrightarrow{OH^-} (\quad)$

(10) $C_2H_5I + CH_3O^- \xrightarrow{CH_3OH} (\quad)$

(11) $CH_3Cl + KCN \xrightarrow{CH_3OH} (\quad)$

(12) $CH_2{=}CHCH_2Br + CH_3COONa \longrightarrow (\quad)$

(13) $PhCH_2Br + OH^- \xrightarrow{H_2O} (\quad)$

Vocabulary

nucleophilic [ˌnjuːkliəu'filik] 亲核的
nucleofuge [njuːklɪəfjuːdʒ] 离核试剂
nucleophile ['njuːkliːəˌfaɪl] 亲核试剂
solvolysis [sɒl'vɒləsɪs] 溶剂解
electrophilic [ɪlektroʊ'fɪlɪk] 亲电的
proportional [prə'pɔːrʃənl] 成比例的
first-order rate equation 一级速率方程
rearrangement ['riːə'reɪndʒmənt] 重排
transition state 过渡态
displacement [dɪs'pleɪsmənt] 置换，位移
internal [ɪn'tɜːrnl] 内部的
retention of configuration 构型反转
rate-determining step 决速步骤
shielding effect 屏蔽效应

Walden inversion 瓦尔登转化
transform 变形，改变
polarizability ['poʊləraɪzə'bɪlətɪ] 极化性
protic 质子的
strip off 剥去，脱离
reinforce 加强，增强
aprotic 非质子的
steric hindrance 位阻
dielectric constant 介电常数
sulfonate ['sʌlfəˌneɪt] 磺酸酯（盐）
tosylate [təʊzɪ'leɪt] 甲苯磺酸盐
sulfonyl ['sʌlfəˌnɪl] 磺酰基
protonation 质子化反应
benzylic ['benzɪlɪk] 苄型
allylic ['æləˌlɪk] 烯丙基的
delocalization 离域
threo [θri'ːəʊ] 苏型
erythro [ərɪθ'rəʊ] 赤型
acetoxy [əs'tɒksɪ] 醋酸基(乙酰氧基)
acetolysis [æsɪ'tɒlɪsɪs] 醋酸解
nonclassical carbocation 非经典碳正离子
exo 外型
brosylate [brəʊsɪ'leɪt] 对溴苯磺酸盐(或酯)
endo 内型
anchimeric assistance 邻位促进
phenonium 苯鎓离子

Chapter 5

Addition and Elimination Reactions（加成与消除反应）

Addition and elimination process are the reverse of one another. Generally, the two process follow a similar mechanistic path in opposite directions, the final product depends on the conditions of the system.

5.1 Electrophilic addition reactions（亲电加成反应）

Electrophilic *addition* can be described in three ways:

(1) $\mathrm{E{-}Y \longrightarrow E^+ + Y^-}$

$$\mathrm{E^+ + {>}C{=}C{<} \longrightarrow {>}C(E){-}C^+{<} \xrightarrow{Y^-} {>}C(E){-}C(Y){<}}$$

(2) $$\mathrm{E{-}Y + {>}C{=}C{<} \longrightarrow {>}C(E){-}C^+{<} + Y^- \longrightarrow {>}C(E){-}C(Y){<}}$$

(3) $$\mathrm{2E{-}Y + {>}C{=}C{<} \longrightarrow [Y{-}E \cdots {>}C{=}C{<} \cdots Y{-}E] \longrightarrow {>}C(E){-}C(Y){<} + E{-}Y}$$

Mechanism (1) implies that a carbocation is generated which is free of the counterion Y^- at its formation. This process involves *prior dissociation* of the electrophilic reagent. Mechanism (2) also involves a carbocation *intermediate*, but it is generated in the presence of an anion and exists initially as an ion pair. Mechanism (3) is a process that has been established for several electrophilic additions. It implies transfer of the *electrophilic* and *nucleophilic* components of the reagent from two separate molecules. *Electron-donating groups* increase the reactivity of a C=C toward electrophilic addition and *electron-withdrawing groups* decrease it. For example, the reactivity toward electrophilic addition of a group of alkenes increased in the order $CCl_3CH{=}CH_2$ < $Cl_2CHCH{=}CH_2$ < $ClCH_2CH{=}CH_2$ < $CH_3CH_2{=}CH_2$.

During the electrophilic addition process, electrophilic reagent may enter from opposite sides for stereospecific *anti* addition or from the same sides for stereospecific *cis* addition.

cis addition   or   *anti* addition

The main electrophilic reagents and their products of a C=C are introduced as follows:

Electrophilic reagent	Product	Electrophilic reagent	Product
H—X	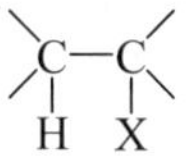 $>C(H)-C(X)<$	$I-N_3$	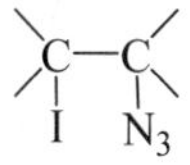 $>C(I)-C(N_3)<$
X—OH	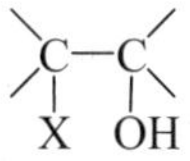 $>C(X)-C(OH)<$	Cl—NO	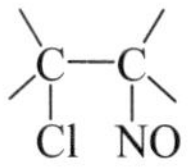 $>C(Cl)-C(NO)<$
X_1-X_2 (ICl, IBr, BrCl)	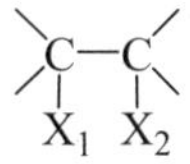 $>C(X_1)-C(X_2)<$	Br—SePh	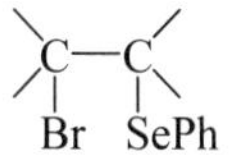 $>C(Br)-C(SePh)<$
X—X	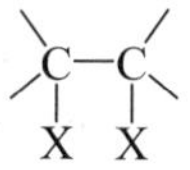 $>C(X)-C(X)<$	HO—SePh	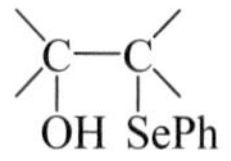 $>C(OH)-C(SePh)<$
$Br-N_3$	 $>C(Br)-C(N_3)<$		

5.1.1 Addition of hydrogen halides to alkenes（氢卤酸对双键的加成）

$$R_2C{=}CHR \xrightarrow{HX} R_2\overset{+}{C}-CH_2R \xrightarrow{X^-} R_2C(X)CH_2R$$

In the addition of hydrogen halides to alkenes, it is generally founded that the positive portion of the reagent goes to the side of the double that has more hydrogens and the *halogen* atom becomes attached to the most-substituted carbon atom of the alkene. This general observation is called Markovnikov's rule. The basis for this *regioselectivity* lies in relative ability of the carbon atoms to accept positive charge. Markovnikov's rule also applies for halogen substituents because the halogen stabilizes the carbocation by *resonance*:

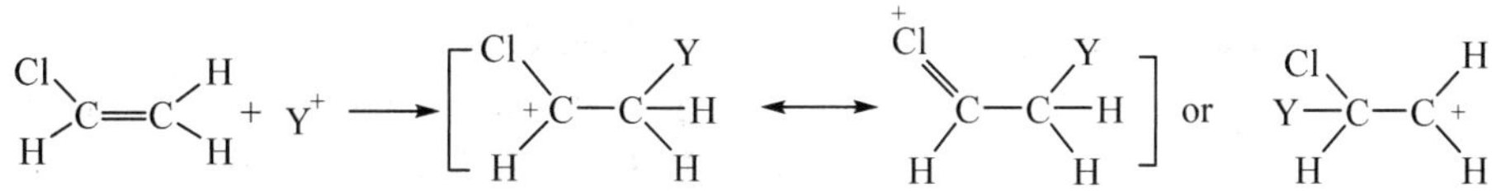

$$(Cl)(H)C{=}CH_2 + Y^+ \longrightarrow \left[(Cl)(H)\overset{+}{C}-CH_2Y \longleftrightarrow (\overset{+}{Cl}{=})(H)C-CH_2Y \right] \text{ or } (Cl)(Y)(H)C-\overset{+}{C}H_2$$

Alkenes containing strong electron-withdrawing groups may violate Markovnikov's rule. For example, the compound $NCCH{=}CH_2$ has been reported to give electrophilic addition with acids in an *anti*-Markovnikov direction.

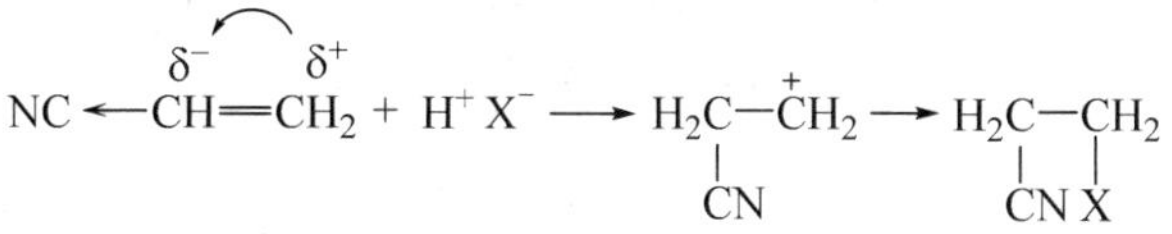

$$NC \leftarrow \overset{\delta^-}{C}H{=}\overset{\delta^+}{C}H_2 + H^+X^- \longrightarrow H_2C(CN)-\overset{+}{C}H_2 \longrightarrow H_2C(CN)-CH_2X$$

The regioselectivity of addition of hydrogen bromide to alkenes can be complicated if a free radical chain addition occur in competition with the ionic addition. The free radical reaction is readily initiated by peroxidic impurities or by light and leads to the *anti*-Markovnikov addition product.

$$CH_3CH{=}CH_2 + HBr \xrightarrow{ROOR} CH_3-CH_2-CH_2Br$$

在氢卤酸对不饱和键的加成反应过程中，氢加在不饱和键电荷偏负的碳原子上，对于脂肪烃基取代的双键，最后的加成产物为“反式”加成，当取代基为芳香基时，得到顺式加成产物。在加成过程中，除了得到正常加成产物以外，还会有重排产物。碳正离子都有重排成更稳定结构的趋势。

5.1.2 Acid-catalyzed hydration of alkenes（烯的酸催化水合反应）

Alkenes can be hydrated by treatment with water and an acid catalyst. The reaction is general acid-catalyzed, implying rate-determining proton transfer from the acid to the double bond. It is a classic organic reaction to synthesis alcohols.

The mechanism explains the observed formation of more highly substituted alcohol (Markovnikov's). The initial carbocation occasionally rearranges to a more stable one.

Alkenes can be hydrated quickly under mild conditions in high yields without rearrangement products by the use of oxymercuration (addition of oxygen and mercury) followed by in situ treatment with sodium borohydride.

(1) $Hg(OAc)_2$, H_2O
(2) $NaBH_4$
90 %

This method, which is applicable to mono-, di-, tri-, and tetraalkyl as well as phenyl-substituted alkenes, gives almost complete Markovnikov addition. Hydroxy, methoxy, acetoxy, halo, and other groups may be present in the substrate without causing difficulties.

烯烃的酸催化水合反应与加卤化氢机理相同，首先是酸提供质子与烯烃加成生成碳正离子，然后水与碳正离子结合，并脱去质子生成醇。水合反应立体选择性较差，加成产物“顺式”和“反式”都有。烯烃的酸催化水合反应过程中也有碳正离子重排产物。

5.1.3 Addition of halogens to alkenes（卤素与烯烃的加成）

During the addition reaction of halogens, interpretations of stereochemistry have focused attention on the role played by cyclic halonium ions. If the addition of Br^+ to the alkene results in a bromonium ion, the *anti* stereochemistry can be readily explained. *Nucleophilic* ring opening by bromide ion would occur by back-side attack at carbon, with rupture of one of C—Br bonds, giving overall *anti* addition. On the other hand, a rotating open of carbocation would be expected to give both *syn* and *anti* addition. If the principal intermediate was an ion pair that collapsed faster than rotation about the C—C bond, *syn* addition could predominate.

syn and *anti* addition

anti addition (X=Br)

syn addition

卤素与烯烃的加成是通过卤桥正离子进行的，溴正离子比氯正离子中间体稳定，溴与烯烃加成的立体选择性较强，加成产物为“反式”。

5.1.4 Hydroboration（硼氢化反应）

When alkenes are treated with borane in ether solvents, BH_3 adds across the double bond. For most substrates, the addition in *hydroboration* is stereospecific and *syn*, with attack taking place from the less-hindered side. The mechanism may be a cyclic four-center one. Thus the product remains the configuration of the starting material with no rearrangement.

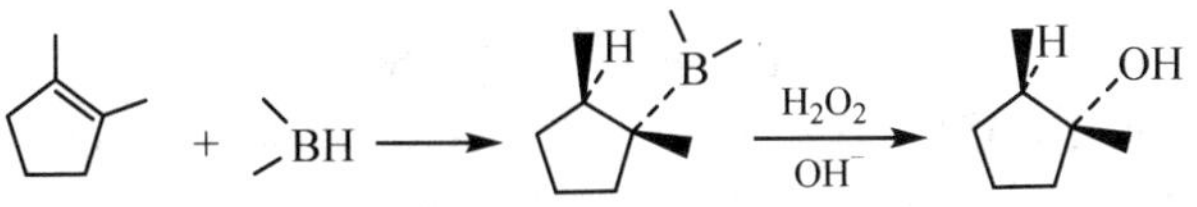

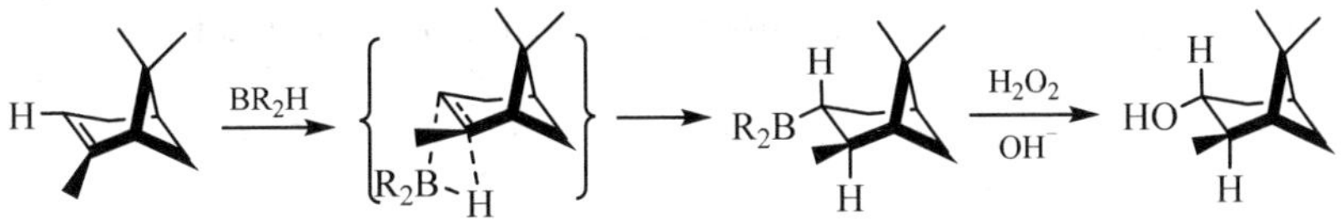

乙硼烷和烷基硼与烯烃迅速反应生成三烷基硼，然后用碱性双氧水氧化，同时水解得伯醇。在硼氢化时，硼总是加到较少取代的不饱和碳原子上。硼氢化反应总是生成“顺式”加成产物。

5.1.5 Addition of carbenes to Alkenes（卡宾对双键的加成）

The generation processes of carbenes can be in the following equations:

$$CHCl_3 + (CH_3)_3COK \longrightarrow :CCl_2 + (CH_3)_3COH + KCl$$

$$CH_2N_2 \xrightarrow[\text{or } h\nu]{\text{heat}} :CH_2 + N_2$$

$$CH_2{=}C{=}O \xrightarrow{h\nu} H_2C: + CO$$

Most carbenes are electrophilic. Electron-donating substituents on the alkene increase the rate of the reaction, and electron-withdrawing groups decrease it. The reactivity of some carbenes can be listed in the order:

$H_2C: > ROOCCH: > PhCH: > BrCH: > ClCH: > Cl_2C: > Br_2C:$

$:CH_2 > :CR_2 > :CAr_2 > :CX_2$

Carbenes in the singlet state (which is the most common state) react with alkenes stereospecifically and *syn*, probably by a one-step mechanism. The products of these cyclopropanations retain any *cis* or *trans* stereochemistry of the reactants.

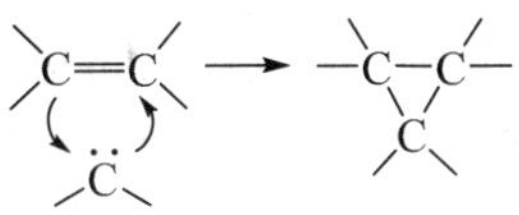

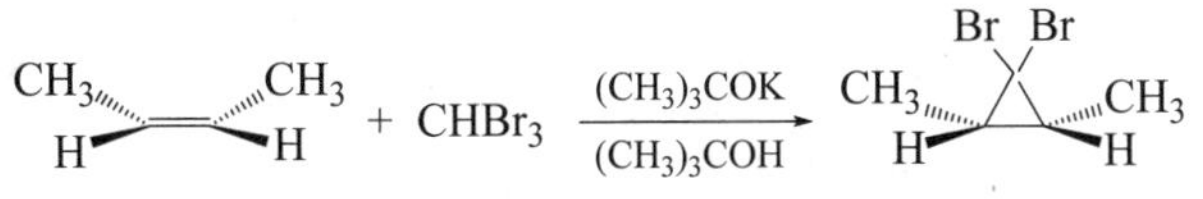

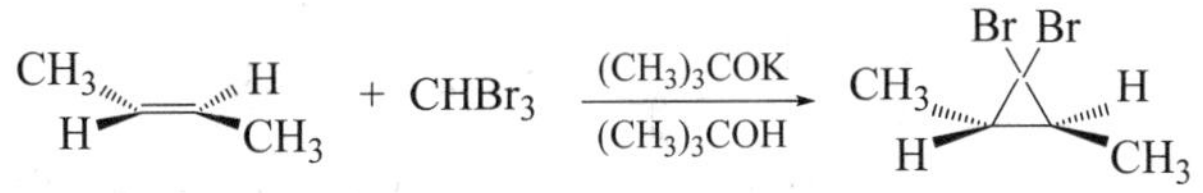

Carbenes in the triplet state react with alkenes nonstereospecifically, probably by a diradical mechanism.

For carbenes or *carbenoids* of the type R—C—R[1] there is another aspect of stereochemistry. When these species are added to all but symmetrical alkenes, two isomers are possible, even if the four groups originally on the double–bond carbons maintain their configurations:

Carbenes are so reactive that they add to the "double bonds" of aromatic rings. The products are usually unstable and rearrange to give ring expansion. Carbene reacts with benzene to give cycloheptatriene.

The Simmons–Smith procedure accomplishes the same result without a free carbene intermediate and without insertion side products. This procedure involves treatment of the double bond compound with CH_2I_2 and a Zn–Cu couple and leads to cyclopropane derivatives in good yields. Many groups such as —X, —NH_2, —OH, —C=O would not interfere the addition of Simmons–Smith reagent.

$$CH_2=CH-COOCH_3 + CH_2I_2 + Zn(Cu) \longrightarrow \triangleright\!-COOCH_3$$

A nitrene (R—N:) is the nitrogen analogue of a carbene. The nitrogen atom has only 6 valence electrons and is therefore considered an *electrophile*. A nitrene is a reactive intermediate and is involved in many chemical reactions.

Because nitrenes are so reactive, they are not isolated. Instead, they are formed as reactive intermediates during a reaction. There are two common ways to generate nitrenes:

$$R-C(=O)-NH_2 \xrightarrow{Br_2/NaOH} R-C(=O)-\ddot{N}:$$

Nitrenes react with alkenes similar to carbenes. A singlet nitrene reacts with retention of configuration. Nitrenes in the triplet state react with alkenes nonstereospecifically.

α 消去得卡宾或氮宾，卡宾为 6 电子结构。卡宾又称碳烯，它是一个中性二价碳的活性中间体。卡宾有单线态和三线态两种形式，卡宾对双键加成时，单线态的碳烯进行顺式加成，具有立体专一性，而三线态的碳烯加成时，没有立体专一性。一般来说，在液相中所得的主要是单线态的卡宾，若在溶液中加入惰性溶剂（如 C_6F_6）稀释，则主要产生三线态的碳烯。在气相中直接光照，若有氩气等惰性气体存在，以三线态碳烯为主，如果加入 O_2 等双游离基，这些双游离基与三线态碳烯结合，留下单线态碳烯。

5.2 Elimination reactions（消除反应）

Usually there are three ways of elimination reactions:

(1) α-elimination（α-消除反应）

In an α elimination, both groups are lost from the same atom to give a carbene (or a nitrene):

$$R_2CH{-}X \longrightarrow R_2C{:} + HX$$

(2) β-elimination（β-消除反应）

When two groups are lost from adjacent atoms so that a new double (or triple) bond is formed, the reaction is called β-elimination. Depending on the reagents and conditions involved, β-elimination might be a first-order (E1), second-order (E2), $E1_{CB}$ and *pyrolytic elimination* process.

$$RCH_2{-}CHR{-}X \longrightarrow RCH{=}CHR + HX$$

(3) γ-elimination（γ-消除反应）

In γ elimination, a three-membered ring is formed:

消除反应分为α 消除、β 消除和γ 消除，其中β 消除又分为 E1 消除、E2 消除、$E1_{CB}$ 消除、热消除等。

5.2.1 The E2 mechanism（E2 消除）

In the E2 mechanism (elimination, bimolecular), the two groups depart *simultaneously*, with the proton being pulled off by a base:

$$R—CH_2—\underset{L}{\overset{H}{C}}—R' + B \longrightarrow [\text{transition state}] \longrightarrow RCH{=}CHR' + HB^+ + L^-$$

The process of E2 elimination takes place in one step and the reaction kinetics is second order: first order in substrate and first order in base. This is a bimolecular process with both the base and the substrate participating in the transition state. A base can be H_2O, $N(CH_3)_3$, CH_3COO^-, $O_2NC_6H_4O^-$, $C_6H_5O^-$, HO^-, $^-OC_2H_5$, $^-NH_2$ or CO_3^{2-}. A leaving group can be $^+NR_3$, $^+PR_3$, $^+SR_2$, SO_2R, —Cl, —Br, —I, —OSO_2R or —OCOR. In most E2 eliminations where there are two or more possible elimination products (*orientation*), the product with the most highly substituted double bond will predominate (*Zaitsev's rule*).

The elimination of *anti-periplanar* conformation, in which H and X depart in opposite directions, is called *anti*-elimination. The elimination of *syn*-periplanar conformation, in which H and X depart in the same directions, is called *syn*-elimination. In the absence of special effects, *anti*-elimination is usually greatly favored over *syn*-elimination.

anti-periplanar
staggered conformation-lower energy

syn-periplanar
eclipsed conformation-higher energy

trans ← KI, acetone — meso — KOH → cis

cis ← KI, acetone — *dl* pair — KOH, alcohol → trans

In open-chain compounds, the molecule can usually adopt anti-periplanar conformation. Elimination of HBr to the right from *meso*-1,2-dibromo-1,2-diphenylethane gives *cis*-2-bromostilbene, while elimination of HBr from the (+) or (−) isomer gives the trans alkene. The reaction is a dehydrohalogenation induced by treatment with KOH in alcohol. Elimination of Br_2 to the left is a dehalogenation (the reverse of halogen addition to an alkene), caused by treatment with *iodide anion*. Zinc dust leads to the same reaction, but with a lower degree of stereospecificity. The mechanism of the iodide anion reaction is shown. The stereospecific results demonstrate that in these cases elimination is *anti*-elimination.

Cyclic compounds generally restrict the spatial orientation of ring substituents to relatively few arrangements. Consequently, reactions conducted on such substrates often provide us with information about the preferred orientation of reactant species in the transition state. Adjacent *trans* groups on a six-membered ring can be *diaxial* or *diequatorial* and the molecule is generally free to adopt either conformation, although one may have a higher energy than the other. *Anti*-periplanarity of the leaving groups requires that they must be diaxial, even if this is the conformation of higher energy. The results with *menthyl* and neomenthyl chlorides are easily interpretable on this basis. Menthyl chloride has two chair conformations: conformation **2,** in which the three substituents are all equatorial, is more stable than conformation **1**, in which the three substituents are all *axial*. The more stable chair conformation of neomenthyl chloride is **3**, in which the chlorine is axial; there are axial hydrogens on both adjacent carbons. The results are: neomenthyl chloride **3** gives rapid E2 elimination and the alkene produced is predominantly **5** (**5**/**4** ratio is 3:1) in accord with *Zaitsev's rule*. Since axial hydrogens are available on both sides, this factor does not control the direction of elimination and *Zaitsev's rule* is free to operate. However, for menthyl chloride, elimination is much slower and the product is entirely the *anti*-Zaitsev, **4**. It is much slow because the unfavorable conformation **1** has to be achieved before elimination can take place, and the product is **4** because only on this side there is an axial hydrogen.

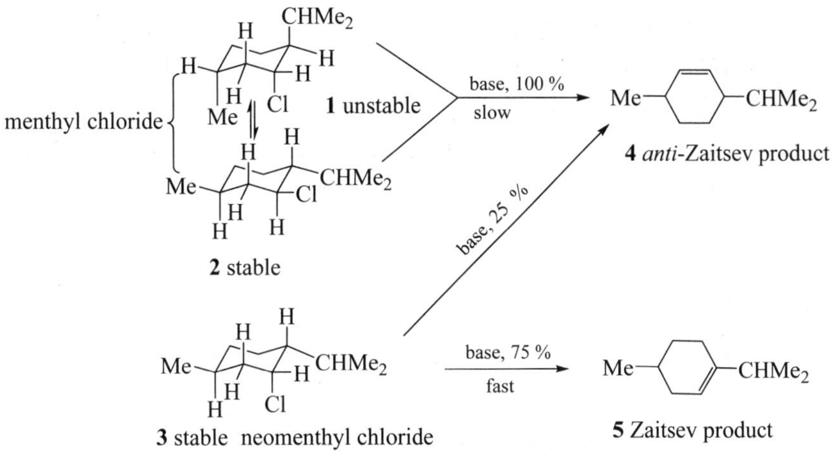

menthyl chloride

1 unstable

2 stable

base, 100 % slow

4 *anti*-Zaitsev product

base, 25 %

base, 75 % fast

3 stable neomenthyl chloride

5 Zaitsev product

Anti-elimination also occurs in the formation of triple bonds. The products in elimination of *cis*-2-chlorofumaric acid and *trans*- 2-chlorofumaric acid are both HOOCC≡CCOOH, but the *trans* isomer reacts ~50 times faster than the *cis* compound.

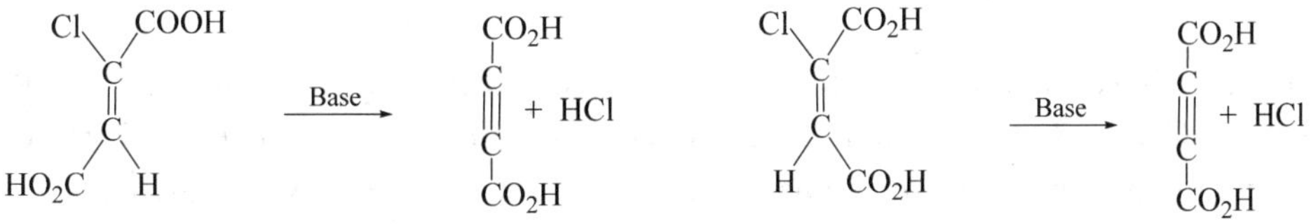

Base → + HCl

trans-2-chlorofumaric acid

Base → + HCl

cis-2-chlorofumaric acid

One example of *syn*-elimination has been found in molecules where H and X could not achieve an *anti*-periplanar conformation. The deuterated norbornyl bromide (X=Br, $^{+}NR_3$ etc) gave the product containing no deuterium. The exo X group cannot achieve a *dihedral* angle of 180° with the endo β hydrogen because of the rigid structure of the molecule. The dihedral angle here is ~120°. These leaving groups prefer *syn*-elimination with a dihedral angle of ~0° to *anti*-elimination with an angle of ~120°.

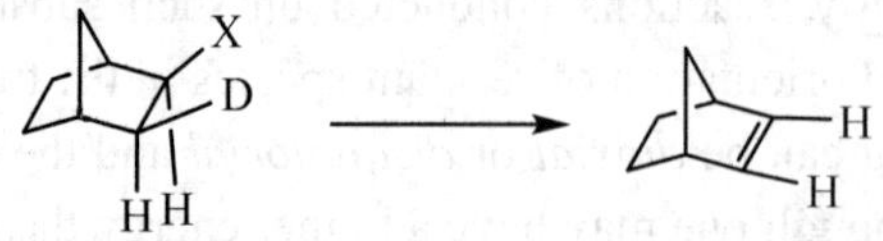

The examples given above illustrate two points: ①*Anti*-elimination requires a dihedral angle of 180°. When this angle cannot be achieved, *anti*-elimination is greatly slowed or prevented entirely. ②*Syn*-elimination is not found to any significant extent unless *anti* elimination is greatly diminished by failure to achieve the 180° dihedral angle.

A double bond does not go to a bridgehead carbon unless the ring sizes are large enough. A substrate with the leaving group on the bridgehead carbon does not undergo elimination.

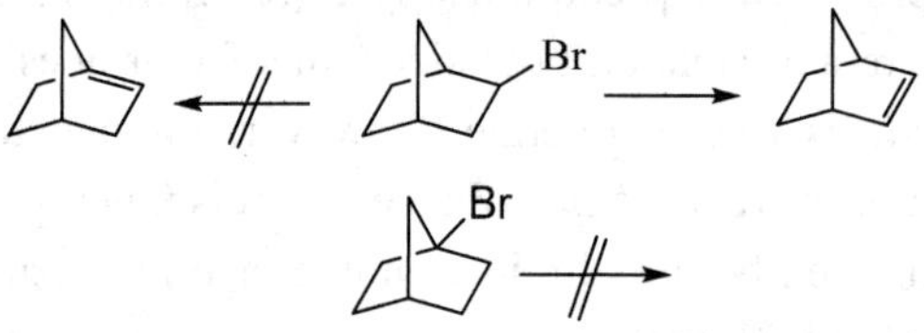

In most cases, compounds containing uncharged *nucleofuges* (those that come off as negative ions) follow *Zaitsev's rule*, just as they do in E1 elimination, no matter what the structure of the substrate. However, elimination from compounds with charged nucleofuges, for example, $\overset{+}{N}R_3$, $\overset{+}{S}R_2$(those that come off as neutral molecules), follow Hofmann's rule if the substrate is acyclic. Hofmann orientation is caused by the fact that the acidity of the β hydrogen is decreased by the presence of the electron-donating alkyl groups. The more acidic hydrogen is removed by the base.

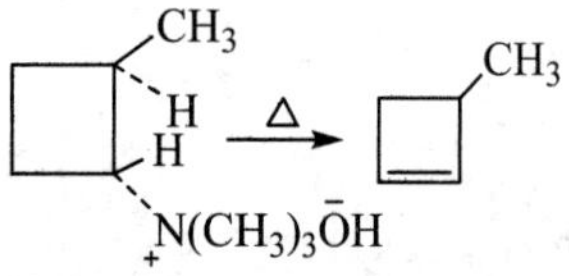

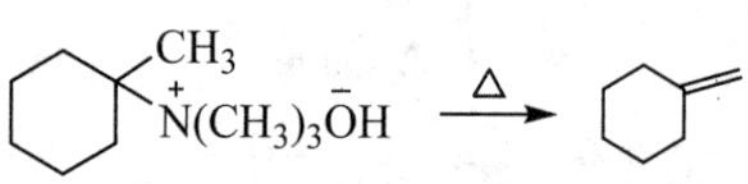

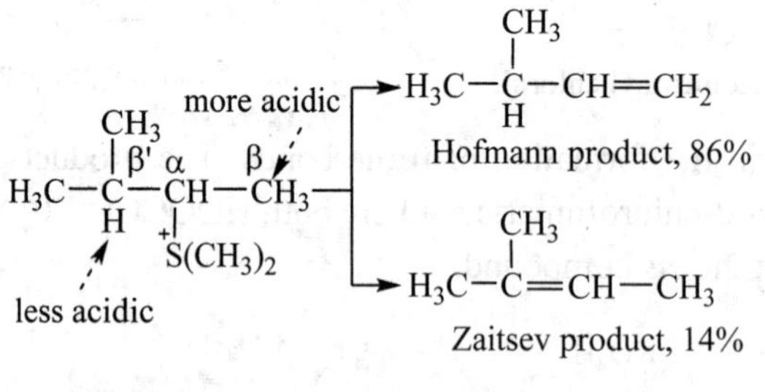

$H_3C-CH(CH_3)-CH(\overset{+}{S}(CH_3)_2)-CH_3$ (β' H: less acidic; β H: more acidic) → $H_3C-CH(CH_3)-CH=CH_2$ Hofmann product, 86% + $H_3C-C(CH_3)=CH-CH_3$ Zaitsev product, 14%

However, if one of the β carbon contains an electron-withdrawing group, such as —COR, —NO_2, —SO_2R, —CN, —C_6H_5 that can be in conjugation with the new double bond, the conjugated product usually predominates, sometimes even when the stereochemistry is unfavorable.

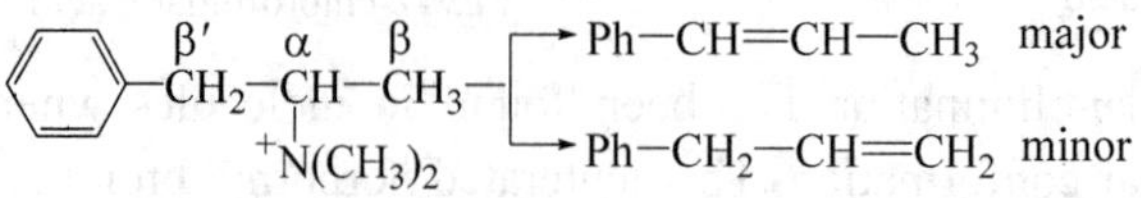

$C_6H_5-CH_2-CH(\overset{+}{N}(CH_3)_2)-CH_3$ (β' α β) → $Ph-CH=CH-CH_3$ major + $Ph-CH_2-CH=CH_2$ minor

按照 E2 历程进行的消除反应，是协同的一步反应，不经过中间体，一般按反式共平面的过渡态进行消除。中性底物的 E2 消除按 Saytzeff 规则进行。而带电荷的底物如季铵碱和锍鎓按 Hofmann 消除形成双键上连较少烷基的烯烃。底物分子α碳上支链越多，消去反应的活性越大。强碱有利于 E2，极性小的溶剂有利于 E2。开链化合物的消除一般以反式共平面为主。六元环化合物主要为反式消除。桥环化合物往往为顺式消除。邻二卤代物可用 Zn 或 NaI/丙酮脱去卤素形成双键。

5.2.2 The E1 mechanism（E1 消除）

The E1 mechanism is nearly identical to the S_N1 mechanism, differing only in the course of reaction taken by the carbocation intermediate. The E1 mechanism is a two-step process in which the rate-determining step is ionization of the substrate to give a carbocation that rapidly loses a β proton to a base.

$$R-CR(H)-CR_2-X \xrightarrow{slow} R-CR(H)-CR_2^+ + X^-$$

$$B^- + R-CR(H)-CR_2^+ \xrightarrow{fast} RRC=CRR + HB^+$$

$$CH_3-\overset{\alpha}{C}(CH_3)(\overset{\beta}{C}H_3)-Br \xrightarrow{slow} CH_3-\overset{\alpha}{C}^+(CH_3)(\overset{\beta}{C}H_2-H) \xrightarrow{OH^-} \begin{cases} \xrightarrow{i} (CH_3)_3COH + Br^- \\ \xrightarrow{ii} CH_3-C(CH_3)=CH_2 + H_2O + Br^- \end{cases}$$

Hydrolysis of *tert*-butyl bromide in a mixed solvent of water and acetonitrile gives a mixture of 2-methyl-2-propanol and 2-methylpropene at a rate independent of the water concentration. The alcohol is the product of an S_N1 reaction and the alkene is the product of the E1 reaction. The characteristics of these two reaction mechanisms are similar, as expected. They both show first order kinetics; neither is much influenced by a change in the nucleophile/base; and both are relatively *non-stereospecific*.

The comparison of E1 and E2 elimination mechanisms of alkyl halides were briefly listed in Table 5. 1.

Table 5.1 comparison of E1 and E2 elimination mechanisms

	E1	E2
promoting factors		
base	weak base work	strong base required
solvent	good ionizing solvent	wide variety of solvents
substrate	3° > 2° >1°	3° > 2° >1°
leaving group	good one required	good one required
characteristics		
kinetics	first order, k_r[RX]	second order, k_r[RX][B^-]
orientation	most highly substituted alkene	most highly substituted alkene
stereochemistry	no special geometry	*anti*-periplanar transition state required
rearrangements	common	impossible

按照 E1 历程进行的消除反应，与 S_N1 相似，为一级反应，可发生重排，不同处在于快步骤中失去β-H。主要生成在双键上连有烷基最多的烯烃，服从查依采夫（Saytzeff）规则。底物分子α碳上支链越多，消去反应的活性越大。弱碱或无碱性的溶剂有利于 E1，极性大的溶剂有利于 E1，好的离去基有利于 E1 和 E2。

5.2.3 The $E1_{CB}$ mechanism（$E1_{CB}$消除）

In the E1 mechanism, the leaving group X leaves first and then β-H leaves to produce an alkene. In the E2 mechanism, the leaving group X and β-H leave at the same time. There is a third possibility: The β-H leaves first to generate a conjugated base, and then the leaving group X leaves to give an alkene. This is a two-step process, called the $E1_{CB}$ mechanism, or the carbanion mechanism, since the intermediate is a carbanion:

$$R-CH_2-\underset{X}{\overset{H}{C}}-R' + B^- \longrightarrow R-\bar{C}H-\underset{X}{\overset{H}{C}}-R' \xrightarrow{-X^-} RCH{=}CHR'$$

The $E1_{CB}$ mechanism would most likely be found with substrates containing acidic hydrogens in the β carbon and poor leaving groups (such as —Oph, —OAc, —F). Compounds with an electron-withdrawing group (e.g., $-NO_2$, $-\overset{+}{S}Me_2$, $ArSO_2-$, —CN, —COOR) in the β carbon would increase the acidity of the β-H and stabilize the conjugated base.

$$\text{(OAc, Ph, H, NO}_2\text{)} + CH_3O^- \longrightarrow \text{(OAc, Ph, }^-\text{, NO}_2\text{)} \xrightarrow{-CH_3COO^-} \text{(Ph, NO}_2\text{)}$$

$$\text{(H, CF}_3\text{)} \underset{}{\overset{B^-}{\rightleftharpoons}} \text{(}\ddot{\ }\text{, CF}_3\text{)} \xrightarrow{-F^-} \text{(=CF}_2\text{)}$$

$$CH_3CH_2CH_2CH_2\underset{F}{C}HCH_3 \xrightarrow{t\text{-}BuO^-} \underset{\text{Hofmann product}}{CH_3CH_2CH_2CH_2CH{=}CH_2}$$

按照 $E1_{CB}$ 历程进行的消除反应，β-C 上连强吸电子基，离去基离核性能较差的底物主要发生 $E1_{CB}$ 消除。碱首先夺取β-H，中间体为 C^-离子，该反应为一级反应。主要生成在双键上连有烷基较少的烯烃，服从 Hofmann 消除规则。不好的离去基如 F^-、R_4N^+有利于 $E1_{CB}$。

5.2.4 Mechanisms of pyrolytic eliminations(热消除机理)

Several types of compounds undergo elimination upon heating, with no other reagent present. Reactions of this type are often run in the gas phase. The mechanisms are obviously different from E1, E2 and $E1_{CB}$, since all these three eliminations require a base in one of the steps, and there is no base or solvent present in pyrolytic elimination. Pyrolytic elimination involves a cyclic transition state, which may be four, five, or six membered. Examples of each size are as follows:

$$-\underset{H}{C}-\underset{Br}{C}- \longrightarrow \;>C{=}C<\; + H-Br$$

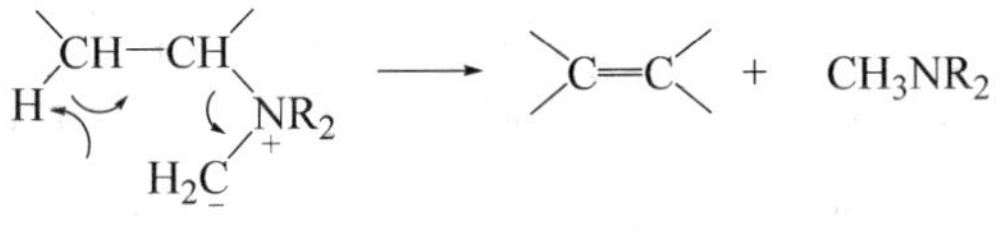

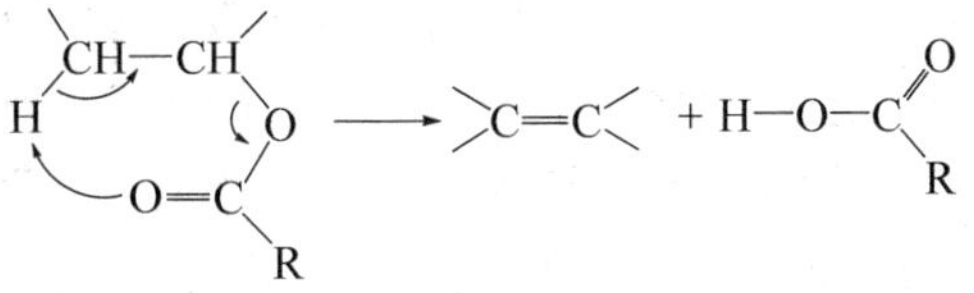

The elimination must be *syn* for the transition states. And for the four and five-membered transition states, the four or five atoms making up the ring must be coplanar and a *cis* β hydrogen is required. Therefore in cyclic systems, if there is a *cis* hydrogen on only one side, the double bond will go that way.

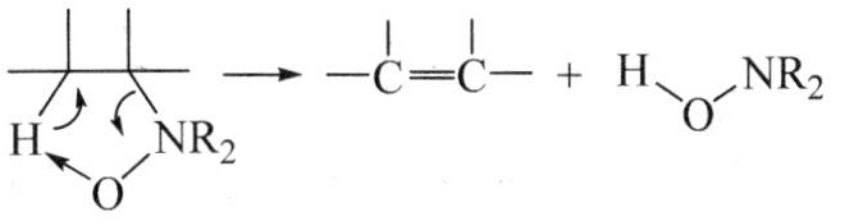

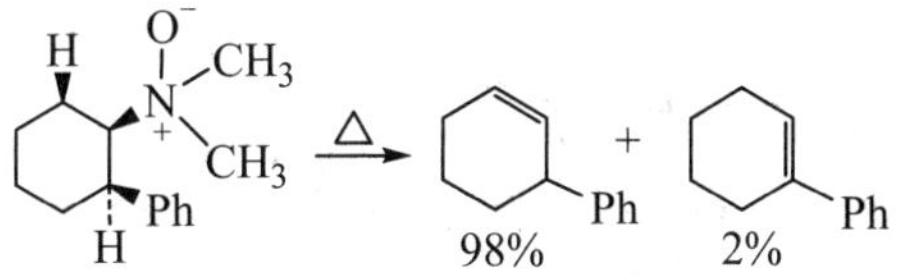

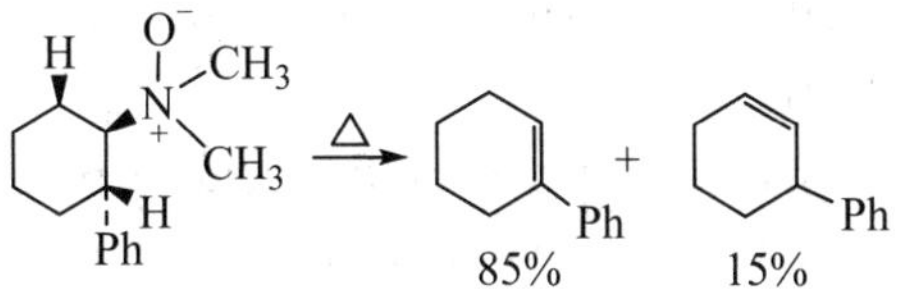

However, when there is a six-membered transition state, this does not necessarily mean that the leaving groups must be *cis* to each other, since such transition states need not be completely coplanar. If the leaving group is axial, then the hydrogen obviously must be equatorial (and consequently *cis* to the leaving group), because the transition state cannot be realized when the groups are both axial. But if the leaving group is equatorial, it can form a transition state with a β hydrogen that is either axial (hence, *cis*) or equatorial (hence, *trans*).

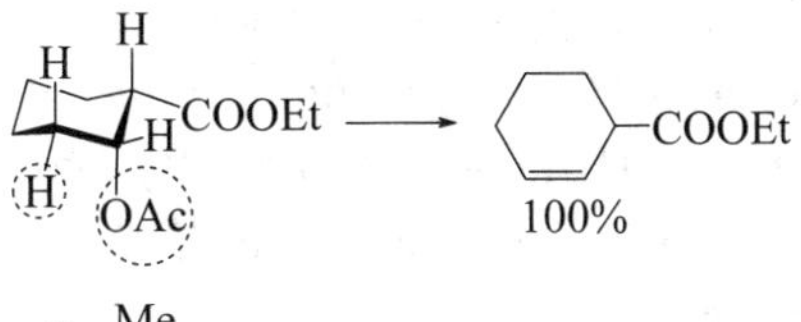

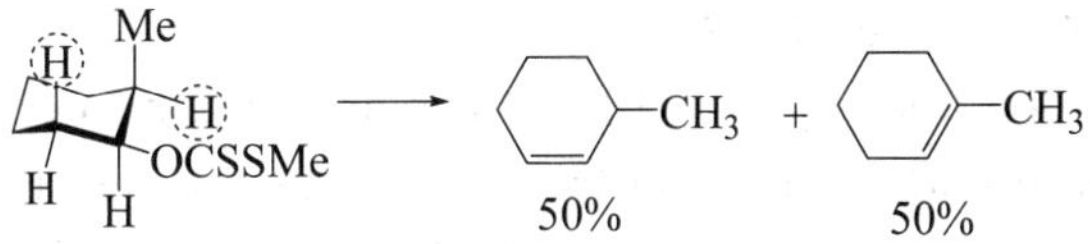

Orientation of pyrolytic elimination is statistical and is determined by the number of β hydrogens available (therefore Hofmann's rule is followed). For example, *sec*-butyl acetate gives 55%～62% yield of 1-butene and 38%～45% yield of 2-butene, which is close to the 3:2 distribution predicted by the number of hydrogens available.

$$CH_3CH_2\underset{\substack{| \\ OCOCH_3}}{C}HCH_3 \xrightarrow{\triangle} \underset{\text{major}}{CH_3CH_2CH{=}CH_2} + \underset{\text{minor}}{CH_3CH{=}CHCH_3}$$

sec-butyl acetate

热消除（分子内的消除反应）即通过四元环、五元环或六元环的环状过渡态进行的消除反应，不需要酸或碱催化，为单分子反应。通过四元环、五元环的环状过渡态进行的消除反应为顺式消除。通过六元环的环状过渡态进行的消除反应既有顺式、也有反式消除。例如羧酸酯的热解可避免醇、酸催化脱水的重排与异构化反应，发生 Hofmann 消除，主要用来合成末端烯烃。

5.3 Competition between elimination and substitution(消除反应与取代反应的竞争)

Elimination reactions are usually accompanied by substitution reaction. When the reagent is a good base it accepts a proton to yield elimination product and if it is a good nucleophile then it attacks the carbon to give substitution product. The proportion of elimination and substitution depends upon the following:

5.3.1 Structure of the substrate（底物结构）

The proportion of elimination increases from $1^\circ \rightarrow 2^\circ \rightarrow 3^\circ$ substrates. The reason is that alkenes formed on elimination are stabilized by hyperconjucation. The steric strain is relieved on the formation of alkene, whereas on substitution the strain is reintroduced.

5.3.2 Nature of the base（碱的性质）

Strong base promotes elimination over substitution and in particular E2 over E1. Alcoholic KOH favors elimination and aqueous KOH favors substitution. Strong nucleophiles but weak bases promote substitution over elimination whereas strong base but weak nucleophile promotes elimination over substitution. Though pyridine and R_3N are not strong bases they are poor nucleophile because the branching at the nitrogen atom causes steric hindrance to nucleophilic attack on carbon. Hence, they act as base to accept the more exposed hydrogens of the substituent groups to afford alkene. A similar steric effect is observed with the size of the base or nucleophile. Elimination increases with increase in the size of the nucleophile.

5.3.3 Nature of solvent（溶剂性质）

A less polar solvent not only favors bimolecular reaction but also E2 over S_N2. Change of protic solvents to aprotic solvents increases the base strength as the solvents layer around the base by hydrogen bonding is absent. Thus Cl^-, OH^-, OR^- etc., are very strong bases in DMF or DMSO. The use of aprotic solvent may change the pathway from E1 to E2.

5.3.4 Effect of temperature（温度影响）

In elimination reaction a strong C—H bond has to break, hence a high activation energy is required for elimination reaction rather than for substitution reaction. In general, the proportion of elimination increases on using a strong base with high concentration and a solvent with low polarity. On the other hand the proportional substitution increases by using a weak base with low concentration and a solvent with high polarity.

亲核取代反应与消除反应的取向与卤代烃的结构有直接关系。直链一级卤代烃易进行 S_N2 反应，不易发生消除反应，只有在强碱条件下才可发生消除反应。二级卤代烃及 β 碳上有支链的一级卤代烃进行 S_N2 反应较慢。在低极性溶剂、强亲核试剂条件下有利于 S_N2 反应；在低极性溶剂、强碱性试剂条件下，有利于 E2 反应。三级卤代烃难以进行 S_N2 反应，在强碱条件下有利于 E2 反应。三级卤代烃在无强碱存在时，得到 S_N1 和 E1 混合产物，生成物的相对比例依条件而定，逐渐增大碱的浓度，将增加消除反应的比例。高温有利于消除反应。

Problems

1. *Trans*-2-methyl-1-chlorocyclohexane reacts with alcoholic KOH at a much slower rate than does its *cis*-isomer. Furthermore, the product from elimination of the *trans*-isomer is 3-methylcyclohexene (not predicted by the *Zaitsev rule*), whereas the *cis*-isomer gives the predicted 1-methylcyclohexene as the chief product. Explain the results.

CH_3, Cl —KOH / alcohol→ CH_3 (slow)

CH_3, Cl —KOH / alcohol→ CH_3 + CH_3 (fast)

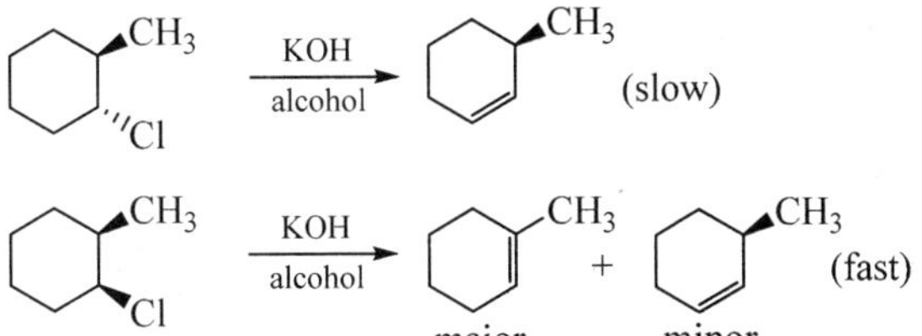

major minor

2. When the first compound shown below is treated with sodium methoxide, the only elimination product is the *trans* isomer. The second diastereomer gives only the *cis* product. Explain the results.

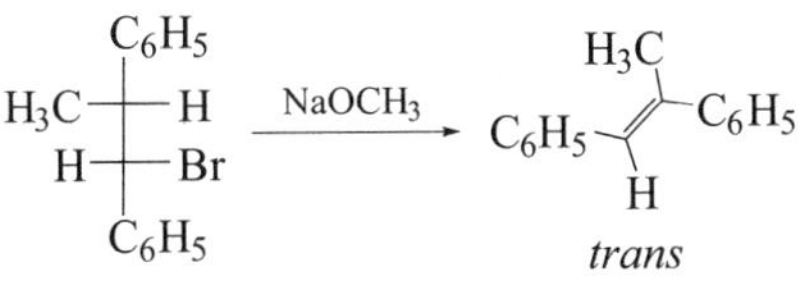

C_6H_5, H_3C—H, H—Br, C_6H_5 —$NaOCH_3$→ H_3C, C_6H_5, C_6H_5, H *trans*

threo *dl* pair

C_6H_5, H_3C—H, Br—H, C_6H_5 —$NaOCH_3$→ H_3C, C_6H_5, H, C_6H_5 *cis*

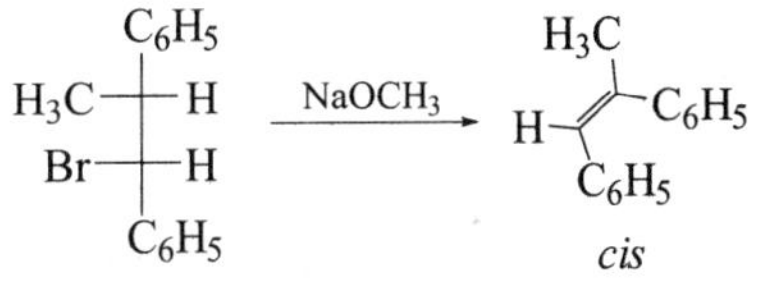

erythro *dl* pair

3. Complete the following reactions.

(1) 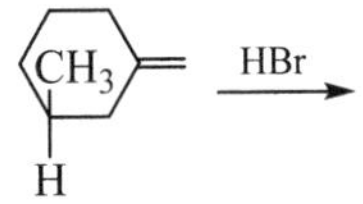 CH_3, H —HBr→

(2) 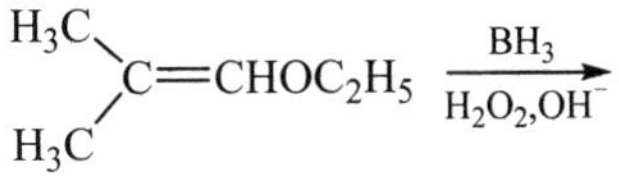 $(H_3C)_2C{=}CHOC_2H_5$ —BH_3 / H_2O_2,OH^-→

(3) 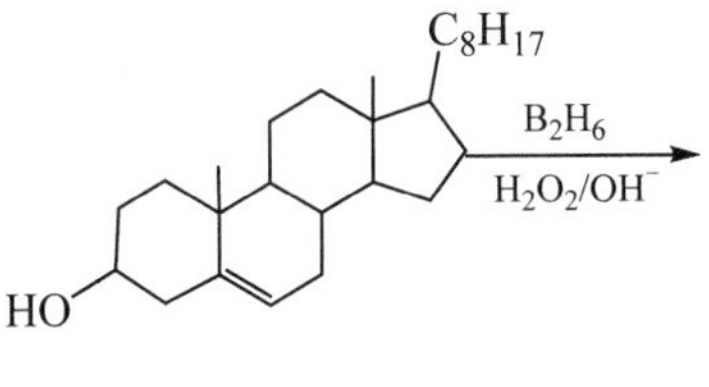 C_8H_{17}, HO —B_2H_6 / H_2O_2/OH^-→

(4) 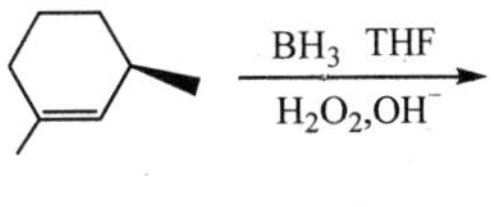 —BH_3 THF / H_2O_2,OH^-→

(5) 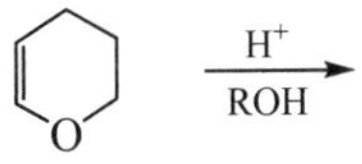 O —H^+ / ROH→

(6) —BH_3 THF / H_2O_2,OH^-→

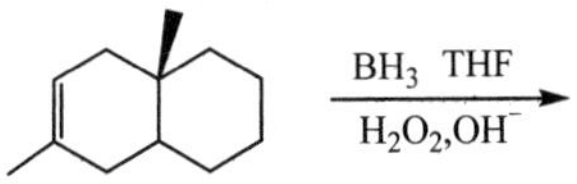

(7) 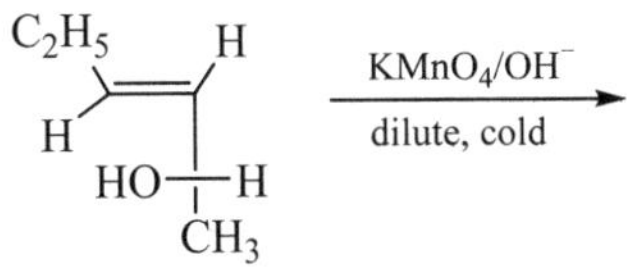 C_2H_5, H, H, HO—H, CH_3 —$KMnO_4/OH^-$ / dilute, cold→

(8) 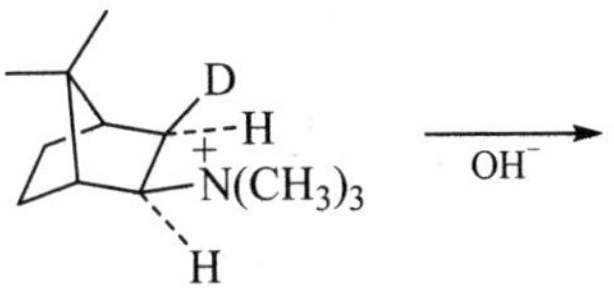 D, H, $\overset{+}{N}(CH_3)_3$, H —OH^-→

(9) 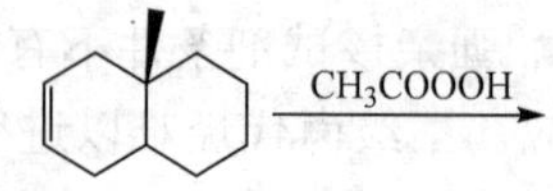 [bicyclic alkene] $\xrightarrow{CH_3COOOH}$ (10) 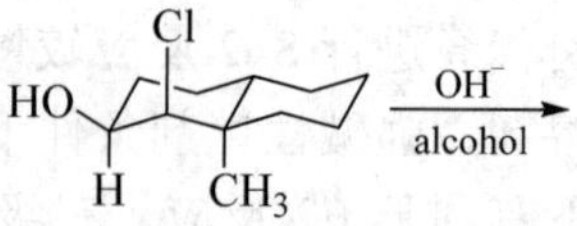 [Cl, HO, H, CH_3 decalin] $\xrightarrow[\text{alcohol}]{OH^-}$

(11) 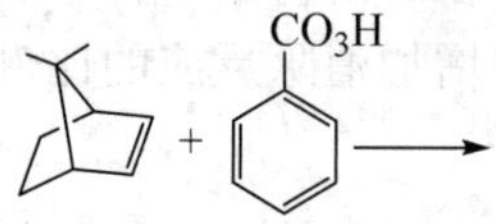 [bicyclic alkene] + [$C_6H_5CO_3H$] $\longrightarrow$ (12) 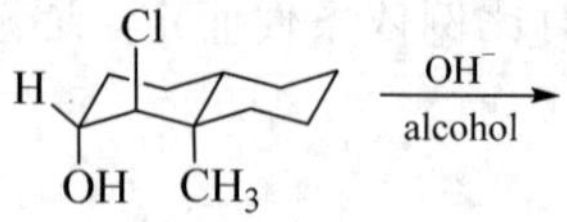 [Cl, H, OH, CH_3 decalin] $\xrightarrow[\text{alcohol}]{OH^-}$

(13) 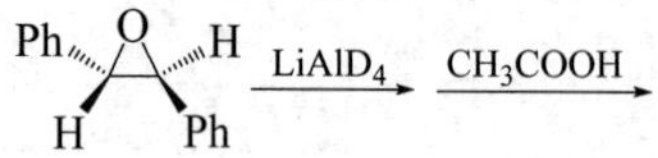 [Ph, H, H, Ph epoxide] $\xrightarrow{LiAlD_4}$ $\xrightarrow{CH_3COOH}$ (14) 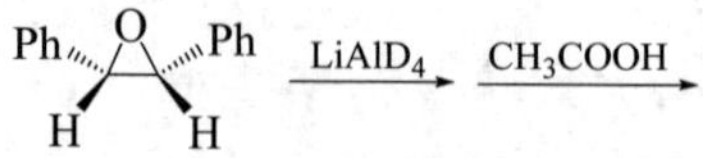 [Ph, Ph, H, H epoxide] $\xrightarrow{LiAlD_4}$ $\xrightarrow{CH_3COOH}$

(15) 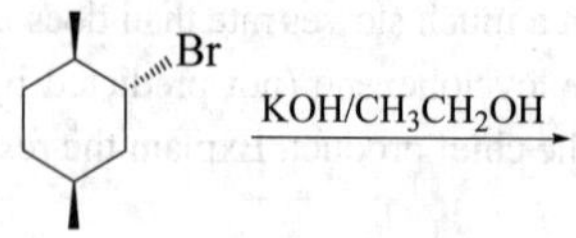 [Br cyclohexane] $\xrightarrow{KOH/CH_3CH_2OH}$ (16) 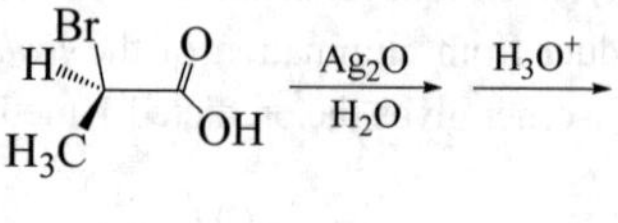 [Br, H, H_3C, COOH] $\xrightarrow[H_2O]{Ag_2O}$ $\xrightarrow{H_3O^+}$

(17) 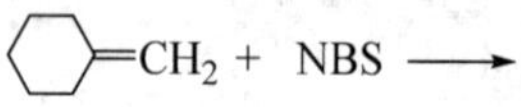 [cyclohexylidene]$=CH_2$ + NBS $\longrightarrow$ (18) $H_2C{=}CH{-}COOH$ + HI $\longrightarrow$

(19) 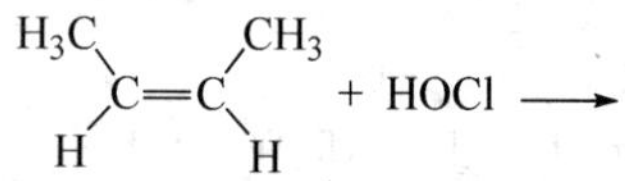 $H_3C(H)C{=}C(H)CH_3$ + HOCl $\longrightarrow$ (20) $(CH_3)_2C{=}CH_2$ + PhSCl $\longrightarrow$

4. Write down the possible mechanisms of the following reactions.

(1) $CH_2{=}CHCH(CH_3)_2 + HBr \longrightarrow CH_3CHBrCH(CH_3)_2 + CH_3CH_2CBr(CH_3)_2$

(2) $CH_2{=}CHC(CH_3)_3 + HBr \longrightarrow CH_3CHBrC(CH_3)_3 + CH_3CH(CH_3)CBr(CH_3)_2$

(3) $2(CH_3)_2C{=}CH_2 \xrightarrow{H^+} (CH_3)_3CCH{=}C(CH_3)_2 + (CH_3)_3CCH_2C(CH_3){=}CH_2$

(4) $CH_2{=}CH_2 + Br_2 \xrightarrow[H_2O]{NaCl} BrCH_2CH_2Br + BrCH_2CH_2OH + BrCH_2CH_2Cl$

(5) $CH_2{=}CH{-}CH{=}CH_2 + Cl_2 \xrightarrow{CH_3OH} CH_3OCH_2CH{=}CHCH_2Cl + CH_2{=}CHCH(OCH_3)CH_2Cl$

(6) 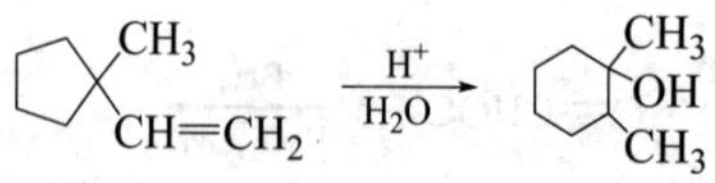 [1-methyl-1-vinylcyclopentane] $\xrightarrow[H_2O]{H^+}$ [CH_3, OH, CH_3 cyclohexane]

5. Complete the following equations.

(1) [1-ethyl... CH_3 methylenecyclopentane $=CH_2$] $\xrightarrow[500℃]{Cl_2}$ ( ) $\xrightarrow[\text{peroxide}]{HBr}$ ( )

(2) [cyclohexyl $CH{=}CH_2$] $\xrightarrow{RCOOOH}$ ( ) $\xrightarrow[OH^-]{H_2O}$ ( )

(3) [octahydronaphthalene] $\xrightarrow[(2)\ Zn/H_2O]{(1)\ O_3}$ 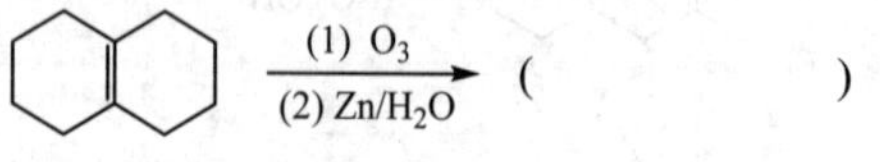 ( )

(4) 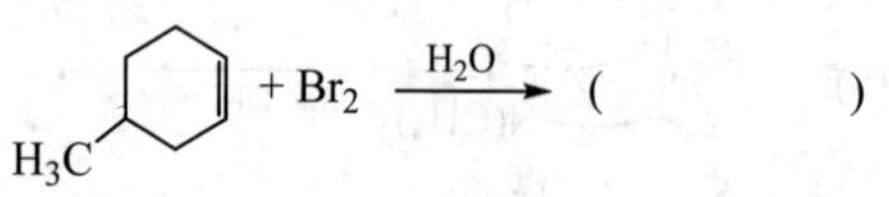 [H_3C cyclohexene] + Br_2 $\xrightarrow{H_2O}$ ( )

Vocabulary

elimination reactions 消除反应
addition 加成
prior ['praiə] 优先的，在前的
dissociation [di,səuʃi'eiʃən] 分离，离解
intermediate 中间体
electrophilic additions 亲电加成
nucleophilic 亲核的
electron-withdrawing groups 吸电子基
electron-donating groups 供电子基
regioselectivity [ri:dʒi:əusilek'tiviti] 区域选择性
resonance ['rezənəns] 共振
halogen ['hælədʒən] 卤族
hydroboration 硼氢化作用
carbenoids [['kɑːbenɔid] 类卡宾
electrophile [i'lektrəfail] 亲电试剂
simultaneously [saiməl'teiniəsli] 同时地
iodide anion 碘负离子
diaxial 双轴向的
dihedral angle 二面角
axial ['æksi:əl] 轴的，轴向的
anti-periplanar 反式共平面
diequatorial 双平伏的
menthyl ['menθɪl] 薄荷基
nucleofuges ['njuːklɪəfjuːdʒ] 离核试剂
non-stereospecific 非立体专一性的
pyrolytic elimination 热解消除，热消除
orientation 取向

Chapter 6

Reactions of Carbonyl Compounds

（羰基化合物的反应）

6.1 The nucleophilic addition reaction mechanisms（亲核加成反应机理）

The most common reaction of *aldehyde*s and *ketone*s is nucleophilic addition.

$$R^1R^2C^{\delta+}=O^{\delta-}:$$

The carbonyl group is polar and the partial positive charge on the carbonyl carbon causes *carbonyl compounds* to be attacked by nucleophiles. In most cases, it is the nucleophile that forms the first new bond to carbon, and these reactions are regarded as nucleophilic additions, which can be represented as:

$$RC(=O)R' + Y^- \xrightarrow{\text{slow}} RR'C(Y)O^- \xrightarrow{H^+} RR'C(Y)OH$$

It is also possible for the oxygen atom to react as a base, attacking the electrophilic species. This species is most often a proton or Lewis acid and the mechanism is:

$$RC(=O)R' + H^+ \longrightarrow R\overset{+}{C}(OH)R' \xrightarrow[Y^-]{\text{slow}} RR'C(Y)OH$$

Whether the nucleophile attacks the carbon or the oxygen atom attacks the electrophilic species, the rate-determining step is usually the one involving nucleophilic attack. Many of these reactions can be catalyzed by both acids and bases.

亲核加成反应机理：羰基是醛、酮的特征官能团，羰基中的碳原子及氧原子均为 sp^2 杂化，碳氧双键上的电子云偏向氧原子，使羰基碳原子的电子云密度显著减少，容易受亲核试剂进攻而发生亲核加成反应。

碱催化：

$$RC(=O)R' + Y^- \xrightarrow{慢} RR'C(Y)O^- \longrightarrow RR'C(Y)OH$$

碱催化可以提高亲核试剂的亲核性。

酸催化：

$$R_2C{=}O + H^+ \longrightarrow R(OH)C^+R' \xrightarrow[Y^-]{slow} R(Y)(OH)CR'$$

酸催化可提高羰基的亲电活性。

决速步骤为亲核试剂对羰基的进攻，得到具有四面体结构的中间体。羰基化合物的结构与亲核试剂的性质对亲核加成的难易程度都有影响。羰基化合物的结构对亲核加成的难易程度有如下影响。①电子效应：羰基碳连有$-I$、$-C$基团将使羰基碳的正电性增加，有利于亲核试剂的进攻；反之，连有$+I$、$+C$基团，将使羰基碳的正电性下降，不利于亲核试剂的进攻。②空间效应：羰基碳连有基团的体积增加，空间位阻增加，不利于亲核试剂进攻，达到过渡状态所需活化能增加，故反应活性相对下降。

6.2 The nucleophilic addition reactions to the carbonyl groups（羰基的亲核加成反应）

6.2.1 The addition of H_2O（与水的加成）

The adduct formed upon addition of water to an aldehyde or ketone is called a hydrate or *gem-diol*. The reaction is subject to both general-acid and general-base catalysis. The products are usually stable only in water solution and decompose on distillation; that is, the equilibrium shifts back toward the carbonyl compound.

$$R_2C{=}O + H_2O \rightleftharpoons R{-}C(OH)(R){-}OH$$

$$R_2C{=}\ddot{O}: \overset{H_3O^+}{\rightleftharpoons} R_2C{=}\overset{+}{O}H \rightleftharpoons R_2C(OH){-}\overset{+}{O}H_2 \rightleftharpoons R_2C(OH)_2$$

In Acid

Step 1: Protonation　Step 2: Water adds　Step 3: Deprotonation

$$HO^- + R_2C{=}\ddot{O}: \rightleftharpoons R_2C(O^-)(OH) \overset{H_2O}{\rightleftharpoons} R_2C(OH)_2$$

In base

Step 1: Hydroxide adds　Step 2: Protonation

The position of the equilibrium is greatly dependent on the structure of the hydrate. Strongly electron-attracting groups on the alkyl group of a ketone or aldehyde favor the hydrate. The hydrate of *chloral* (trichloroacetaldehyde) is a stable crystalline substance.

6.2.2 The addition of alcohols（与醇的加成）

In a similar fashion to the formation of hydrates with water, aldehydes and ketones form *acetals* and *ketal*s through reaction with alcohols in the presence of acid catalysts. In the formation of an acetal and ketal, two molecules of alcohol add to the carbonyl group, and one mole of water is eliminated.

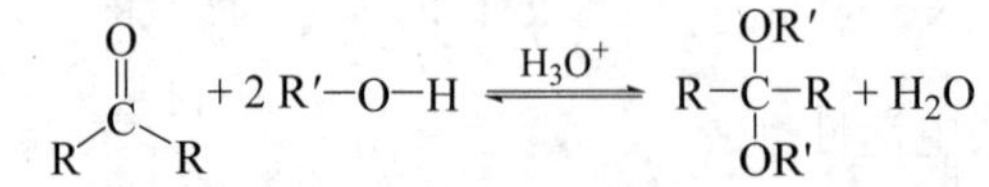

$$R-CO-R + 2\,R'-O-H \xrightleftharpoons{H_3O^+} R-C(OR')_2-R + H_2O$$

Acetals and ketals formation only occurs with acid catalysis. The first step is the typical acid catalyzed addition to the carbonyl group. The *hemiacetal* reacts further to produce the more stable acetals and ketals. The second half of the mechanism starts with protonation of the hydroxyl group, followed by loss of water, gives a resonance-stabilized carbocation. Attack on the carbocation by methanol, followed by loss of a proton, gives the acetal.

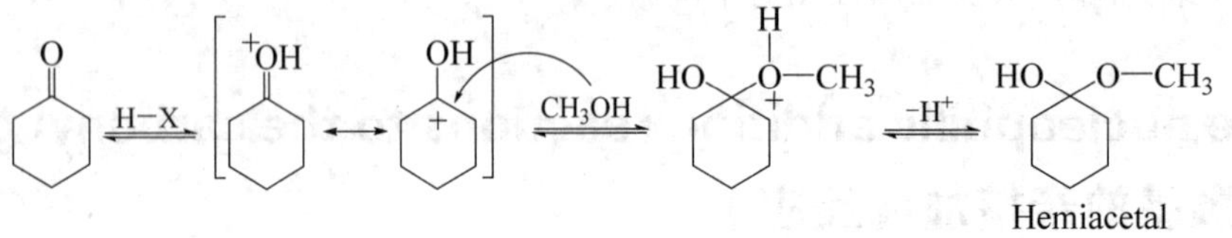

H–X, CH_3OH, $-H^+$

Hemiacetal

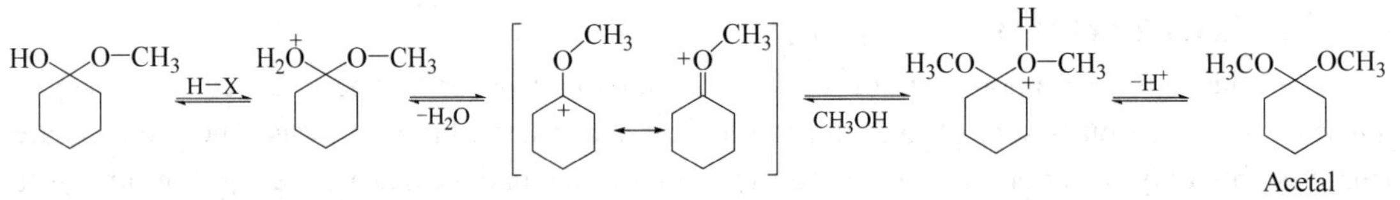

H–X, $-H_2O$, CH_3OH, $-H^+$

Acetal

More commonly, instead of two molecules of alcohols being used, a diol is used (*entropically* more favorable). This produces cyclic acetals and ketals.

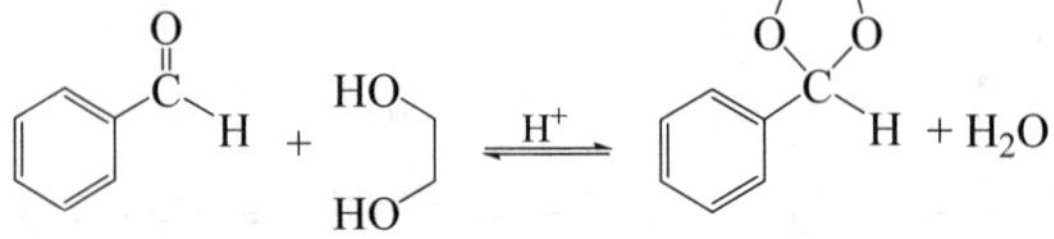

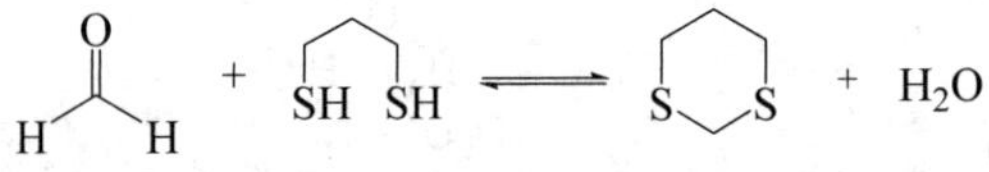

H^+ + H_2O

+ H_2O

This reaction is reversible, and acetals and ketals can be hydrolyzed by treatment with acid. The large excess of water drives the equilibrium toward the ketone or aldehyde. The mechanism is the reverse of the acetal formation. The reaction is not catalyzed in either direction by bases, so most acetals and ketals are quite stable to bases, though they are easily hydrolyzed by acids. This reaction is therefore a useful method of protection of aldehyde or ketone functions from attacking by strong bases and nucleophiles.

Consider the proposed synthesis below. The necessary Grignard reagent could not be made because the carbonyl group would react with the nucleophilic organometallic group.

Proposed synthesis

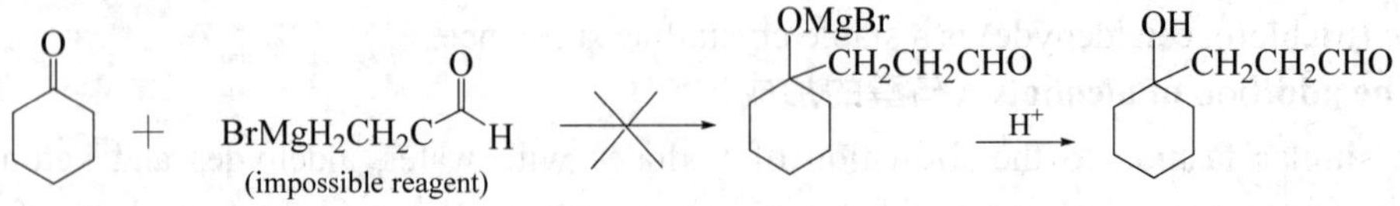

$BrMgH_2CH_2C(O)H$ (impossible reagent); OMgBr, CH_2CH_2CHO; H^+; OH, CH_2CH_2CHO

If the aldehyde is protected as an acetal, it is unreactive toward a Grignard reagent. The protected aldehyde is converted to the Grignard reagent, which can be allowed to react with *cyclohexanone*.

Actual synthesis

$$BrH_2CH_2C(=O)H \xrightarrow{HOCH_2CH_2OH} BrH_2CH_2C(OCH_2CH_2O)H \xrightarrow[ether]{Mg} BrMgH_2CH_2C(OCH_2CH_2O)H \xrightarrow{\text{cyclohexanone}} \text{1-(OMgBr)-1-}(CH_2CH_2CH(OCH_2CH_2O))\text{cyclohexane} \xrightarrow{H^+} \text{1-(OH)-1-}(CH_2CH_2CHO)\text{cyclohexane}$$

Most aldehydes are easily converted to acetals. For ketones the process is more difficult, presumably for steric reasons.

Hemiacetals can gain stability by being cyclic—when the carbonyl group and the attacking hydroxyl group are part of the same molecule. The reaction is now an intramolecular (within the same molecule) addition, as opposed to the intermolecular (between two molecules) ones.

$$OHC(CH_2)_3CH_2OH \underset{}{\overset{HCl}{\rightleftharpoons}} \text{2-hydroxytetrahydropyran (cyclic hemiacetal, OH)}$$

6.2.3 The addition of hydrogen cyanide（与氢氰酸的加成）

Cyanide reacts rapidly with carbonyl compounds producing *cyanohydrins*, via the base catalyzed nucleophilic addition mechanism.

$$R_2C{=}O + {}^{-}C{\equiv}N \longrightarrow R{-}C(O^-)(CN){-}R \xrightarrow{H{-}CN} R{-}C(OH)(CN){-}R$$

Step 1: Cyanide adds　　　　Step 2: Protonation

Like hydrate formation, cyanohydrin formation is an equilibrium governed reaction (i.e. reversible reaction), and corresponding aldehydes are more reactive than ketone.

6.2.4 The addition of organometallic reagents（与金属有机试剂的加成）

Two most commonly used organometallic reagents are alkyl lithium reagents and Grignard reagents. These reagents are powerful nucleophiles and very strong bases, so they bond readily to carbonyl carbon atoms, giving *alkoxide* salts of lithium or magnesium. Reactions of this kind are among the most important synthetic methods available to chemists, because they permit simple starting compounds to be joined to form more complex structures.

Because they are so reactive, organolithiums are usually reacted at low temperature, often –78℃, in aprotic solvents such as Et_2O or THF. Organolithiums also react with oxygen, so they have to be handled under a dry, inert atmosphere of nitrogen or argon.

Organomagnesium reagents known as Grignard reagents (RMgX) react in a similar way. The addition of Grignard reagents to aldehydes and ketones is known as the Grignard reaction. The initial product is a magnesium alkoxide, requiring a hydrolysis step to generate the final alcohol product. *Formaldehyde* reacting with Grignard reagents gives primary alcohols; other aldehydes reacting with Grignard reagents give secondary alcohols; and ketones reacting with Grignard reagents give tertiary alcohols. The reaction is of very broad scope.

$$\text{(CH}_3)_2\text{CHCHO} + \text{C}_5\text{H}_9\text{Li} \longrightarrow \longrightarrow (\text{CH}_3)_2\text{CH}-\text{CH(OH)}-\text{C}_5\text{H}_9$$

$$\text{HCHO} + (\text{CH}_3)_3\text{C}-\text{MgCl} \longrightarrow \longrightarrow (\text{CH}_3)_3\text{C}-\text{CH}_2\text{OH}$$

$$\text{C}_6\text{H}_5\text{COCH}_3 + \text{C}_2\text{H}_5\text{MgBr} \longrightarrow \longrightarrow \text{C}_6\text{H}_5\text{C(OH)(C}_2\text{H}_5)\text{CH}_3$$

$$\text{furan-2-CHO} + \text{CH}_2{=}\text{CH}-\text{CH}_2\text{MgBr} \longrightarrow \longrightarrow \text{furan-2-CH(OH)CH}_2\text{CH}{=}\text{CH}_2$$

What's more, the preparation of the same tertiary alcohol, different RMgX and ketones can be chosen.

$$\left.\begin{array}{l} CH_3-MgX + C_2H_5COC_3H_7 \\ C_2H_5-MgX + CH_3COC_3H_7 \\ C_3H_7-MgX + CH_3COC_2H_5 \end{array}\right\} \longrightarrow CH_3CH_2C(OH)(CH_3)-CH_2CH_2CH_3$$

The reaction of RMgX or RLi with α,β-unsaturated aldehydes or ketones can proceed via 1,4-addition as well as normal 1,2-addition. In general, alkyllithium reagents give less 1,4-addition products than the corresponding Grignard reagents.

Two additional examples of the addition of organometallic reagents to carbonyl compounds are informative. The first demonstrates that active metal derivatives of terminal alkynes function in the same fashion as alkyl lithium and Grignard reagents. The second example illustrates the use of acetal protective groups in reactions with powerful nucleophiles.

$$\text{cyclohexanone} \xrightarrow[\text{(2) } H_3O^+,\ 70\%]{\text{(1) } HC{\equiv}C^-\ Na^+,\ NH_3(l),\ -33℃} \text{1-ethynylcyclohexanol } (\text{C(OH)}-C{\equiv}CH)$$

$$\text{2-(2-bromoethyl)-1,3-dioxolane} \xrightarrow{CH_3CH_2C{\equiv}CLi} \text{dioxolane}-CH_2CH_2C{\equiv}CCH_2CH_3 \xrightarrow{H_3^+O} CH_2CH_3C{\equiv}CCH_2CH_2CHO$$

6.2.5 The addition of *bisulfite* （与亚硫酸氢盐的加成）

$$\underset{(R')H}{\overset{R}{}}\!\!>C{=}O + NaHSO_3 \rightleftharpoons \underset{(R')H}{\overset{R}{}}\!\!>C<\underset{SO_3Na}{\overset{OH}{}}$$

Bisulfite addition products are formed from aldehydes, methyl ketones, cyclic ketones (generally seven-membered and smaller rings), and α-keto esters, upon treatment with sodium bisulfite. Most other ketones do not undergo the reaction, probably for steric reasons. The reaction is reversible (by treatment of the addition product with either acid or base) and is useful for the purification of the starting compounds, since the addition products are soluble in water and many of the impurities are not.

6.2.6 The Reformatsky reaction（瑞佛马斯基反应）

Organozinc compounds are prepared from *α-halogenesters* in the same manner as Grignard Reagents. This reaction is possible due to the stability of esters against organozincs. Due to the very low basicity of zinc *enolate*s, there is hardly any competition from proton transfer, and the scope of carbonyl addition partners is quite broad. In presence of ketones or aldehydes, the organozinc compounds react as the nucleophilic partner in an addition to give β-hydroxy esters.

$$\mathrm{R(R)HC{=}O + Br{-}CR'R''{-}CO_2R + Zn \longrightarrow R{-}C(OZnBr)((R)H){-}CR'R''{-}CO_2R}$$

$$\left[\mathrm{BrZn{-}CR'R''{-}CO_2R}\right]$$

$$\xrightarrow{H_3O^+} \mathrm{R{-}C(OH)((R)H){-}CR'R''{-}CO_2R}$$

Compared to organolithiums and organomagnesium halides (Grignard reagents), the organozinc halide reagents used in the Reformatsky reaction are relatively stable, and many are available commercially.

亲核加成反应

加水：羰基水化反应既受酸催化，又受碱催化。产物为同碳二醇。水的亲核性较小，反应后空间位阻增大，故不利于水化。若带吸电基，有利于水化反应。

加醇：醇与醛或酮进行可逆的加成反应，醛或酮在酸催化下，与一分子醇反应得到半缩醛或半缩酮，半缩醛（酮）不稳定。与二分子醇反应得缩醛或缩酮，缩醛（酮）对碱、氧化剂、还原剂、碱金属等稳定。在酸性环境中分解为原来的醛或酮，利用此法可以在合成中保护醛或酮羰基。

加 HCN: 反应为碱催化。羰基与 HCN 加成，生成α-羟基腈，这个反应可以使碳链中增加一个碳原子。

与有机金属试剂反应：甲醛与 RMgX 反应，再水解得到 RCH_2OH，其他醛与 RMgX 反应，得到仲醇，与酮反应得到叔醇。有机锂的活性比 RMgX 高，大位阻酮不能与格氏试剂反应，但可以与有机锂反应得到三级醇。醛或酮还可与炔钠等有机金属试剂发生亲核加成反应，实现碳链增长。醛或酮与α-卤代酯制得的有机锌试剂反应生成β-羟基酯，叫做 Reformatsky 反应。

与亚硫酸氢钠的加成：醛、脂肪族甲基酮及少于 8 个碳原子的环酮与过量的 $NaHSO_3$ 饱和溶液（40%）在室温下反应，生成 α-羟基磺酸钠，溶于水，不溶于亚硫酸氢钠饱和溶液，形成白色结晶沉淀析出。α-羟基磺酸钠用稀酸或弱碱（如碳酸钠）处理，又可得到原来的醛或脂肪族甲基酮。

6.3 The addition and elimination reactions（加成与消除反应）

The reaction of aldehydes and ketones with ammonia or primary amines forms imine derivatives, also known as Schiff bases.

$$\mathrm{R((H)R')C{=}O + R''NH_2 \longrightarrow R((H)R')C(OH)(NHR'') \longrightarrow R((H)R')C{=}NR''}$$

According to the pattern followed by analogous nucleophiles, the initial products of the addition of ammonia or primary amines to aldehydes or ketones would be expected to be *hemiaminal*s, but these compounds are generally unstable. The initial N-substituted hemiaminals lose water to give the stable Schiff bases.

Aldehydes and ketones also condense with other ammonia derivatives, such as *hydroxylamine* and *hydrazine*s.

$$\mathrm{R(H)R'C{=}O} + \begin{cases} \mathrm{NH_2{-}OH} \\ \mathrm{NH_2{-}NH_2} \\ \mathrm{NH_2{-}NH{-}C_6H_5} \\ \mathrm{NH_2{-}NH{-}C_6H_3(NO_2)_2} \\ \mathrm{NH_2{-}NHCONH_2} \end{cases} \longrightarrow \begin{cases} \mathrm{R(H)R'C{=}N{-}OH} \\ \mathrm{R R'(H)C{=}N{-}NH_2} \\ \mathrm{R(H)R'C{=}N{-}NH{-}C_6H_5} \\ \mathrm{R(H)R'C{=}N{-}NH{-}C_6H_3(NO_2)_2} \\ \mathrm{R(H)R'C{=}N{-}NHCONH_2} \end{cases}$$

醛或酮与胺的衍生物（氨、羟胺、肼、苯肼、缩氨脲等）先加成、后消除一分子水得到含 C═N 的产物。产物具有特殊的颜色、固定的熔点，不溶于水，可用来鉴定羰基化合物。仲胺与醛酮加成后不能形成亚胺，如果α-碳上有氢则可脱水得烯胺。烯胺的β-碳原子具有较强的亲核性能，可与卤代烃，卤代酮，卤代酸酯等反应，在烯胺β-碳上引入烃基或与酰卤作用可在羰基的β-碳上引入酰基，产物经水解可得到相应的醛酮。

6.4 The reactivity and stereoselectivity of nucleophilic additions（亲核加成反应的活性和立体选择性）

6.4.1 Reactivity（反应活性）

Aldehydes are more reactive than ketones. This stems from two factors: electronics effect and steric effect.

$$\mathrm{HCHO} > \mathrm{CH_3CHO} > \mathrm{(CH_3)_2C{=}O} > \mathrm{RCOCH_3} > \mathrm{ArCOCH_3}$$

Ketones have two alkyl substituents whereas aldehydes only have one. Carbonyl compounds undergo reaction with nucleophiles because of the polarization of the C═O bond. Alkyl groups are electron donating, and so ketones have their effective partial positive charge reduced more than aldehydes (two alkyl substituents vs. one alkyl substituent).

The electrophilic carbon is the site that the nucleophile must approach for reaction to occur. In ketones the two alkyl substituents create more steric hindrance than the single substituent that aldehydes have. Therefore ketones offer more steric resistance to nucleophilic attack.

6.4.2 Stereoselectivity（立体选择性）

Almost without exception, every stereoselective reaction involves in a double bond (usually C=C; sometimes C=O) with *diastereotopic* faces. When an acyclic carbonyl compound is attacked by a nucleophile, the reaction involves the creation of a new, tetrahedral chiral centre at a carbon that was planar and *trigonal*. The carbonyl carbon atom that is not chiral center but can be made into a chiral center is called *prochiral*.

If the carbon atoms adjacent to the carbonyl group are not chiral centers, nucleophile attacks planar carbonyl from either side to yield a pair of enantiomers and the products will be *racemic*.

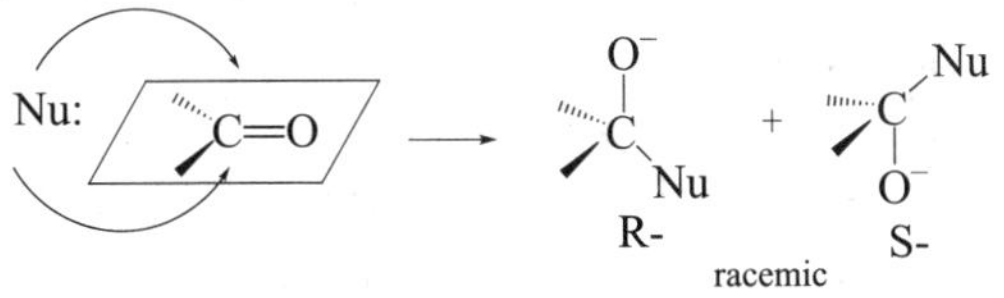

Nu: C=O → R- + S-
racemic

However, if the carbonyl compound with a chiral center adjacent to the carbonyl group, the matter has a greater influence over reactions of the carbonyl group. The most important conformation of the carbonyl compound is the one which place the largest group *perpendicular* to the carbonyl group. There are no eclipsing interactions, and the large group is held satisfactorily far away from the carbonyl group.

Due to free rotation of single bonds, sterically most demanding group is perpendicular to the plane of the carbonyl, *anti* to incoming nucleophile. The large group L *eclipse*s with R and not the carbonyl, the incoming nucleophile would try to avoid it and approach from side of small (S) group. The primary interaction is now between the nucleophile and the small or medium substituent and stereoselectivity observed is usually modest. This is Cram's rule.

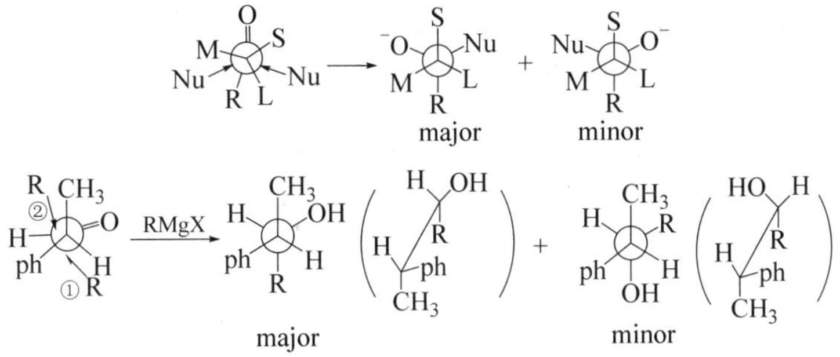

major　　minor

RMgX

major　　minor

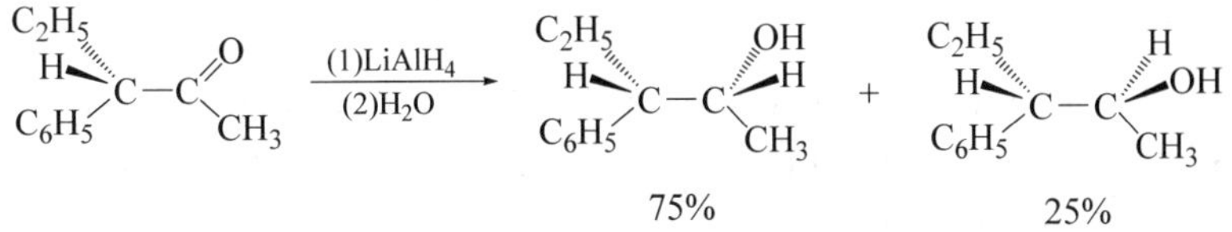

(1)$LiAlH_4$
(2)H_2O

75%　　25%

羰基化合物加成的立体选择性

α-碳不具手性的羰基化合物［前（潜）手性分子］的亲核加成，亲核试剂从羰基平面两侧进攻机会均等，得到外消旋化产物。

含 *α*-手性中心的羰基化合物的亲核加成和还原（$LiAlH_4$或 $NaBH_4$）反应，遵循 Cram 规则，得立体选择性的非对称异构体产物。进攻含*α*-手性碳的醛酮时，手性碳上两个体积较小的基团（S 与 M）分别处于羰基两侧，体积最大的基团 L 则与羰基呈反位交叉构象，进攻剂从体积较小的 S 与 M 之间进攻羰基占优势，这就是 Cram 规则。

6.5 Condensation reactions（缩合反应）

6.5.1 Aldol condensation（羟醛缩合反应）

In the *aldol* reaction, the α carbon of one aldehyde or ketone molecule adds to the carbonyl carbon of another. The fundamental transformation in this reaction is a *dimerization* of an aldehyde (or ketone) to a β-hydroxy aldehyde (or ketone) by α C–H addition of one reactant molecule to the carbonyl group of a second reactant molecule. The products of aldol reactions often undergo a subsequent elimination of water, made up of an α-hydrogen and the β-hydroxyl group.

Base– catalyzed aldol condensations:

$$CH_3{-}CHO + OH^- \longrightarrow {}^-CH_2CHO \xrightarrow{CH_3-CHO} CH_3{-}\overset{O^-}{\overset{|}{C}H}{-}CH_2CHO \longrightarrow CH_3\overset{OH}{\overset{|}{C}H}CH_2CHO \longrightarrow CH_3CH{=}CHCHO$$

Under basic conditions, the aldol condensation occurs by a nucleophilic addition of the enolate ion (a strong nucleophile) to a carbonyl group. Protonation gives the aldol product.

Acid-catalyzed aldol condensations:

$$CH_3\overset{O}{\overset{\|}{C}}CH_3 \rightleftharpoons CH_3\overset{OH}{\overset{|}{C}}{=}CH_2 \xrightarrow{CH_3\overset{^+OH}{\overset{\|}{C}}CH_3} CH_3{-}\overset{^+OH}{\overset{\|}{C}}{-}CH_2{-}\overset{OH}{\overset{|}{C}}(CH_3)_2$$

$$\xrightarrow{-H^+} CH_3{-}\overset{O}{\overset{\|}{C}}{-}CH_2{-}\overset{OH}{\overset{|}{C}}(CH_3)_2 \xrightarrow{-H_2O} CH_3{-}\overset{O}{\overset{\|}{C}}{-}CH{=}C(CH_3)_2$$

Aldol condensations also take place under acidic conditions. The first step is form of the enol by acid-catalyzed aldol *keto-enol tautomerism*. The enol serves as a weak nucleophile to attack an activated (protonated) carbonyl group in the second step. Loss of the enol proton gives the aldol product.

Heating an aldol product leads to dehydration of the hydroxyl group. Thus, an aldol condensation, followed by dehydration forms a conjugated *α, β*-unsaturated aldehyde or ketone. When the enolate of one aldehyde (or ketone) adds to the carbonyl group of another, the result is called a *crossed aldol condensation*. The compounds used in the reaction must be selected carefully, or a mixture of several products will be formed. If only one of the reactants has an α hydrogen, only one enolate will be present in the solution. If the other reactant contains a particularly electrophilic carbonyl group, it is more likely to be attacked by the enolate ion.

$$HCHO + CH_3CHO \xrightarrow[B.\ \triangle,\ -H_2O]{A.\ dil.\ OH^-} H_2C{=}\underset{H}{C}{-}CHO$$

Intramolecular aldol reactions of dikeones are useful for making five and six-membered rings. Aldol cyclizations of rings smaller than five or larger than six are less common because larger or smaller rings are less favored by their energy and entropy. The carbonyl group of the product may be outside the ring in some cases.

R, H, R′, O, O　$\overset{OH^-}{\rightleftharpoons}$　R, H, R′, OH, O　$\xrightarrow{OH^-}$　R, R′, O

O, O, CH_3　$\overset{OH^-}{\rightleftharpoons}$　O, CH_3, OH　$\xrightarrow{OH^-}$　O, CH_3

$$\text{2-acetyl-2-(3-oxobutyl)... } \xrightleftharpoons{OH^-} \xrightarrow{OH^-}$$

6.5.2 Knoevenagel condensation（Knoevenagel 缩合）

A Knoevenagel condensation is a nucleophilic addition of an active hydrogen compound to a carbonyl group under the catalysis of weakly basic amine (such as pyridine or amines) followed by a dehydration reaction in which a molecule of water is eliminated (hence condensation). The product is often an α,β conjugated enone.

$$\text{>C=O} + H_2C(X)(Y) \xrightarrow{\text{weak base}} \text{>C=C}(X)(Y)$$

X= —COOH, —COR, —CN, —NO_2, —H

Y= —COOH, —COOR, —NO_2, etc

$$CH_2(COOEt)_2 + B \rightleftharpoons \bar{C}H(COOEt)_2 \xrightarrow{RCH=O} RCH(O^-)CH(COOEt)_2 \xrightarrow{BH^+} RCH(OH)CH(COOEt)_2$$

$$\xrightleftharpoons{B} R-CH(OH)-\bar{C}(COOEt)_2 \longrightarrow RCH=C(COOEt)_2$$

$$PhCHO + H_2C(CO_2Et)_2 \xrightarrow{Py} PhCH=C(CO_2Et)_2 \xrightarrow{H_3O^+} \xrightarrow{\triangle} PhCH=CHCOOH$$

6.5.3 Stobbe condensation（Stobbe 缩合）

The Stobbe condensation is a modification specific for the diethyl ester of *succinic acid* requiring less strong bases. An example is its reaction with *benzophenone.*

$$Ph-CO-Ph + (CH_2CO_2Et)_2 \rightleftharpoons Ph_2C=C(CO_2Et)CH_2COO^-$$

$$CR_2=O + H\bar{C}(CO_2Et)CH_2CO_2Et \rightleftharpoons R_2C(O^-)CH(CO_2Et)CH_2C(=O)OEt \rightleftharpoons \text{cyclic intermediate } (O^-, OEt)$$

$$\xrightleftharpoons{-EtO^-} \text{lactone } R_2C\text{–}CH(CO_2Et) \xrightleftharpoons[-H^+]{EtO^-} \text{lactone carbanion} \rightleftharpoons CR_2=C(CO_2Et)-CH_2COO^-$$

6.5.4 Mannich reaction（Mannich 反应）

The Mannich reaction is an organic reaction which consists of an amino alkylation of an acidic proton placed next to a carbonyl functional group with formaldehyde and ammonia or any primary or secondary amine. The final product is a β-amino-carbonyl compound also known as a Mannich base.

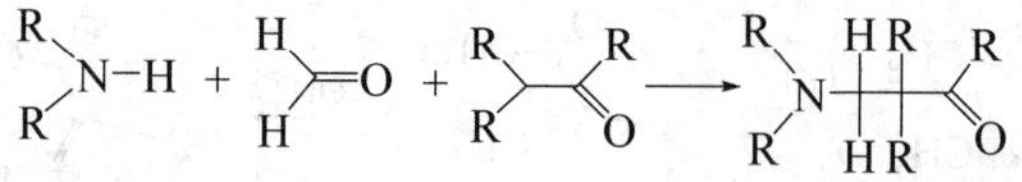

$$CH_3COCH_3 + HCHO + CH_3NH_2 \rightleftharpoons H_3C-\overset{O}{\overset{\|}{C}}CH_2CH_2NHCH_3$$

$$\xrightarrow{HCHO,\ CH_3COCH_3} H_3C-\overset{O}{\overset{\|}{C}}CH_2CH_2\underset{CH_3}{\underset{|}{N}}CH_2CH_2COCH_3$$

The Mannich reaction is an example of nucleophilic addition of an amine to a carbonyl group followed by dehydration to the Schiff base. The Schiff base is an electrophile which reacts in the second step in an electrophilic addition with a compound containing an acidic proton (which is, or had became an enol). The Mannich reaction is also considered as a condensation reaction. In the Mannich reaction, ammonia or primary or secondary amines are employed for the activation of formaldehyde. Tertiary amines lack a N—H proton to form the intermediate imine. α-CH-acidic compounds (nucleophiles) include carbonyl compounds, nitriles, acetylenes, aliphatic nitro compounds, α-alkyl-pyridines or imines. It is also possible to use activated phenyl groups and electron-rich heterocycles such as furan, pyrrole, and thiophene.

The mechanism of the Mannich reaction starts with the formation of an *iminium ion* from the amine and the formaldehyde. The compound with the carbonyl functional group (in this case a ketone) can *tautomerize* to the enol form, after which it can attack the iminium ion.

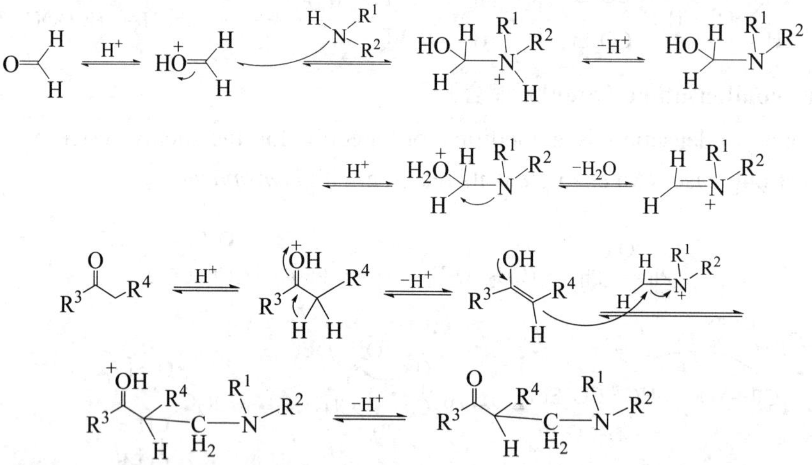

6.5.5 Claisen condensation（Claisen 缩合）

The Claisen condensation (not to be confused with the Claisen rearrangement) is a carbon–carbon bond forming reaction that occurs between two esters or one ester and another carbonyl compound in the presence of a strong base, resulting in a β-keto ester or a β-diketone.

At least one of the reagents must be enolized (have an α-proton and be able to undergo deprotonation to form the enolate anion). The base used must not interfere with the reaction by undergoing nucleophilic substitution or addition with a carbonyl carbon. For this reason, sodium alkoxide is often used. In mixed Claisen condensations, a non-nucleophilic base such as *lithium diisopropylamide* (LDA), may be used, since only one compound is enolized. The alkoxy portion of the ester must be a relatively good leaving group. Methyl and ethyl esters, which yield methoxide and ethoxide, respectively, are commonly used.

$$RCH_2COOR' \xrightleftharpoons{EtONa} RCH_2-\underset{}{\overset{ONa}{C}}=\underset{R}{C}COOR' \xrightarrow{H^+} RCH_2-\overset{O}{\overset{\|}{C}}-\underset{R}{C}COOR'$$

$$C_2H_5O^- + H-CH_2-\overset{O}{\overset{\|}{C}}-OC_2H_5 \rightleftharpoons H_2C=\overset{O^-}{C}-OC_2H_5 \xrightleftharpoons{CH_3\overset{O}{\overset{\|}{C}}-OC_2H_5} CH_3-\overset{O^-}{\underset{CH_2COOC_2H_5}{C}}-OC_2H_5$$

$$\rightleftharpoons CH_3\overset{O}{\overset{\|}{C}}CH_2\overset{O}{\overset{\|}{C}}-OC_2H_5 \rightleftharpoons CH_3\overset{O}{\overset{\|}{C}}\overset{-}{C}H\overset{O}{\overset{\|}{C}}-OC_2H_5 \xrightarrow{H_3\overset{+}{O}} CH_3\overset{O}{\overset{\|}{C}}CH_2\overset{O}{\overset{\|}{C}}-OC_2H_5$$

In the first step of the mechanism, an α-proton is removed by a strong base, resulting in the formation of an enolate anion, which is relatively stable by the delocalization of electrons. Next, the carbonyl carbon of the (other) ester is nucleophilic attacked by the enolate anion. The alkoxy group is then eliminated (resulting in (re)generation of the alkoxide), and the alkoxide removes the newly-formed α-proton to form a new, highly resonance-stabilized enolate anion. Aqueous acid (e.g. sulfuric acid or phosphoric acid) is added in the final step to neutralize the enolate and any base still present. The newly-formed β-keto ester or β-diketone is then isolated. The last step is irreversible, which drives the reaction forward. If there is not a proton between the ketone and an ester in the product, the product will not be formed. Therefore, two hydrogens on the α-carbon are necessary.

For example, the self-condensation of ethyl acetate gives ethyl acetoacetate. Ethoxide is used as the base to avoid *transesterification* or hydrolysis of the ethyl ester.

$$2\ H_3C\overset{O}{\overset{\|}{C}}OCH_2CH_3 \xrightarrow{C_2H_5ONa} H_3C\overset{O}{\overset{\|}{C}}\underset{H}{\overset{-}{C}}\overset{O}{\overset{\|}{C}}OCH_2CH_3 \xrightarrow{H^+} H_3C\overset{O}{\overset{\|}{C}}\underset{H_2}{C}\overset{O}{\overset{\|}{C}}OCH_2CH_3$$

In a crossed Claisen condensation, an ester without α-protons serves as the electrophilic component. Some useful esters without α-protons are benzoate, formate, carbonate and *oxalate* esters.

$H\overset{O}{\overset{\|}{C}}OC_2H_5$	$C_6H_5COOC_2H_5$	$COOC_2H_5$–$COOC_2H_5$	$O=C(OC_2H_5)_2$
ethyl formate	ethyl benzoate	diethyl oxalate	diethyl carbonate

$$EtOOC-CH_2-CH_2-CH_2-COOEt + EtO-\overset{O}{\overset{\|}{C}}-\overset{O}{\overset{\|}{C}}-OEt \xrightarrow{C_2H_5ONa} \xrightarrow{H_3\overset{+}{O}} \text{(EtOOC)}_2\text{-cyclopentane-dione}$$

$$HCOOC_2H_5 + PhCH_2CO_2C_2H_5 \longrightarrow HCO\underset{Ph}{C}HCO_2C_2H_5$$

Crossed Claisen condensation between ketones and esters are also possible. Ketones are more acidic than esters, and the ketone component is more likely to deprotonate and serves as the enolate component in the condensation.

$$C_6H_5COOC_2H_5 + C_6H_5COCH_3 \longrightarrow C_6H_5COCH_2COC_6H_5$$

The Dieckmann condensation, where a molecule with two ester groups reacts intramolecularly, forms a cyclic β-keto ester. In this case, the ring formed must not be strained, usually a 5- or

6-membered ring.

$$CH_3OOC(CH_2)_4COOCH_3 \xrightarrow{CH_3ONa} [\text{intermediate}] \longrightarrow \text{methyl 2-oxocyclohexanecarboxylate}$$

$$CH_3OOC(CH_2)_3COOCH_3 \xrightarrow{CH_3ONa} [\text{intermediate}] \longrightarrow \text{methyl 2-oxocyclopentanecarboxylate}$$

6.5.6 Darzens reaction（Darzens 反应）

The Darzens reaction (also known as the Darzens condensation or *glycidic* ester condensation) is the chemical reaction of a ketone or aldehyde with an α-haloester in the presence of a base to form an α,β-*epoxy* ester.

$$RCOR' \xrightarrow[NaOEt]{ClCH(R'')COOEt} \text{epoxy ester}(R, R', R'', COOEt)$$

The reaction process begins when a strong base is used to form a carbanion at the halogenated position. Because of the ester, this carbanion is a resonance-stabilized enolate, which makes it relatively easy to form. This nucleophilic structure attacks another carbonyl component, forming a new carbon–carbon bond. These first two steps are similar to a base-catalyzed aldol reaction. The oxygen anion in this aldol-like product then does an intramolecular S_N2 attack on the formerly-nucleophilic halide-bearing position, displacing the halide to form an *epoxide*. This reaction sequence is thus a condensation reaction, since there is a net loss of HCl when two reactant molecules join.

$$ClCH_2COOC_2H_5 \xrightarrow{RONa} Cl\bar{C}HCOOC_2H_5 \xrightarrow{RCOR'} RR'C(O^-)-CH(Cl)-COOC_2H_5 \xrightarrow{-Cl^-} RR'C(O)CH-COOC_2H_5$$

The product of Darzens reaction can be reacted further to form various types of compounds. Hydrolysis of the ester can lead to decarboxylation, which triggers a rearrangement of the epoxide into a carbonyl.

$$PhCOCH_3 + ClCH_2COOC_2H_5 \xrightarrow{C_2H_5ONa} Ph(CH_3)C(O)CH-CO_2C_2H_5 \xrightarrow{NaOH} Ph(CH_3)C(O)CH-CO_2^-$$

$$\xrightarrow[-CO_2]{HCl} (H_3C)(Ph)C{=}CHOH \longrightarrow (H_3C)(Ph)CH-CHO$$

$$\beta\text{-ionone} \xrightarrow[MeONa,\ Py]{ClCH_2CO_2Et} \text{glycidic ester}-CO_2Et \xrightarrow[HCl]{NaOH} \text{glycidic acid}-COOH \longrightarrow \text{enol}-OH \longrightarrow \text{aldehyde}-CHO$$

6.5.7 Michael addition（Michael 加成）

α,β-Unsaturated carbonyl compounds have unusually electrophilic double bonds. The β carbon is electrophilic because it shares the partial positive charge of the carbonyl carbon through resonance. The Michael reaction or Michael addition is the nucleophilic addition of a carbanion or another nucleophile to an α,β-unsaturated carbonyl compound. It belongs to the larger class of conjugate additions. This is one of the most useful methods for the mild formation of C—C bonds.

$$Nu^- + H_2C{=}CH{-}C(O){-}R \longrightarrow H_2C(Nu){-}CH{=}C(O^-){-}R \xrightarrow{H^+} H_2C(Nu){-}CH{=}C(OH){-}R \rightleftharpoons H_2C(Nu){-}CH_2{-}C(O){-}R$$

The electrophile (the α, β-unsaturated carbonyl compound) accepts a pair of electrons; it is called the Michael acceptor. The attacking nucleophile donates a pair of electrons; it is the Michael donor. Common donors are enolate ions that are stabilized by strong electron withdrawing groups such as carbonyl groups, cyano groups or nitro groups. Common acceptors contain a double bond conjugated with a carbonyl group, a cyano group, or a nitro group (Table 6.1).

Table 6.1 Some common Michael donors and Michael acceptors

Michael donors		Michael acceptors	
$RCO\bar{C}HCOR'$	β-diketone	$H_2C{=}CH{-}C(O){-}H$	conjugated aldehyde
$RCO\bar{C}HCOOR'$	β-keto ester	$H_2C{=}CH{-}C(O){-}R$	conjugated ketone
R_2CuLi	dialkyl cuprate	$H_2C{=}CH{-}C(O){-}OR$	conjugated ester
$-N{-}C{=}C$	enamine	$H_2C{=}CH{-}C(O){-}NH_2$	conjugated amide
$RCO\bar{C}H{-}C{\equiv}N$	β-keto nitrile	$H_2C{=}CH{-}C{\equiv}N$	conjugated nitrile
$RCO\bar{C}H{-}NO_2$	α-nitro ketone	$H_2C{=}CH{-}NO_2$	nitroethene

For example in the diagram below, in the Michael reaction, the ketone is deprotonated by a base to form an enolate nucleophile which attacks the Michael acceptor in next step. The aldol condensation is an intramolecular process that creates the namesake ring of the Robinson *annulation* product.

$$\text{2-methylcyclohexanone} + CH_3\overset{O}{\overset{\|}{C}}CH{=}CHCH_3 \xrightarrow{KOH/DMSO} \text{Michael adduct} \longrightarrow \text{Robinson annulation product}$$

6.5.8 Perkin reaction（Perkin 反应）

The Perkin reaction is an organic reaction developed by William Henry Perkin that can be used to make *cinnamic acids* i.e. α, β-unsaturated aromatic acid by the aldol condensation of aromatic aldehydes and acid anhydrides in the presence of an alkali salt of the acid.

$$PhCHO + (CH_3CO)_2O \xrightarrow{CH_3COONa} Ph{-}CH{=}CHCOOH$$

6.5.9 Benzoin condensation（安息香缩合）

The *Benzoin* condensation is a reaction (often called a condensation reaction, for historical reasons) between two aromatic aldehydes, particularly benzaldehyde. The reaction is catalyzed by a nucleophile such as a cyanide anion or an *N*-heterocyclic carbene. The reaction product is an aromatic acyloin with benzoin as the parent compound.

In the first step, the cyanide anion (as sodium cyanide) reacts with the aldehyde in a nucleophilic addition. Rearrangement of the intermediate results in polarity reversal of the carbonyl group, which then adds to the second carbonyl group in a second nucleophilic addition. Proton transfer and elimination of the cyanide ion affords benzoin as the product. This is a reversible reaction.

The cyanide ion serves three different purposes in the course of this reaction. It acts as a nucleophile, facilitates proton abstraction, and is also the leaving group in the final step. The benzoin condensation is in effect a dimerization and not a condensation because a small molecule like water is not released in this reaction. For this reason the reaction is also called a benzoin addition. In this reaction, the two aldehydes serve different purposes; one aldehyde donates a proton and one aldehyde accepts a proton. 4-dimethylaminobenzaldehyde is an efficient proton donor while benzaldehyde is both a proton acceptor and donor. In this way it is possible to synthesise mixed benzoins, i.e. products with different groups on each half of the product.

RCH_2X（X 为—CHO、—COR、—COOR、$—NO_2$、—CN 等）在酸或碱作用下形成亲核试剂，再与羰基化合物加成。

羟醛缩合：在酸或碱催化剂作用下，含有 α-H 的醛或酮相互作用，生成 β-羟基醛（酮）（反应为可逆反应）或进一步失水得到 α,β-不饱和醛酮的反应叫羟醛缩合。无 α-H 的醛酮，自身不能发生羟醛缩合，但可以与另一分子含 α-H 的醛酮发生交叉羟醛缩合。两种酮的交叉缩合中，至少有一种为甲基酮或环酮，位阻小的甲基酮或环酮在反应中作为羰基组分。

分子内羟醛缩合产物一般为五元环或六元环的 α,β-不饱和醛(酮)。

Knoevenagel 缩合：醛酮与含活泼亚甲基的化合物（如丙二酸、丙二酸酯，β-羰基酸酯、β-氰基酸酯等）在温和的碱作用下的 α,β-不饱和酸酯的反应叫 Knoevenagel 缩合反应。

Stobbe 缩合：醛酮与 1,4-丁二酸酯在碱性催化条件下，生成亚甲基丁二酸单酯的反应叫 Stobbe 缩合反应。

Mannich 反应：具有 α-H 的醛、酮或其他含活泼氢的化合物与醛及胺或氨作用，生成 β-氨基酮的反应叫 Mannich 反应。含活泼氢的化合物有：RCH_2COR、RCH_2COOR、RCH_2COOH、RCH_2CN 和 RCH_2NO_2 等。

Claisen 缩合：含 α-H 的羧酸酯在强碱作用下得 β-酮酸酯的反应叫 Claisen 缩合反应。不含 α-H 的羧酸酯不发生自身 Claisen 酯缩合，但可与另一含 α-H 的酯发生交叉 Claisen 酯缩合。1,6-或 1,7-二元羧酸酯发生分子内酯缩合得五元环或六元环的 β-环酮酸酯的反应叫 Dieckmann 缩合。

Darzens 反应：由 α-卤代羧酸酯（酰胺）在醇钠催化下与醛酮反应得 α,β-环氧酸酯（酰胺）。

Michael 加成：碳负离子对 α,β-不饱和羰基化合物发生的 1,4-加成反应叫 Michael 加成。α,β-不饱和羰基化合物叫 Michael 受体，活泼亚甲基化合物的碳负离子叫 Michael 供体。有机铜锂与 α,β-不饱和醛酮反应主要发生 1,4-加成。有机锂与 α,β-不饱和醛酮反应主要发生 1,2-加成。经环酮的碳负离子与 α,β-不饱和酮发生 1,4-加成，再发生分子内羟醛缩合，得到增加一个环的 α,β-不饱和酮叫做 Robinson 环化法。

Perkin 反应：芳香醛和乙酐或取代乙酐在相应羧酸的钠盐或钾盐存在下发生缩合反应得 α,β-不饱和芳香酸。

安息香缩合：两分子芳醛在 CN^-存在下，缩合生成 α-羟酮的反应。

6.6 Reaction with ylides（与叶立德的反应）

In 1954 Wittig discovered that the addition of a phosphorus stabilized anion to a carbonyl compound did not generate an alcohol, but an alkene! This reaction proved so useful that Wittig received the Nobel Prize in chemistry in 1979 for this discovery.

$$>C=O + Ph_3P=CR_2 \longrightarrow >C=CR_2 + Ph_3P=O$$

The phosphorus stabilized anion is called an ylide, which is a molecule that is overall neutral, but exists as a carbanion bound to a positively charged heteroatom. Phosphorus ylides are produced from the reaction of *triphenylphosphine* and alkyl halides.

$$Ph_3P: + H-CH(R)-X \longrightarrow Ph_3\overset{+}{P}-CH(R)-H \xrightarrow{n\text{-BuLi}} \left[Ph_3\overset{+}{P}-\overset{-}{C}(R)-H \longleftrightarrow Ph_3P=C(H)R \right]$$

ylide　　ylene

This two step reaction starts with the nucleophilic attack of the phosphorus on the (usually primary) alkyl halide. This generates an alkyl triphenylphosphonium salt. Treatment of this salt with a strong base removes a proton from the carbon bound to the phosphorus, and generates the ylide. The ylide is a resonance form of a C=P double bond. The double bond resonance form requires 10 electrons around the P atom. This is achievable through use of its d electrons (3rd row element), but the π bond between carbon and phosphrous is weak, and this is only a minor contributor. The carbon atom actually bears a partial negative charge, balanced by a corresponding charge on phosphrous. The carbanionic character of the ylide makes it a very powerful nucleophile, and so it reacts rapidly with a carbonyl group.

$$Ph_3\overset{+}{P}-\overset{-}{C}(R)-H + R'_2C{=}O \longrightarrow R-C(H)(\overset{+}{P}Ph_3)-C(R')(O^-)-R' \longrightarrow R-C(H)(PPh_3)-C(R')(O)-R' \xrightarrow{-Ph_3P=O} R(H)C{=}CR'_2$$

This produces a charge-separated intermediate (a *betaine*). Betaines are unusual since they have a negatively charged oxygen and a positively charged phosphorus (a "phosphonium oxide") on adjacent carbon atoms. Phosphorus and oxygen always form strong bonds, and the attraction of opposite charges promotes the generation of a four-membered oxaphosphetane ring. This four-membered ring quickly collapses to generate an alkene and triphenyl phosphine oxide (very stable). The elimination of $Ph_3P{=}O$ is the driving force of this reaction. This is a good general route to make new C=C double bonds starting from carbonyl compounds.

The following examples show the formation of carbon-carbon double bond using the Wittig reaction. Mixtures of *cis* and *trans* isomers often result when geometric isomerism is possible.

$$C_6H_{10}{=}O + \overset{-}{C}H_2-\overset{+}{P}Ph_3 \longrightarrow C_6H_{10}{=}CH_2 + O{=}PPh_3$$

$$Ph_3\overset{+}{P}-\overset{-}{C}HPh + PhCHO \longrightarrow \textit{cis-}PhCH{=}CHPh + \textit{trans-}PhCH{=}CHPh$$

羰基与叶立德的反应

叶立德又称叕内盐，指的是一类在相邻原子上有相反电荷的中性分子。最常见的叶立德是磷叶立德。Wittig 反应中即是用磷叶立德与羰基化合物反应制得烯烃和氧化三苯磷。这是一个非常有价值的合成方法，用于从醛、酮直接合成烯烃。硫叶立德被用于合成环氧化合物。

6.7 The nucleophilic substitutions of carboxylic acids and their derivatives（羧酸及其衍生物的亲核取代反应）

This is the most important reaction of carboxylic acid derivatives. The overall transformation is defined by the following equation, and may be classified either as nucleophilic substitution at an *acyl* group or as acylation of a nucleophile. For certain nucleophilic reagents, the reaction may assume other names as well. If the nucleophile is water, the reaction is often called hydrolysis. If the nucleophile is an alcohol, the reaction is called alcoholysis. And for ammonia and amines, it is called aminolysis.

$$RC(=O)L + Nu^- \longrightarrow R-C(O^-)(L)-Nu \longrightarrow RC(=O)Nu + L^-$$

Acylation reactions generally take place by an addition-elimination process in which a nucleophilic reactant bonds to the electrophilic carbonyl carbon atom to create a tetrahedral intermediate. This tetrahedral intermediate then undergoes an elimination to yield the products. In this two-step mechanism, bond formation occurs before bond cleavage, and the carbonyl carbon atom undergoes a hybridization change from sp^2 to sp^3 and back again.

Different carboxylic acid derivatives have very different reactivities, acyl chlorides and bromides being the most reactive and amides the least reactive, as noted in the following qualitatively ordered list. The change in reactivity is dramatic.

Reactivity: acyl halides > anhydrides >> esters ≈ acids >> amides

However, some nucleophiles are too weak to attack an unactivated carbonyl group. For example, an alcohol attacks the carbonyl group of an acid chloride, but it does not attack an acid. If a strong acid promotes the carbonyl group of the carboxylic acid, the carbonyl group is activated toward attack by alcohol, resulting in Fischer *esterfication.* A similar mechanism is responsible for the acid-catalyzed transesterification of an ester, in which one alkoxy group substitutes for another. Acid catalysis of acylation can be generalized in the following scheme.

$$R-\overset{\overset{O}{\|}}{C}-OH + H^+ \rightleftharpoons R-\underset{HOR'}{\overset{\overset{^+OH}{\|}}{C}}-OH \rightleftharpoons R-\underset{R'\overset{+}{O}H}{\overset{OH}{\overset{|}{C}}}-OH \rightleftharpoons R-\underset{OR'}{\overset{OH}{\overset{|}{C}}}-\overset{+}{O}H_2 \rightleftharpoons R-\overset{\overset{^+OH}{\|}}{C}-OR' \rightleftharpoons R-\overset{\overset{O}{\|}}{C}-OR'$$

Transesterification also occurs under basic conditions, catalyzed by a small amount of alkoxide ion. For example, in the base-catalyzed transesterification of ethyl benzoate, methoxide ion is sufficiently nucleophilic to attack the ester carbonyl group. Ethoxide ion serves as a leaving group in a strongly exothermic second step. Base catalysis of acylation can be generalized in the following scheme.

$$C_6H_5-\overset{\overset{O}{\|}}{C}-OCH_2CH_3 \xrightarrow[CH_3O^-]{} \rightleftharpoons C_6H_5-\underset{OCH_3}{\overset{\overset{O^-}{|}}{C}}-OCH_2CH_3 \rightleftharpoons C_6H_5-\overset{\overset{O}{\|}}{C}-OCH_3$$

The common acid or base-catalyzed mechanisms for ester hydrolysis, which proceed by way of a tetrahedral intermediate, have been classified as $A_{Ac}2$ and $B_{Ac}2$ respectively. This notation refers to the nature of the catalysis (acid or base), the C—O bond which is broken (acyl-oxygen or alkyl-oxygen) and the molecularity of the rate-determining step (k_r) as summarized in the following diagram.

The acid catalyzed process is fully reversible reaction, driven to completion by an excess of water, whilst the base catalyzed reaction is essentially a fully complete process once the carboxylate ion has been formed. The rate of acid-catalyzed hydrolysis of common esters is found to be proportional to both H^+ and the ester and when the hydrolysis is carried out in the presence of $H_2{}^{18}O$, the product alcohol has no ^{18}O content. Thus, it is an $A_{Ac}2$ reaction. Acidic hydrolysis (and esterification) is rather insensitive to polar effects. This appears quite reasonable as the two parts of the process, protonation and subsequent nucleophilic attack by water, require two opposing electronic influences. Electron donating groups accelerate protonation, but they inhibit the subsequent attack by a nucleophile. On the other hand, electron-withdrawing substituents repress protonation but they accelerate the nucleophilic attack. These reactions, however, are subject to steric effects; for instance, esterification of 2-methyl benzoic acid is retarded while that of 2, 6-dimethylbenzoic acid is completely prevented.

$A_{Ac}2$

$$R-\overset{O}{\overset{\|}{C}}-OR' + H^+ \rightleftharpoons R-\overset{^+OH}{\overset{\|}{C}}-OR' \xrightleftharpoons{+H_2O} R-\underset{OH}{\overset{^+OH_2}{\overset{|}{\underset{|}{C}}}}-OR'$$

$$\rightleftharpoons R-\underset{OH}{\overset{H-O}{\overset{|}{\underset{|}{C}}}}-\overset{+}{\underset{H}{O}}R' \rightleftharpoons R-\overset{O}{\overset{\|}{C}}-OH + R'OH + H^+$$

$$HOOCCH_2CH_2\overset{O}{\overset{\|}{C}}OC_2H_5 + H_2^{18}O \xrightleftharpoons{H^+} HOOCCH_2CH_2\overset{O}{\overset{\|}{C}}{}^{18}OH + C_2H_5OH$$

A $B_{Ac}2$ hydrolysis means a bimolecular basic hydrolysis of an ester proceeding through the acyl-oxygen bond cleavage. The hydrolysis of esters by hydroxide ions in aqueous solution is kinetically a second order reaction, first order in ester and first order in hydroxide ion. It is, therefore, a bimolecular reaction. The evidence for the acyl-oxygen bond fission is provided by carrying out the hydrolysis of *n*-amyl acetate in water enriched in ^{18}O and then isolating *n*-amyl alcohol without ^{18}O. Electron-withdrawing substituents in either the acyl or alkyl part of the ester facilitate attack by hydroxide ions.

$B_{Ac}2$

$$R\overset{O}{\overset{\|}{C}}OR' + OH^- \rightleftharpoons R-\underset{OH}{\overset{O^-}{\overset{|}{\underset{|}{C}}}}-OR' \rightleftharpoons R\overset{O}{\overset{\|}{C}}OH + R'O^- \longrightarrow R\overset{O}{\overset{\|}{C}}O^- + R'OH$$

$$H_3C-\overset{O}{\overset{\|}{C}}-OC_5H_{11} + H_2^{18}O \xrightleftharpoons{OH^-} H_3C-\overset{O}{\overset{\|}{C}}{}^{18}O^- + C_5H_{11}OH$$

Equations illustrating uncommon acid-catalyzed or base-catalyzed mechanisms, $A_{Al}1$, $A_{Ac}1$ and $B_{Al}1$, are shown at the bottom.

When groups such as tert-alkyl, benzyl, etc., capable of forming stable carbonium ions are the alkyl moiety in the substrate, hydrolysis and esterification may take place by an alkyl-oxygen bond cleavage. Confirmation of this mechanism is provided by carrying out the hydrolysis in $H_2{}^{18}O$. This type of mechanism has been labeled as $A_{Al}1$ mechanism (under acid condition) or $B_{Al}1$ mechanism (under basic condition). As the intermediate has property of a free carbonium ion, it can undergo racemization and rearrangement, if the structure permits.

$A_{Al}1$

$$R-\overset{O}{\overset{\|}{C}}-OCR'_3 + H^+ \rightleftharpoons R-\overset{^+OH}{\overset{\|}{C}}-O-CR'_3 \longrightarrow RCOOH + R'_3\overset{+}{C} \xrightarrow[-H^+]{H_2O} R'_3COH$$

$$C_6H_5-COO(CH_3)_3 + H_2^{18}O \xrightarrow{H^+} C_6H_5-COOH + H^{18}OC(CH_3)_3$$

$B_{Al}1$

$$o\text{-}HOOC-C_6H_4-COO\overset{*}{C}H(C_6H_5)-C_6H_4-OCH_3 \xrightleftharpoons{NaOH} o\text{-}HOOC-C_6H_4-COO^- + H\overset{+}{C}(C_6H_5)(C_6H_4OCH_3)$$

$$\xrightarrow{OH^-} C_6H_5-\underset{H}{\overset{OH}{\overset{|}{\underset{|}{C}}}}-C_6H_4-OCH_3$$

racemate

Hydrolysis and esterification of sterically hindered compounds, such as derivatives of trialkylacetic acids and ortho disubstituted benzoic acids, are carried out by dissolving them in concentrated sulphuric acid and then pouring this solution into cold water (for hydrolysis) or into an alcohol (for esterification). This type of mechanism has been labeled as $A_{Ac}1$ mechanism. Like $A_{Ac}2$ mechanism, there is an initial formation of an oxonium ion by the addition of a proton to the substrate which then undergoes a rate-controlling heterolytic fission to yield an acylium ion. The acylium ions are linear and their formation is favored primarily because of relief in the steric strains of the sterically hindered compounds. Another factor that favors their formation is the conjugation of the $—C{\equiv}O^+$ group with the unsaturated system thus delocalizing the positive charge over the whole ion. A rapid attack by water or alcohol on this ion results in the formation of acid or ester, respectively.

$$A_{Ac}1 \quad R-\overset{O}{\overset{\|}{C}}-OR' + H^+ \rightleftharpoons R-\overset{^+OH}{\overset{\|}{C}}-OR' \rightleftharpoons R-\overset{O}{\overset{\|}{C}}-\underset{H}{\overset{+}{O}}R' \rightleftharpoons RC{\equiv}O^+ + R'OH$$

$$RC{\equiv}O^+ \xrightarrow{H_2O} RCOOH + H^+$$

2,4,6-Trimethylbenzoate (Me groups, COEt) $\xrightarrow{\text{conc. } H_2SO_4}$ Me–C₆H₂(Me)₂–$C{\equiv}O^+$ $\xrightarrow{H_2O}$ Me–C₆H₂(Me)₂–COOH

羧酸及其衍生物的亲核取代

羧酸及其衍生物发生亲核反应时，先加成，生成四面体中间体，后消除，得到取代产物。反应活性：酰氯 > 酸酐 > 酯 > 酰胺。

酯的水解包括酸催化的双分子酰氧断裂（$A_{Ac}2$）机理和碱催化的双分子酰氧断裂（$B_{Ac}2$）机理。酯分子中的羰基受亲核试剂的进攻生成四面体中间体，再脱去离去基，得到取代产物。反应物空间位阻增大时，不利于反应按上述机理进行。有吸电子基时对 $B_{Ac}2$ 反应有利。羰基碳的正电性增加，有利于分散中间体的负电荷，从而有利于亲核加成。电子效应对 $A_{Ac}2$ 反应几乎没有影响。

叔醇的酯一般按酸催化的单分子烷氧断裂（$A_{Al}1$）或碱催化的单分子烷氧断裂（$B_{Al}1$）历程进行水解。

空间位阻很大的酯在强酸介质(H_2SO_4)中，由于水的亲核性降低，酰基正离子稳定，按酸催化的单分子酰氧断裂（$A_{Ac}1$）机理水解。

Problems

1. The triketone shown below is called “*ninhydrin*” and is used for the detection of amino acids. It exists in aqueous solution as a monohydrate. Which of the three ketones is hydrated and why?

2. This hydroxyketone shows no peaks in its infrared spectrum between 1600 cm^{-1} and 1800 cm^{-1} but it does show a broad absorption at 3000 to 3400 cm^{-1}. In the ^{13}C NMR spectrum, there are no peaks above 150 ppm. But there is a peak at 110 ppm. Suggest an explanation.

ninhydrin

HO (one hydroxyketone structure)

one hydroxyketone

3. Each of these compounds is a hemiacetal and therefore formed from an alcohol and a carbonyl compound. In each case give the structure of these original materials.

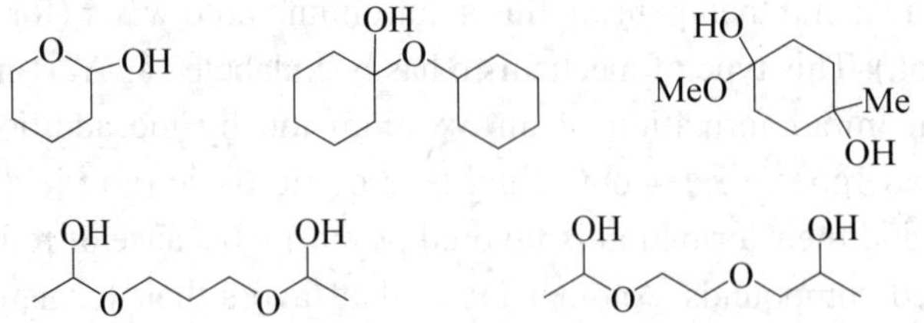

4. What would be the mechanism of the reaction? Give your predictions.

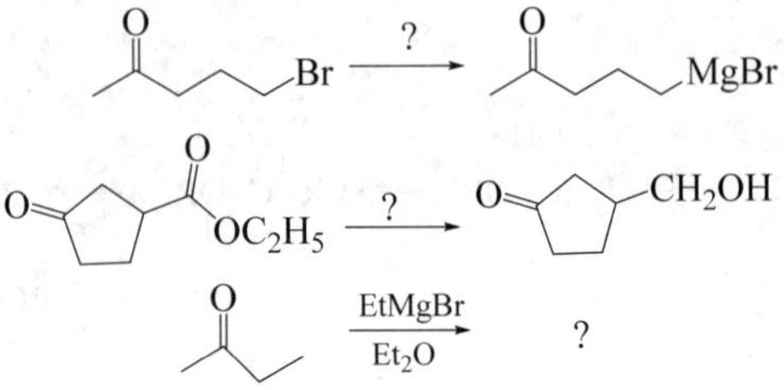

5. A typical Darzens reaction involves the base-catalysed formation of an epoxide from an α-haloketone and an aldehyde. Suggest a mechanism for the Darzens reaction consistent with the results shown below.

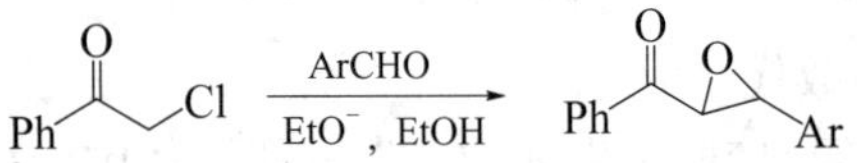

6. The aldehyde and ketone below are self-condensed with aqueous NaOH so that an unsaturated carbonyl compound is the product. Give a structure for each product and explain why you think this product is formed.

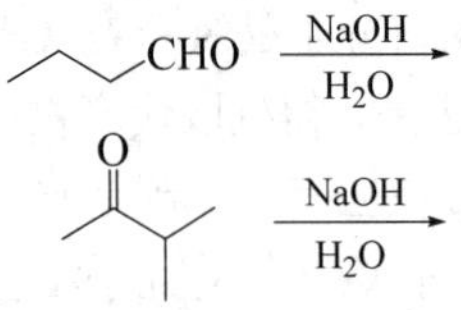

7. How would you synthesize the following compounds?

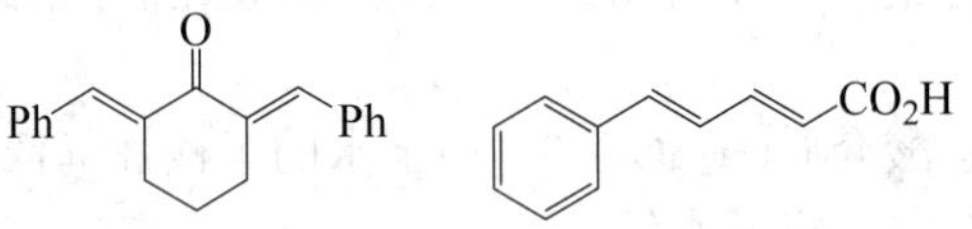

8. Complete the following reactions.

(1) 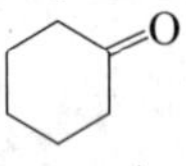 (cyclohexanone) + $(C_6H_5)_3P{=}C(CH_3)_2 \longrightarrow$ (　　　)

(2) 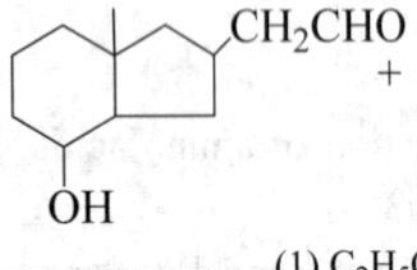 (bicyclic compound bearing CH_2CHO and OH) + $(C_6H_5)_3P{=}CHCH(CH_3)_2 \longrightarrow$ (　　　)

(3) $CH_2(CO_2Et)_2 \xrightarrow[(2)\ C_2H_5Br]{(1)\ C_2H_5ONa}$ (　　　) $\xrightarrow[(2)\ CH_3I]{(1)\ C_2H_5ONa}$ (　　　) $\xrightarrow[(2)\ H^+,\ (3)\ \Delta]{(1)\ OH^-}$ (　　　)

(4) $CH_3COCH_2CO_2Et \xrightarrow[(2)\ n\text{-}C_4H_9Br]{(1)\ C_2H_5ONa}$ (　　　) $\xrightarrow[(2)\ H^+,\ (3)\ \triangle]{(1)\text{conc. NaOH}}$ (　　　)

(5) $CH_3COOEt \xrightarrow[(2)\ H_3O^+]{(1)\ PhMgBr}$ (　　　)

(6) $CH_3CH_2COOH + NH_3 \xrightarrow{\triangle}$ (　　　) $\xrightarrow{P_2O_5}$ (　　　)

(7) $(CH_3)_2CO \xrightarrow[OH^-]{HCN}$ (   ) $\xrightarrow[HCl]{H_2O}$ (   ) $\xrightarrow{\triangle}$ (   )

(8) (cyclohexane-1,2-diyl)COOEt, COOEt $\xrightarrow{C_2H_5ONa}$ (   ) $\xrightarrow[(2)CH_3I]{(1)\ C_2H_5ONa}$ (   )

(9) (butane chain)COCH$_3$, COCH$_3$ $\xrightarrow[H_2O,\triangle]{NaOH}$ (   )

(10) (chain)CO_2Et, CO_2Et $\xrightarrow[(2)\ H^+,\ \triangle]{(1)\ C_2H_5ONa}$ (   )

(11) cyclohex-2-enone + $CH_2(CO_2Et)_2$ $\xrightarrow{C_2H_5ONa}$ (   ) $\xrightarrow[(2)\ ClCH_2CO_2Et]{(1)\ C_2H_5ONa}$ (   )

9. Give the reactivity order of nucleophilic addition reactions of the following compounds.

（1） CH_3CHO、CH_3COCH_3、CF_3CHO、$CH_3CH{=}CHCHO$、$CH_3COCH{=}CH_2$

（2） $ClCH_2CHO$、$BrCH_2CHO$、$CH_2{=}CHCHO$、CH_3CH_2CHO、CH_3CF_2CHO

（3） $HSCH_2CH_2CHO$、$NCCH_2CHO$、CH_3OCH_2CHO、CH_3SeCH_2CHO

Vocabulary

aldehyde ['ældə,haɪd] 醛；乙醛
ketone ['ki:toʊn] 酮
carbonyl compounds 羰基化合物
gem-diol 偕二醇
chloral ['kloʊrəl] 三氯乙醛
ketal ['ki:tæl] 缩酮
hemiacetal [ˌhemi'æsitæl] 半缩醛
entropically 与熵有关的
cyclohexanone 环己酮
cyanohydrins [ˌsaɪənə(ʊ)'haɪdrɪn] 氰醇
alkoxide [æl'kɔksaid, -sid] 醇盐，酚盐
formaldehyde [fɔː'mældɪhaɪd]，甲醛，蚁醛
bisulfite [ˌbaɪ'sʌlfaɪt] 亚硫酸盐
α-halogenester α-卤代酯
enolate ['iːnə,leit] 烯醇化物
hemiaminal [hiːma'ɪæmɪnl] α-氨基醇
hydroxylamine [haɪˌdrɒksɪlə'miːn] 羟胺
hydrazine ['haɪdrəziːn] 肼
diastereotopic [daɪəstɪərɪə'tɒpɪk] 非对映(异构)的
trigonal ['trɪg(ə)n(ə)l] 三角的，三角形的
prochiral 前手性的
racemic [rə'siːmɪk] 外消旋的
perpendicular [ˌpə:pən'dikjulə] 垂直，垂线
eclipse [i'klips] 日蚀，月蚀，重叠式
aldol ['ældɒl] 羟醛
dimerization [ˌdaimərai'zeiʃən] 二聚
keto-enol tautomerism 酮式-烯醇式互变异构
crossed aldol condensation 交叉羟醛缩合
succinic acid 琥珀酸
benzophenone [ˌbenzəʊ'fɪnəʊn] 苯甲醛
iminium ion 亚胺离子
tautomerize [tɔː'tɔməraiz] 使互变异构
lithium diisopropylamide, or LDA 二异丙基胺基锂
transesterification [trænsəsterəfɪ'keɪʃən] 酯交换
oxalate ['ɒksəˌleɪt] 草酸盐（酯）
glycidic [gliː'sɪdɪk] 缩水甘油的
epoxy [ɪ'pɑːksi] 环氧树脂，
epoxide [e'pɒksaɪd] 环氧化物
annulation [ænjʊ'leɪʃən] 环形物
cinnamic acids 肉桂酸
benzoin ['benzəʊɪn] 安息香
ylide 叶立德
triphenylphosphine 三苯基膦
betaine ['biːtəiːn] 内鎓盐，甜菜碱
acyl ['æsəl] 酰基
esterfication [estəfɪ'keɪʃn] 酯化
ninhydrin [nin'haidrin] 茚三酮

Chapter 7

Rearrangements

（重排）

The carbon skeleton or the position of functional group in the molecule changes in *rearrangement* reactions. Most rearrangements are 1,2-*shifts*, but some rearrangements are longer *migrations*, for example, 1,3-shifts, 1,5-shifts, etc. Rearrangements can be divided into *nucleophilic rearrangements*, *electrophilic rearrangements* and *free-radical rearrangements*. When an atom or a group with its electron pair migrates, nucleophilic rearrangement reactions or anionotropic rearrangement reactions occur, for the migrating group can be regarded as a nucleophile. While the migrating atom or group (W) move without its electron pair, electrophilic or cationotropic rearrangement reactions undergo, for the migrating group can be considered as an electrophile. If an atom or a group moves with just one electron, the free-radical rearrangement reactions happen. The atom A can be called the migration origin; B can be called the migration terminus. On the other hand, rearrangement reactions can in principle be divided into intermolecular rearrangement reactions and intramolecular rearrangement reactions. In the former reactions, the group (W) completely detaches from A in a molecule and goes to B of the other molecule; while in the latter reactions, the group (W) goes from A to B in the same molecule. It is usually not difficult to distinguish a given rearrangement in this manner by crossover experiments.

$$\mathrm{W{-}A{-}B} \longrightarrow \mathrm{A{-}B{-}W}$$

分子中的基团或原子从一个原子迁移到另一个原子上，碳链或官能团的位置发生变化的一类反应称为分子重排反应。重排分为三类：亲核重排、亲电重排和自由基重排。迁移基团带着电子对进行的重排为亲核重排。

7.1 Nucleophilic rearrangements（亲核重排）

Nucleophilic rearrangements consist of three steps: ① creation of B with an open *sextet* such as a *carbocation*, a carbine, a *nitrene* and oxonium cation etc; ② migration of the group (W) with one electron pair; ③ stabilization of A with an open sextet by further reaction. There are various ways for the first step, but the formation of a carbocation or a nitrene is the most important. The acid treatment of an alcohol is one of the most common methods for the formation of a carbocation. The formation of a nitrene can be realized by the decomposition of *acyl azides*.

$W{-}A{-}B{-}Y \xrightarrow[\text{step 1}]{-Y^-} W{-}A{-}B^+ \xrightarrow{\text{step 2}} A^+{-}B{-}W \xrightarrow{\text{step 3}}$ product

nucleophilic rearrangement

carbocation　　carbene　　nitrene　　oxonium cation

亲核重排的中间体为碳正离子、碳烯、氮烯、缺电子氧等。

7.1.1 Nucleophilic rearrangements between carbon atoms（碳原子之间的亲核重排）

（1）Wagner-Meerwein rearrangements（Wagner-Meerwein 重排）

The first Wagner-Meerwein rearrangements were found in the bicyclic terpenes. Nowadays 1,2-rearrangements of H atoms or alkyl groups in carbenium ions that do not contain any heteroatoms attached to A or B can be called Wagner-Meerwein rearrangements. There are some reactions having Wagner-Meerwein rearrangement step, such as electrophilic additions of alkenes, nucleophilic *substitutions*, E_1 elimination, Friedel-Crafts *alkylation* reactions, etc. An example of electrophilic additions of alkenes containing Wagner-Meerwein rearrangement step is the formation of bornyl chloride, isobornyl chloride and camphene. Electrophilic addition of C=C affords a *tertiary* carbocation, then migration occurs, and the 3°-carbocation is rearranged to a 2°-carbocation, for the *strain* of small ring can be released, although 3°-carbocation is usually more stable than 2°-carbocation. Further reaction with Cl^- produces bornyl chloride. If steric congestion is released, the other kind of 3°-carbocation is formed by further alkyl rearrangement, and camphene is produced along with the loss of H^+ from the methyl group. The second 3°-carbocation also can be rearranged to 2°-carbocation, and isobornyl chloride is obtained by attraction of Cl^- with 2°-carbocation.

HCl　relief of small-ring strain　bornyl chloride

relief of steric congestion

Camphene　　isobornyl chloride

The following example shows nucleophilic substitutions containing Wagner-Meerwein rearrangement step. 1°-Carbocation is initially formed by the reaction of neopentyl alcohol with H^+ and the loss of water. The migration of methyl with its bonding electrons to the adjacent 1°-carbocation produces a more stable 3°-carbocation. Nucleophilic substitution of Cl^- affords chlorinated compounds.

$$(CH_3)_3C{-}CH_2OH \underset{}{\overset{H^+}{\rightleftharpoons}} (CH_3)_3C{-}CH_2OH_2^+ \xrightarrow{-H_2O} (CH_3)_3C{-}CH_2^+ \longrightarrow (CH_3)_2\overset{+}{C}{-}CH_2CH_3 \xrightarrow{Cl^-} (CH_3)_2CCl{-}CH_2CH_3$$

The direction of rearrangements of alkyl carbocations is usually toward the most stable carbocation, which is 3° >2° >1° . However, the stability of carbocation *intermediates* is not the only factor that leads to molecular rearrangement. The release of angle strain, *torsion* strain or steric crowding in the reactant structure also can drive such a rearrangement by an alkyl or aryl shift to a carbocation site. The following example illustrates a rearrangement induced by the strain in a small ring. The reaction of cyclobutyl tertiary alcohol with HBr or H_2SO_4 leads to ring expansion product, respectively. Although a 3° -carbocation is initially formed and more stable, the release of the angle and torsion strain of the four-membered ring drives a methylene group shift to form a 2° -carbocation. Nucleophilic substitution gives the final product. The nucleophilic substitution reactions also produce elimination products.

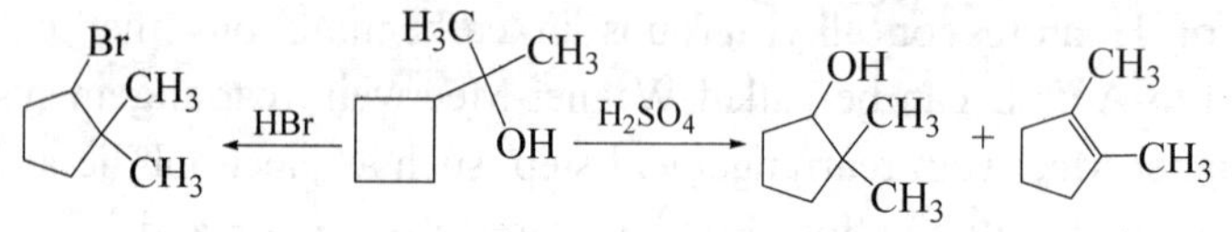

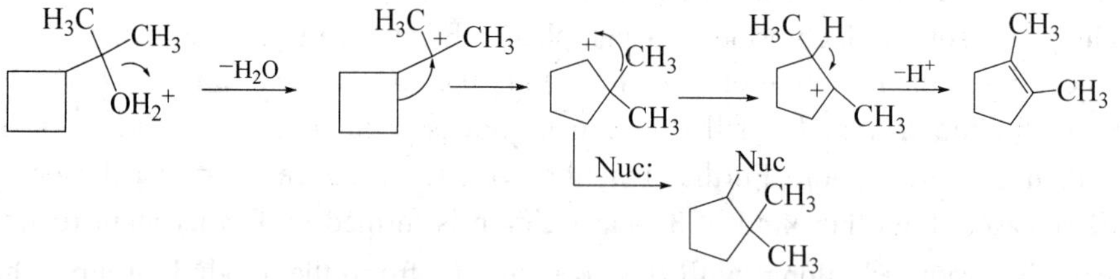

An example of E_1 containing Wagner-Meerwein rearrangement step is the formation of 2-methyl 2-butene from neopentyl alcohol, too. Firstly, 1° -carbocation is formed. Then migration of methyl and the loss of H^+ give E_1 product.

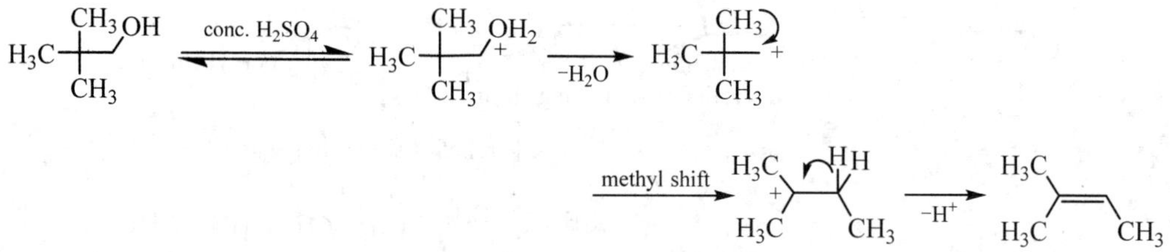

The treatment of primary amine with HNO_2 also can initiate Wagner-Meerwein rearrangements. 1° -Carbocation is formed via the reaction of primary amine with HNO_2. Then methyl migrates; water reacts with the formed 3° -carbocation; H^+ is lost, finally rearrangement product is formed.

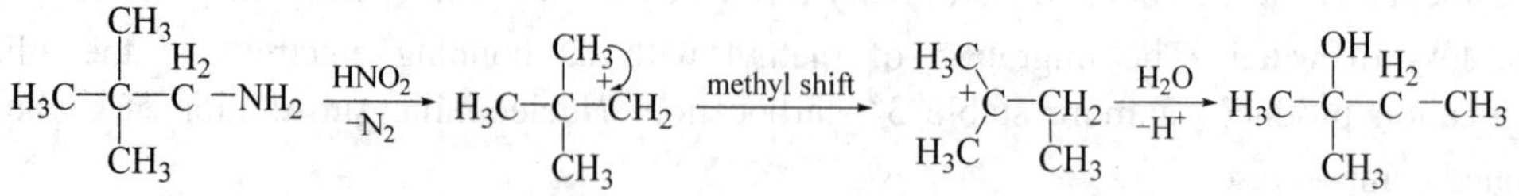

The reaction of alkyl halide with *Lewis acid* is the other method to initiate Wagner-Meerwein rearrangements. 1° -carbocation is rearranged to a more stable 2° -carbocation. Then Friedel-Crafts alkylation reaction affords isopropylbenzene.

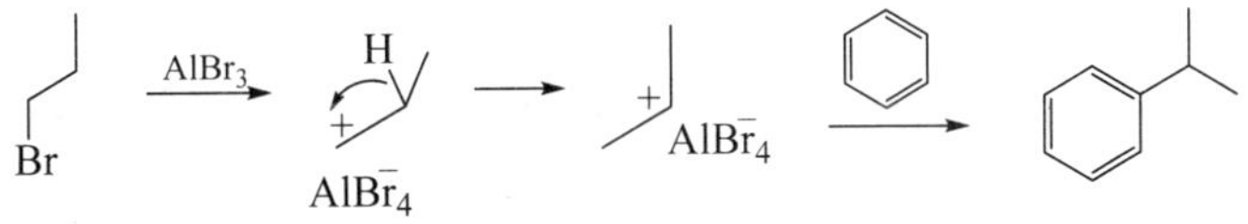

烃基和氢都可以作为迁移基团，重排为更稳定的碳正离子，或通过重排减小环的张力。迁移基团的迁移能力：芳基>烷基 (3°>2°>1°)>H。

（2）The *pinacol* rearrangement（Pinacol 重排）

When glycols react with acids, they can be rearranged to afford aldehydes or ketones. This reaction is called the pinacol rearrangement. The classical example of a pinacol rearrangement is the formation of pinacolone. Since the diol is symmetrical, each hydroxyl has equal chance to react with H^+ to produce 3°-carbocation by the subsequent loss of water. The resulting 3°-carbocation is relatively stable, for it can return to pinacol which can be approved by reaction in the presence of isotopically labeled water. A 1,2-methyl shift generates an even more stable carbocation in which the charge is delocalized by heteroatom resonance. This new cation is the conjugate acid of pinacolone. Each step in this rearrangement reaction is potentially reversible, which is demonstrated by the acid catalyzed dehydration of pinacolone (and pinacol) to 2,3-dimethyl-1,3-butadiene under vigorous conditions.

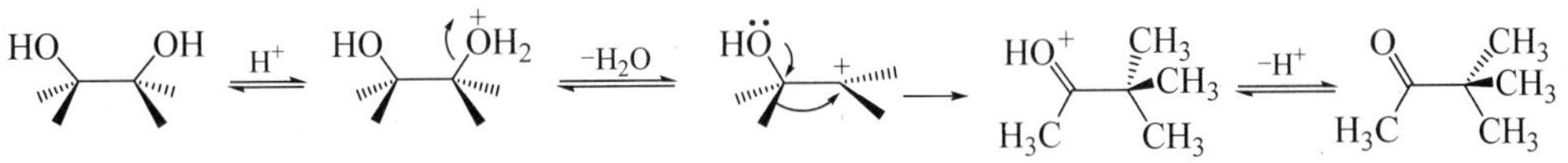

However, if the diol is *unsymmetrical*, there are many factors to be considered when the course of a pincol rearrangement is analyzed. These include: ① which hydroxyl group is lost as water? or which intermediate carbocation is more stable? ② Which is the migrating group? ③ How do steric hindrance and other strain factors influence on the rearrangement? ④ Is it possible for *epoxides* to be intermediates in the pinacol rearrangement? ⑤ Does the product stability decide the outcome of competing rearrangements? ⑥ Do the reaction conditions (i.e. type of acid, concentration, solvent and temperature) influence the course of rearrangement?

The reaction shown below may answer the first question. Diphenyl 3°-carbocation is initially formed, while dimethyl 3°-carbocation is not formed, for the former is double benzyl carbonium ion, more stable than the latter. The 1,2-methyl shift and the loss of H^+ finish the pincol rearrangement. Therefore, the hydroxyl group which is helpful for the formation of more stable carbocation is lost as water.

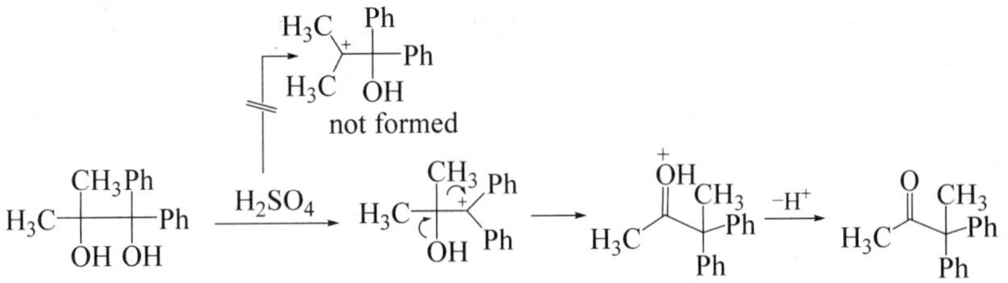

In the following reaction, diphenyl 3°-carbocation is rearranged to its major product, while methylene cation gives its minor product. However, the major product is formed by hydride migration. Phenyl migration gives the minor product. Obviously, diphenyl 3°-carbocation is more stable than methylene cation, thus the stable cation is formed superiorly.

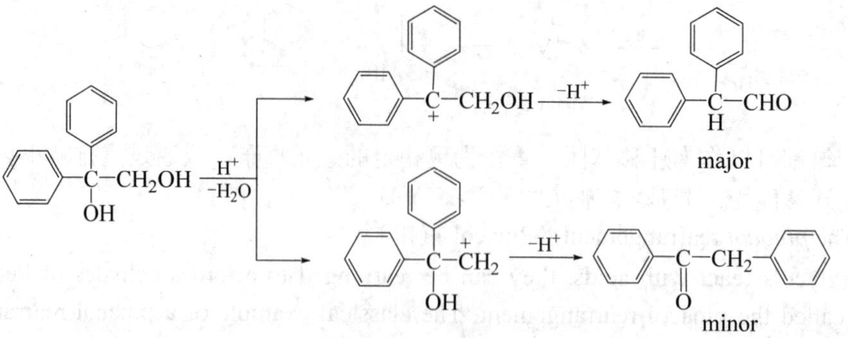

Which group migrates? It is determined not only by which group has a higher inherent migrating ability, but also by whether the group that does not migrate is better at stabilizing the positive charge. The migratory aptitudes are in the order of aryl > alkyl. As to the order of hydrogen, it is usually unpredictable. In some cases, it is hydrogen that migrates, instead of aryl; in other cases, migration of alkyl is preferred to that of hydrogen. The reaction conditions also play important role on the group migration. The following is an example of aryl migration which gives the major product.

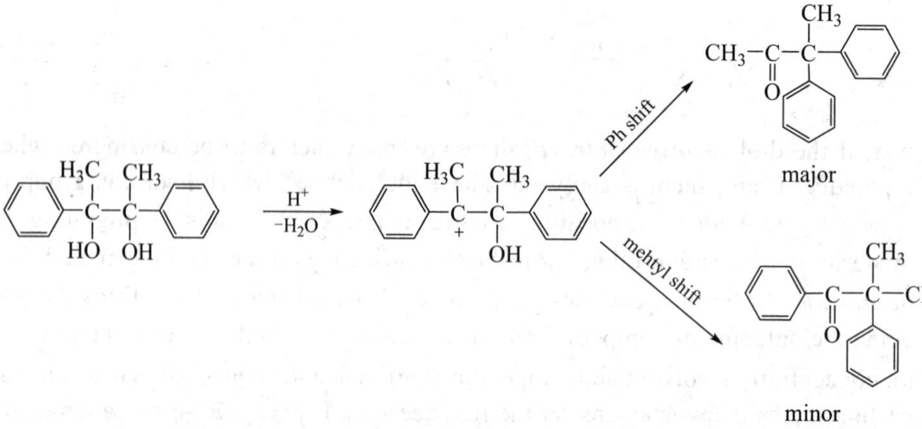

When cyclic diols are treated with acid, the effect of steric hindrance must be considered. The stereochemistry of the diol plays a crucial role in deciding the major product. If an alkyl group, such as methyl, is situated *trans-* to the leaving —OH group alone, it may migrate. If otherwise, *ring contraction* occurs, i.e. the ring carbon itself migrates to the carbocation centre. The following example shows the formation of five-membered ring ketones from cyclohexyl diol. There appears to be a connection between the migration origin and migration terminus throughout the reaction. If the migrating alkyl group has a chiral center as its key atom, the configuration at this center is retained even after migration takes place.

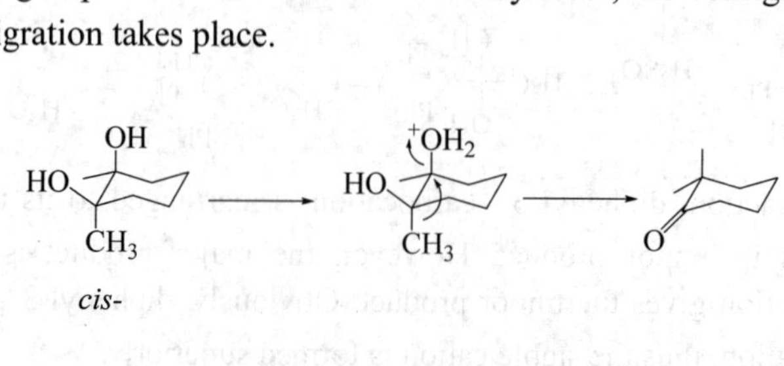

cis-

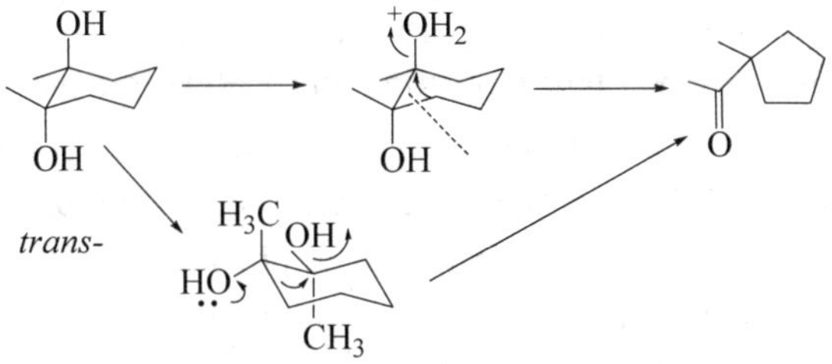

邻二醇在酸作用下的重排产物为醛或酮。非对称的取代乙二醇中，能生成稳定碳正离子的羟基首先质子化，首先离去。质子化以后，亲核性强的基团优先迁移，迁移能力：对甲氧基苯基 > 对甲基苯基>苯基>对溴苯基>烷基>氢。

When β-amino alcohols are treated with nitrous acid, the semipinacol rearrangements occur. As to the chemical reaction of a 1-aminomethyl-cycloalkanol with nitrous acid to form an enlarged cycloketone, it is called Tiffeneau-Demjanov rearrangement. The following is an example of Tiffeneau-Demjanov rearrangement. The initial tertiary cabocation is formed by the loss of nitrogen gas via the reaction of β-amino alcohol with nitrous acid. Then methyl migration gives resource-stabilized carbocation, and final aldehyde is produced by *deprotonation*.

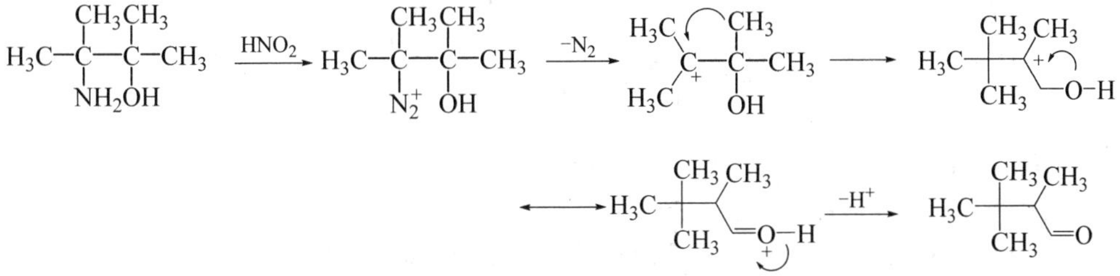

Some cyclic β-amino alcohols involve ring contraction or expansion. Both include the formation of initial carbocation step, migration step and deprotonation step. If cyclic β-amino alcohols bear substituents, the effect of steric hindrance should be considered. Dimethyl substituted cyclic β-amino alcohol gives a mixture, while the amino alcohol bearing *t*-Bu gives only one product.

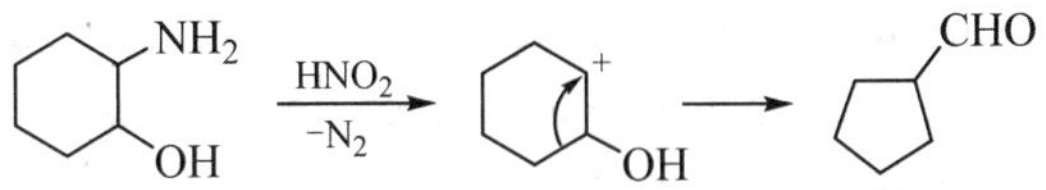

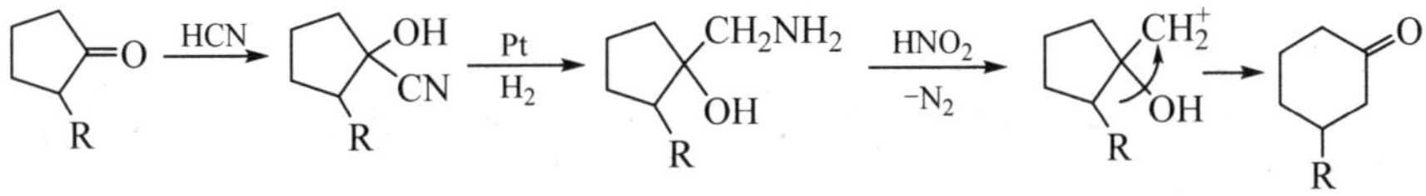

R=H or CH_3

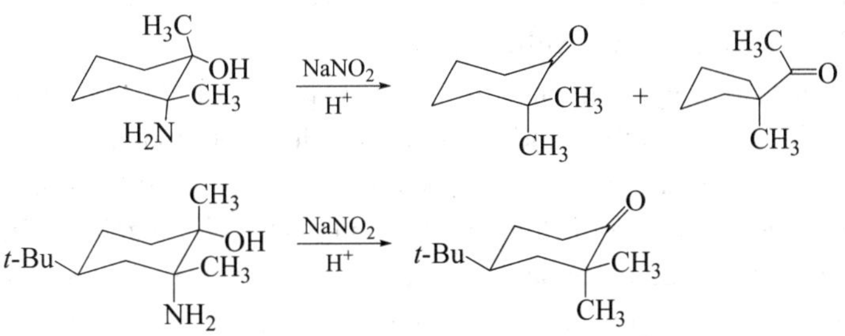

The following example illustrates the importance of substrate configuration on the course of rearrangement. When phenyl group at the α position and —NH_2 are anti-coplanar, the initial stage is the phenyl group shift to an adjacent carbocation site, it may be viewed as an intramolecular electrophilic substitution of the Friedel-Crafts type. Phenyl ring approach from the side opposite to the departing nitrogen of the diazonium ion generates a phenonium ion intermediate, even though the anisyl group (An) has a much greater migratory aptitude than phenyl. Electron pair donation by the hydroxyl substituent then acts to open the three-membered ring of the phenonium ion intermediate, yielding the anisyl ketone product. When anisyl group at the α position and—NH_2 are anti-coplanar, it is the anisyl group that shifts to the adjacent carbocation, instead of the phenyl group. Deprotonation diastereoselectively gives Ph ketone product.

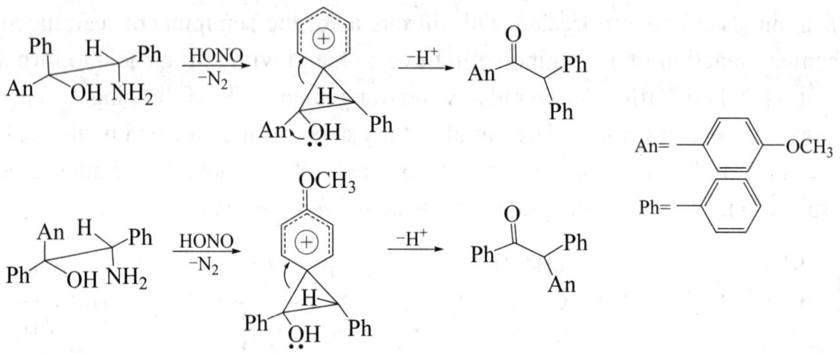

Tiffeneau-Demjanov 重排：卤代醇、氨基醇发生的类似 Pinacol 重排。

（3）The benzilic acid rearrangement（benzilic acid 重排）

The reaction of α-diketones with base affording the salt of α-hydroxyl acid is called the benzilic acid rearrangement. There is a difference between the benzilic acid rearrangement and the pinacol rearrangement. In the benzilic acid rearrangement the initial step is not the formation of a carbocation with an open sextet, but the release of a pair of π-electrons from the C=O to the oxygen by the addition of base to the carbonyl group. Then migration to the adjacent carbon occurs to form the negative oxygen ion. This benzylic acid rearrangement complements a weak electrophilic pull by the adjacent carbonyl carbon with the "push" of the alkoxide anion. Proton shift gives the salt of α-hydroxy acid. It is noticed that aliphatic diketones and α-keto aldehydes also are suitable substrates for the benzilic acid rearrangement.

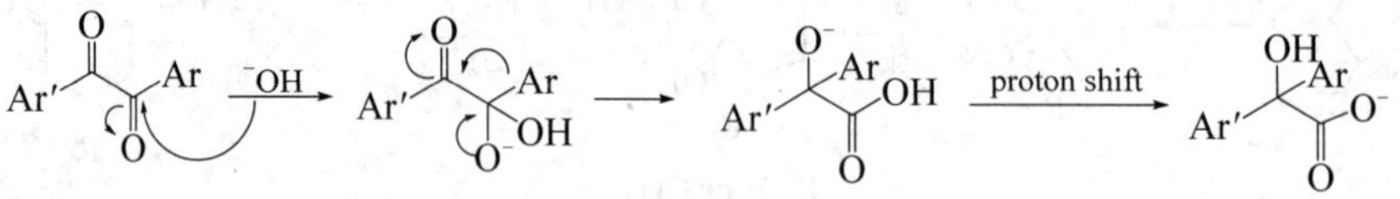

用强碱处理 *α*-二酮，重排得到 *α*-羟乙酸盐（酯）。

（4）The Wolff rearrangement（Wolff 重排）

The rearrangement of acyl carbenes to *ketenes* is called the Wolff rearrangement. The rearrangement is a critical step in the Arndt-Eistert procedure for elongating a *carboxylic acid* by a single methylene unit, as shown in the diagram below. The reaction of the starting acid with $SOCl_2$ gives an acyl chloride derivative. As is known, diazomethane has a nucleophilic methylene group. *Acylation* of the methylene carbon produces an equilibrium mixture of a diazonium species and the diazomethyl ketone plus hydrogen chloride. Thus the following reaction of CH_2N_2 affords a diazomethyl ketone in the presence of base. In the presence of HCl the reaction gives a chloromethyl ketone. The formed diazo ketone is decomposed to an acyl carbene in the presence of a silver catalyst (usually Ag_2O or $AgNO_3$) together with heat or light energy. Then R group shifts to the adjacent carbene to finish the Wolff rearrangement, and a ketene is generated. If the ketene reacts with water, then a carboxylic acid product is generated. When the ketene reacts with alcohol or amine, then an ester or an amide is formed, respectively.

$$H_2C{=}\overset{+}{N}{=}\overset{-}{N} \longleftrightarrow H_2\overset{-}{C}{-}\overset{+}{N}{\equiv}N:$$

diazomethane

$$RCOOH \xrightarrow{SOCl_2} RCOCl \xrightarrow{CH_2N_2} [RCOCH{=}N_2 + HCl] \xrightarrow{base} RCOCH{=}N_2$$

a diazomethyl ketone

$$[RCOCH{=}N_2 + HCl] \xrightarrow{HCl} RCOCH_2Cl$$

a chloromethyl ketone

$$\xrightarrow[\text{heat or light}, -N_2]{Ag^+} RCO\ddot{C}H \xrightarrow{\text{Wolff rearrangement}} R{-}CH{=}C{=}O \begin{cases} \xrightarrow{H_2O} RCH_2COOH \\ \xrightarrow{R'OH} RCH_2COOR' \\ \xrightarrow{R'NH_2} RCH_2CONHR' \end{cases}$$

an acyl carbene; ketene

The first example of the following Arndt-Eistert reactions shows the tolerance for other functional groups, such as nitro. The second example shows that the configuration of the migrating group is retained in the rearrangement.

$$o\text{-}O_2NC_6H_4CO_2H \xrightarrow{SOCl_2} o\text{-}O_2NC_6H_4COCl \xrightarrow[B.\ Ag_2O,\ H_2O]{A.\ CH_2N_2} o\text{-}O_2NC_6H_4CH_2CO_2H$$

$$(H_3C)(C_2H_5)CH{-}CO_2H \xrightarrow{SOCl_2} (H_3C)(C_2H_5)CH{-}COCl \xrightarrow[B.\ Ag_2O,\ H_2O]{A.\ CH_2N_2} (H_3C)(C_2H_5)CH{-}CH_2CO_2CH_3$$

α-重氮酮在 Ag_2O 存在或者光照射下，失去 N_2，经酰基碳烯重排得烯酮，烯酮经水解得到羧酸，烯酮也可与醇或氨反应得到酯或酰胺。

7.1.2 Rearrangement by atom or group shifts from a carbon atom to a nitrogen atom（原子或基团从碳原子迁移到氮原子的重排）

（1）The Hoffmann rearrangement（Hofmann 重排）

In the Hoffmann rearrangement, the reaction of an unsubstituted amide with sodium hypobromite or sodium hydroxide and bromine gives a primary amine. Firstly, *N*-bromo amide is formed. Then a proton on the nitrogen atom is lost by the interaction of OH^-, for the proton is acidic in the presence of two electron-withdrawing groups (acyl and bromo). The loss of Br^- generates an acyl nitrene. The R group migrates from a carbon atom to the adjacent nitrogen atom to form an *isocyanate*. The addition reaction of water and the loss of carbon dioxide give a primary amine. The R group may be alkyl or aryl.

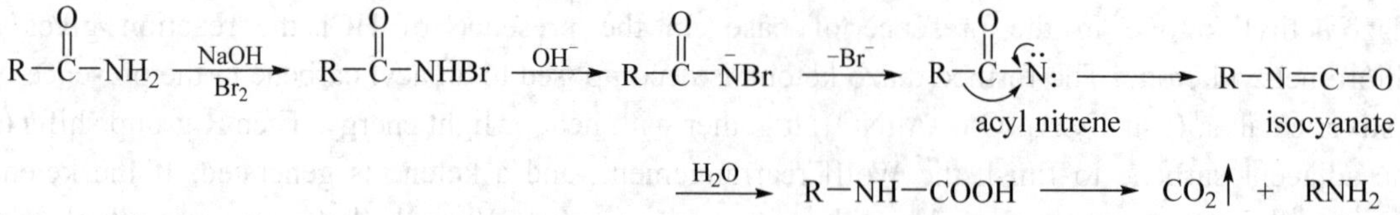

$$R-\overset{O}{\overset{\|}{C}}-NH_2 \xrightarrow[Br_2]{NaOH} R-\overset{O}{\overset{\|}{C}}-NHBr \xrightarrow{OH^-} R-\overset{O}{\overset{\|}{C}}-\overset{-}{N}Br \xrightarrow{-Br^-} R-\overset{O}{\overset{\|}{C}}-\ddot{N}: \longrightarrow R-N=C=O$$

acyl nitrene　　isocyanate

$$\xrightarrow{H_2O} R-NH-COOH \longrightarrow CO_2\uparrow + RNH_2$$

The examples of the Hoffmann rearrangement are illustrated below. The first example shows the stereostructure of chiral center will be kept during the migration course in the Hoffmann rearrangement. The second example shows the configurations of the migrating group is retained in the rearrangement.

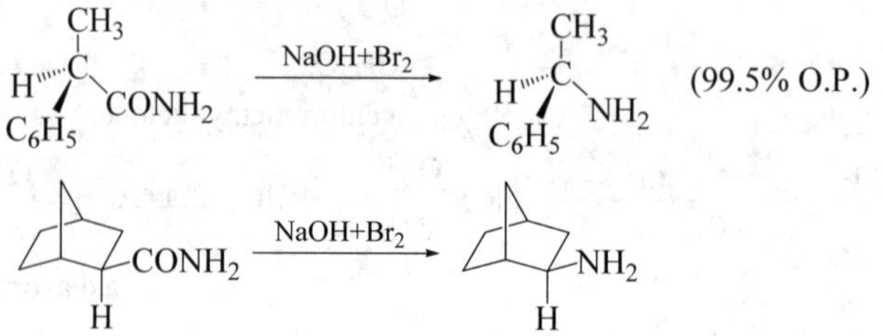

$$H\text{''''}C(CH_3)(C_6H_5)CONH_2 \xrightarrow{NaOH+Br_2} H\text{''''}C(CH_3)(C_6H_5)NH_2 \quad (99.5\%\ O.P.)$$

$$\text{(bicyclic)}-CONH_2 \xrightarrow{NaOH+Br_2} \text{(bicyclic)}-NH_2$$

酰胺与 $Br_2(Cl_2)$ 在碱性条件下经过异氰酸酯中间体重排成胺。如果酰胺的 α-碳是手性碳，反应后手性碳的构型保持不变。

（2）The Curtius rearrangement（Curtius 重排）

In the Curtius rearrangement, the *pyrolysis* of acyl azide generates isocyanate with good yield due to the fact that no water is present to hydrolyze it to the amine. The formed isocyanate also can be subsequently hydrolyzed in water to produce the amine, as in the Hoffmann rearrangement. The acyl azide ($RCON_3$) can be prepared by the reaction of an acyl chloride with sodium azide or by the reaction of an ester with excess hydrazine, followed by reaction of the acylhydrazide product ($RCONHNH_2$) with cold nitrous acid.

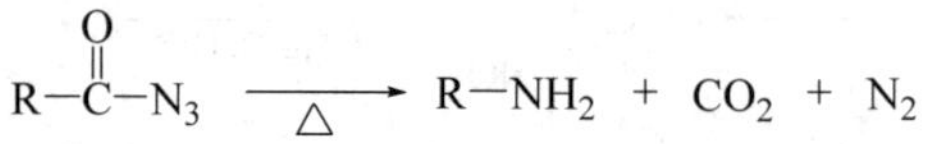

$$R-\overset{O}{\overset{\|}{C}}-N_3 \xrightarrow[\triangle]{} R-NH_2 + CO_2 + N_2$$

（3）The Schmidt rearrangement（Schmidt 重排）

The Schmidt rearrangement involves the addition of hydrazoic acid (HN_3) to a carboxylic acid, as illustrated below. It is also a rearrangement of acyl nitrenes to isocyanates. The Schmidt rearrangement can be regarded as a variant of the Curtius rearrangement.

$$R-\overset{O}{\overset{\|}{C}}-OH + HN_3 \xrightarrow[C_6H_6]{H_2SO_4} R-NH_2 + CO_2 + N_2$$

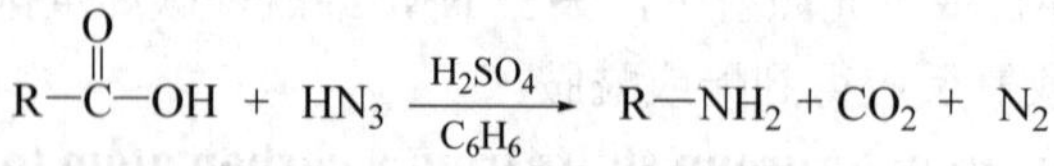

（4）The Lossen rearrangement（Lossen 重排）

When a hydroxamic acid, which can be prepared by the reaction of an ester with hydroxyl amine, is *O*-acylated, the formed *O*-acyl derivatives give isocyanates just by heating or in the presence of P_2O_5, Ac_2O, and $SOCl_2$ etc. The rearrangement is called the Lossen rearrangement, as is shown below.

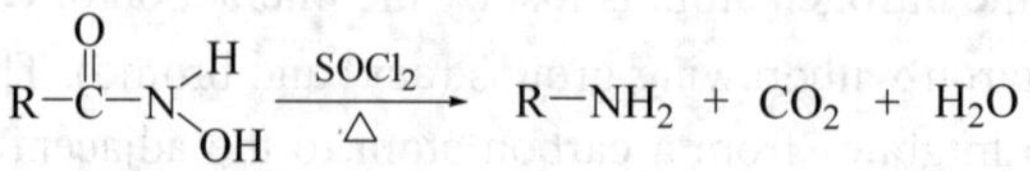

$$R-\overset{O}{\overset{\|}{C}}-N(H)(OH) \xrightarrow[\triangle]{SOCl_2} R-NH_2 + CO_2 + H_2O$$

From the Hoffmann rearrangement, the Curtius rearrangement and the Lossen rearrangement, we can see that there are some general procedures for obtaining the acyl nitrene precursors, which are listed below.

$$R-\overset{O}{\overset{\|}{C}}-\ddot{\underset{..}{N}}-Z \xrightarrow{-Z} R-\overset{O}{\overset{\|}{C}}-\ddot{N} \longrightarrow R-\ddot{N}=C=\ddot{O} \xrightarrow{H_2O} R-\underset{H}{\ddot{N}}-\overset{O}{\overset{\|}{C}}-O-H \longrightarrow R-NH_2 + CO_2$$

an acyl nitrene　an isocyanate　a carbamic acid

Z = Br or Cl　Hoffmann Rearrangement
　= N_2^+　Curtius Rearrangement
　= OCOR′　Lossen Rearrangement

经过异氰酸酯中间体重排还有 Lossen 重排、Curtius 重排、Schimidt 重排。

（5）The Beckmann rearrangement（Beckmann 重排）

When a ketoxime is treated with a strong acid, the hydroxyl group on the ketoxime is removed. Then a group on the carbon atom migrates to the adjacent nitrogen atom. After hydrolysis, an amide is formed. If the protonated hydroxyl group and its bonding electron pair were completely removed, a divalent sp-hybridized azacation would be generated. If it was true, both carbon substituents (R & R′) would be candidates for the subsequent 1,2-shift. However, in practice, it is always the group *anti* to the departing OH that migrates to the adjacent nitrogen atom. This *stereospecificity* indicates that the 1,2-shift is concerted with N—O cleavage, as shown below. The resulting *N*-alkylated nitrilium intermediate will react with nucleophiles, such as water, at the electrophilic carbon atom adjacent to the "onium" nitrogen. The loss of H^+ generates the initial product, an iminol, which immediately *tautomerizes* to the more stable amide.

$$\underset{R'}{\overset{R}{}}C=N\text{-}OH \xrightarrow{H^+} \underset{R'}{\overset{R}{}}C=N\text{-}\overset{+}{O}H_2 \longrightarrow \left[R-\overset{+}{C}=N-R' \longleftrightarrow R-C\equiv\overset{+}{N}-R'\right] \xrightarrow{H_2O} R-\underset{\overset{+}{O}H_2}{C}=N-R'$$

$$\xrightarrow{-H^+} R-\underset{OH}{C}=N-R' \rightleftharpoons R-\underset{O}{\overset{\|}{C}}-NH-R'$$

(Fluorenone oxime with NO_2 $\xrightarrow{PCl_5}$ phenanthridine with N=C–OH and NO_2 $\longrightarrow$ phenanthridinone, HN–C=O, with NO_2)

If a ketoxime reacts with phosphorous pentachloride, an iminochloride is probably formed. Then stereospecific rearrangement occurs, and the respective amide is produced after hydrolyzing.

$$\text{HO-N}=C(R^1)(R^2) \xrightarrow{PCl_5} \text{ClnPO-N}=C(R^1)(R^2) \longrightarrow R^2-\overset{+}{N}\equiv C-R^1 \longrightarrow R^2-N=C(OH)R^1 \rightleftharpoons R^1C(=O)NHR_2$$

$$\text{HO-N}=C(R^1)(R^2) \xrightarrow{PCl_5} \text{OPCln-N}=C(R^1)(R^2) \longrightarrow R^1-\overset{+}{N}\equiv C-R^2 \longrightarrow R^1-N=C(OH)R^2 \rightleftharpoons R^2C(=O)NHR^1$$

syn & *anti* isomers　stereospecific rearrangement　iminol tautomer　amide tautomer

When a ketoxime reacts with acyl chloride, the *O*-acyl is removed. Hydrolysis gives the amide, too.

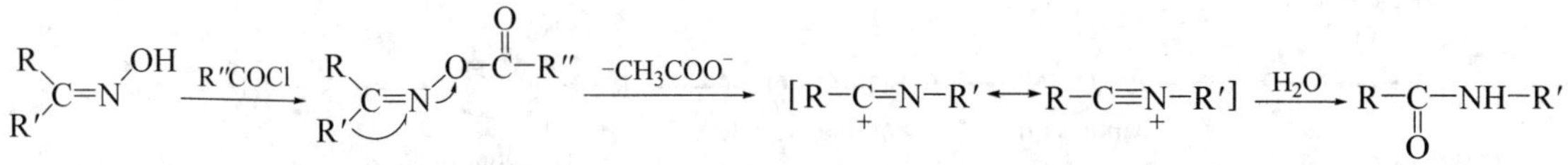

$$\mathrm{RR'C{=}N{-}OH} \xrightarrow{R''COCl} \mathrm{RR'C{=}N{-}O{-}C(O){-}R''} \xrightarrow{-CH_3COO^-} [\mathrm{R{-}\overset{+}{C}{=}N{-}R'} \longleftrightarrow \mathrm{R{-}C{\equiv}\overset{+}{N}{-}R'}] \xrightarrow{H_2O} \mathrm{R{-}C(O){-}NH{-}R'}$$

In the Beckmann rearrangement, the stereostructure of chiral center will be kept during the migration course, as is shown below.

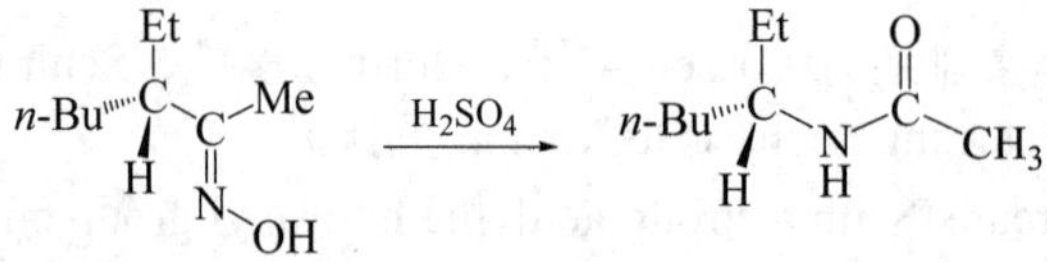

n-Bu, Et, H, Me, N–OH $\xrightarrow{H_2SO_4}$ n-Bu, Et, H, N–H, C(O)CH_3

Beckmann 重排指酮肟在磷酸、硫酸、多磷酸、五氯化磷等试剂催化下重排形成酰胺。反应中间体是乃春（Nitrene, 氮烯）。迁移基（R）从离去基（OH）的背后进攻缺电的氮。因此，反应是立体专属的反应迁移。迁移基团中心碳原子重排前后构型不变。

7.1.3 Rearrangement by atom or group shifts from a carbon atom to an oxygen atom（原子或基团从碳原子迁移到氧原子的重排）

（1）The Baeyer-Villiger rearrangement（Baeyer-Villiger 重排）

The acid-catalyzed reaction of ketones with hydroperoxide derivatives is known as the Baeyer-Villiger reaction. In the reaction, the carboxylic esters or lactones are formed by "*insertion*" of oxygen between R′ and carbonyl, thus it is a kind of oxidation reaction. R′ group migration from the carbon atom of carbonyl group to the oxygen atom of hydroperoxide derivatives occurs in the reaction, as is shown below. Therefore, the reaction also can be called the Baeyer-Villiger rearrangement. The rearrangement is more likely concerted. Once the peracid has added to the carbonyl group, the rearrangement may be facilitated by an intramolecular hydrogen bond.

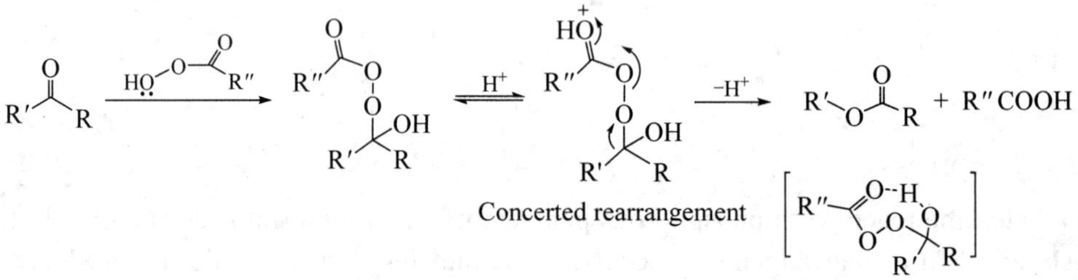

$\mathrm{R'C(O)R} \xrightarrow{HO{-}O{-}C(O)R''}$ adduct $\underset{}{\overset{H^+}{\rightleftharpoons}}$ protonated adduct $\xrightarrow{-H^+}$ $\mathrm{R'{-}O{-}C(O){-}R}$ + $R''COOH$

Concerted rearrangement

If the ketones are unsymmetrical, the approximate ability of the migration is 3°-alkyl>2°-alkyl~benzyl~phenyl>1°-alkyl>methyl. The position of the insertion of the oxygen is between the easier migrating group and the carbonyl. The following example shows that the migration of p-MeO-C_6H_4-is easier than Ph group, thus the oxygen is inserted between the carbonyl and the p-MeO-C_6H_4- group.

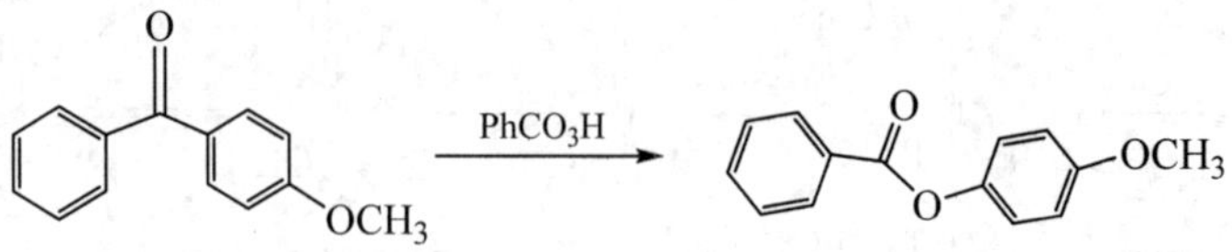

$\mathrm{PhC(O)C_6H_4OCH_3} \xrightarrow{PhCO_3H} \mathrm{PhC(O)OC_6H_4OCH_3}$

The stereostructure of chiral center will be kept during the migration course in the Baeyer-Villiger reaction.

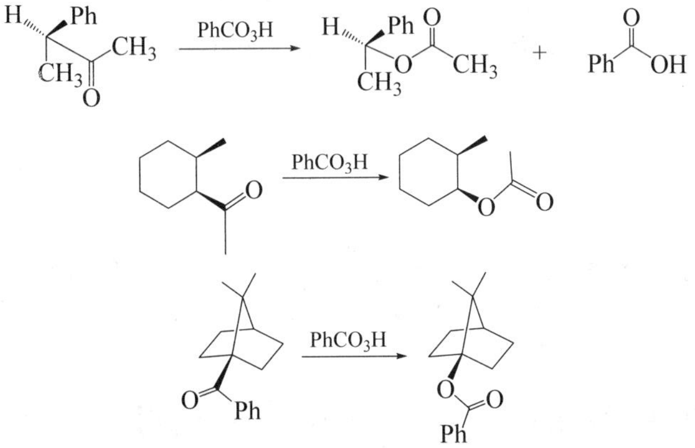

The cyclic ketones in the following reaction give ring expansion products. If aldehydes are substrates in the Baeyer-Villiger rearrangement, carboxylic acids are produced.

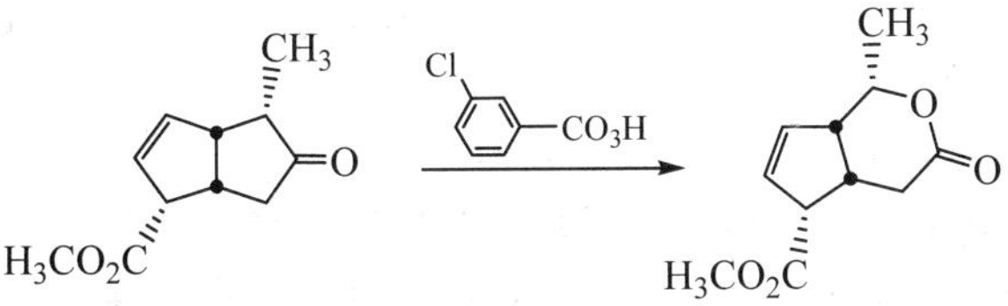

Baeyer-Villiger 重排指醛和酮被过氧酸或过氧化氢氧化，与羰基直接相连的碳链断裂，插入一个氧生成酯的反应。顺序：3°-烷基>2°-烷基～苄基～苯基>1°-烷基>甲基。迁移基团中心碳原子重排前后构型不变。

（2）Rearrangement of hydroperoxides（氢过氧化物重排）

Hydroperoxides can be cleaved by proton or Lewis acids. In the reaction, the leaving group becomes water by protonating the hydroxyl group. It is possible to initially form an oxacation, then the relatively unstable oxacation is transformed into a more stable carbocation by a 1,2-alkyl or aryl shift, as is shown below. The formed carbocation intermediate reacts with water to produce an unstable *hemiacetal*, which is hydrolyzed to generate the ketone product. The rearrangement reaction also takes place when the alcohols are treated with H_2O_2 and acids. Peroxy esters are also suitable substrates for the rearrangement reaction.

rearrangement

a hemiacetal

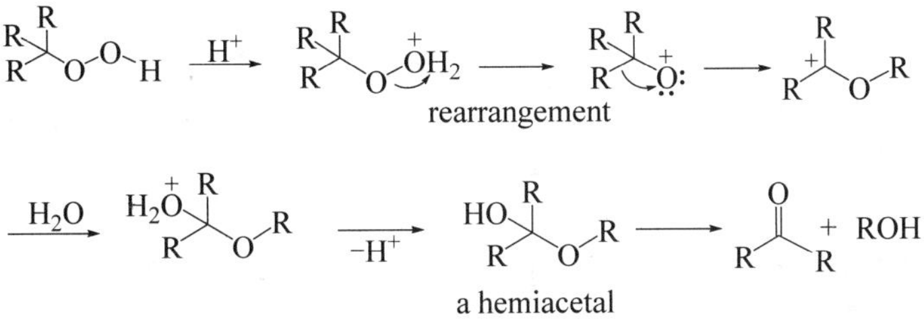

When the carbon in the hydroperoxides bears different groups, such as alkyl group and aryl group, the migration of the aryl group dominates. A useful industrial procedure for preparing phenol (and acetone) is based on this strategy. Interestingly, a hydroperoxide is formed by the oxidation of cumene, instead of the treatment of alcohols and H_2O_2 and acids.

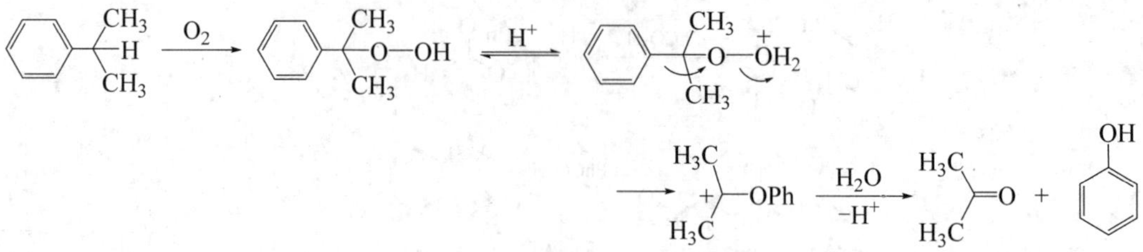

在酸的作用下，过氧键断裂，得到缺电子氧正离子，烷基迁移后产生更稳定的碳正离子，碳正离子遇水产生不稳定的半缩醛，经过水解后得到产物酮。

7.2 Electrophilic rearrangements（亲电重排）

Electrophilic rearrangements also consist of three steps. The first step is the generation of a *carbanion* (or other negative ion); the second step is the migration of the group without its electron pair; the third step is the formation of the product by further reaction. Electrophilic rearrangement is not as common as nucleophilic rearrangement.

亲电重排中迁移基团不带电子对进行重排。

7.2.1 The Stevens rearrangement and the Wittig rearrangement（Stevens 重排和 Wittig 重排）

If one of the carbons attached to the nitrogen in a *quaternary ammonium salt* has an electron-withdrawing group Z, and the ammonium salt is treated with a strong base, such as NaOR or $NaNH_2$, then a rearrangement tertiary amine is formed. The reaction is called the Stevens rearrangement. When Z is an aryl group, a competing reaction which is called the Sommelet-Hauser rearrangement occurs. In the Stevens rearrangement, the electron-withdrawing group may be RCO—, ROOC—, phenyl, vinyl, ethynyl. The most common migrating groups are allylic, benzylic, benzhydryl, 3-phenylpropargyl, and phenacyl. The following scheme indicates how the Stevens rearrangement finishes. The initial carbanion is formed by the treatment of a quaternary ammonium salt with a base, then the migrating group without an electron pair shifts from the nitrogen atom to the adjacent carbanion to generate the final product. During the migration process, the configuration of the migrating group is retained.

Stevens rearrangement

a strong base

HO^- / $-H_2O$

t-BuOK

PhLi

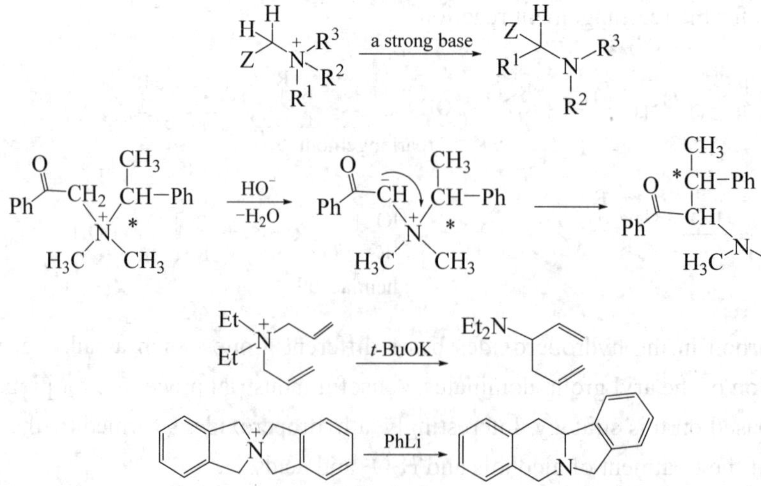

If a sulfonium salt containing an electron-withdrawing group is treated with a base, a thio ether is produced. The reaction is also Stevens rearrangement. In this reaction, $PhCH_2$—group migrates from the sulfur atom to the adjacent carbanion.

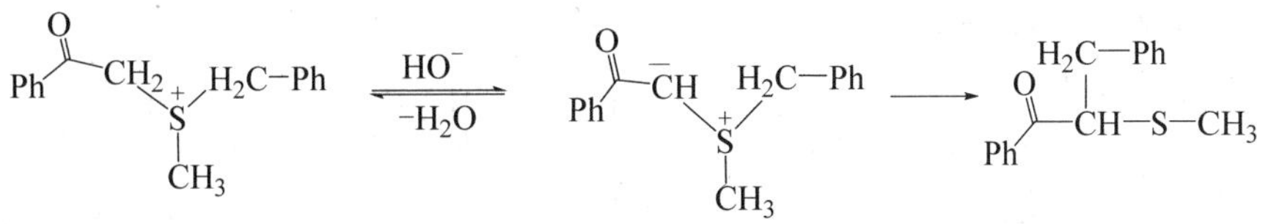

α-位上含吸电子基的季铵盐或季锍鎓，在强碱作用下，发生分子内烷基的[1,2]-迁移，烷基从氮原子或硫原子迁移至邻近的碳负离子上，生成叔胺或硫醚。吸电子基为酰基、酯基、芳基、乙烯基和乙炔基等。迁移顺序：烯丙基>苄基>乙基>甲基>苯基。迁移基团迁移前后中心原子的构型保持。

When benzylic quaternary ammonium salts are treated with alkali hydroxides, the Sommelet-Hauser rearrangement occurs to produce a benzylic tertiary amine. The reaction is most often carried out with three methyl groups on the nitrogen atom. The Sommelet-Hauser rearrangement often competes with the Stevens rearrangement. If the two rearrangements are possible, the former is favored at lower temperatures, while the latter is favored at higher temperatures. The Sommelet-Hauser rearrangement may be considered as an addition-elimination process. The benzylic hydrogen is acidic, thus it can be lost by the strong base to form a *nitrogen ylide*. The nitrogen ylide can be transformed to a carbanion, and then the carbanion undergoes a nucleophilic addition on the phenyl ring to generate another carbanion. Further steps finish the rearrangement. If cyclic quaternary ammonium salt is treated with sodium amide, then a ring expansion product is formed.

Sommelet-Hauser rearrangement

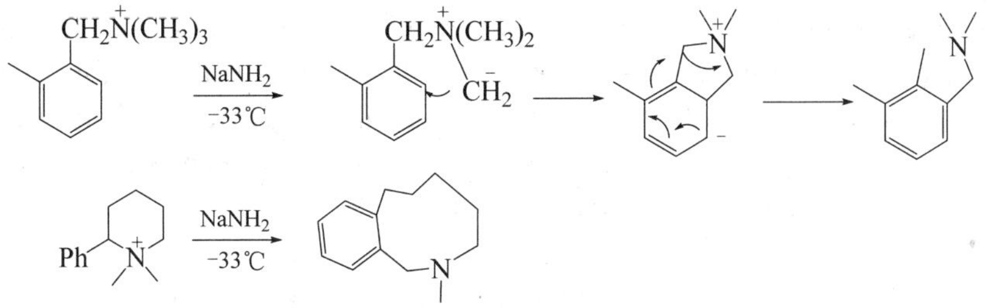

苄基季铵盐在强碱作用下，重排生成邻位烷基取代的苄基叔胺。

If a sulfonium salt contains a benzylic group, the Sommelet-Hauser rearrangement occurs, too.

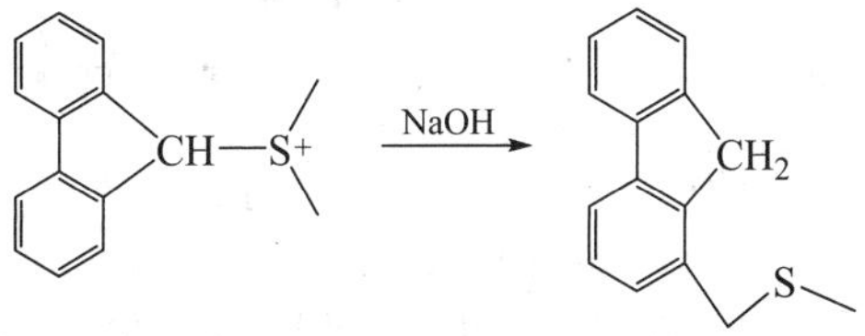

When an ether is treated with a strong base, such as *alkyllithium*, *phenyllithium* or sodium amide, the Wittig rearrangement occurs to produce the salt of a tertiary alcohol. In the Wittig rearrangement, a reactive carbanion α to an ether is formed by a powerful base. An intramolecular shift of an alkyl or aryl group from oxygen to carbon then creates a much more stable alkoxide anion. After hydrolysis, a tertiary alcohol is produced. Thus the Wittig rearrangement is a useful method for preparation of multi substituted alcohols. The R and R′ groups may be alkyl, aryl or vinyl. Migratory aptitudes are allylic, benzylic > ethyl > methyl > phenyl.

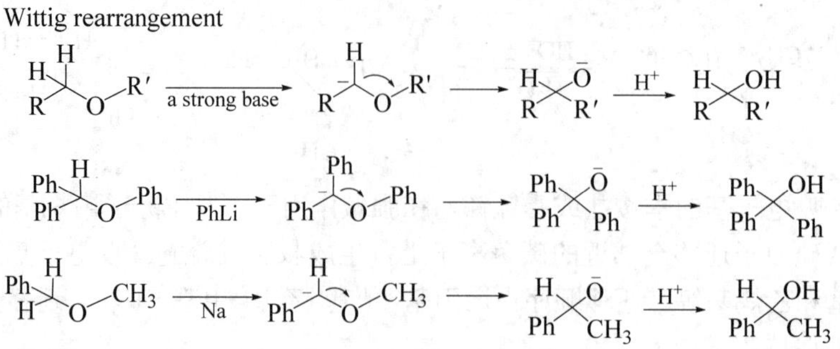

Wittig rearrangement

醚类化合物经强碱(烷基锂或氨基钠等)处理，分子中的一个烷基迁移生成醇。中间经由碳负离子重排生成稳定的氧负离子。迁移顺序：烯丙基，苄基>乙基>甲基>苯基。

7.2.2 The Favorskii rearrangement（Favorskii 重排）

In the Favorskii rearrangement, α-halo ketones react with alkoxide ions to give esters. If hydroxide ions or amines are used as bases, the carboxylic acid (salt) or amide is produced, respectively, instead of the ester.

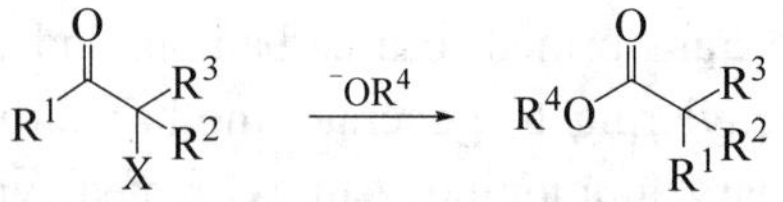

First, the α hydrogen on the non-halogenated side of the carbonyl is removed by an alkoxide ion to form a carbanion. Then a cyclopropanone is formed via the intramolecular nucleophilic substitution of the formed carbanion. The nucleophilic addition of the alkoxide ion to the carbon atom of the carbonyl group leads to the formation of a negative oxygen ion. If the formed cyclopropanone is a symmetric compound, the three-membered ring can be opened with equal probability on either side of the carbonyl. However, in the general case, the three-membered ring is unsymmetrical and should open on the side that gives the more stable carbanion. The formed carbanion reacts with ROH to generate an ester.

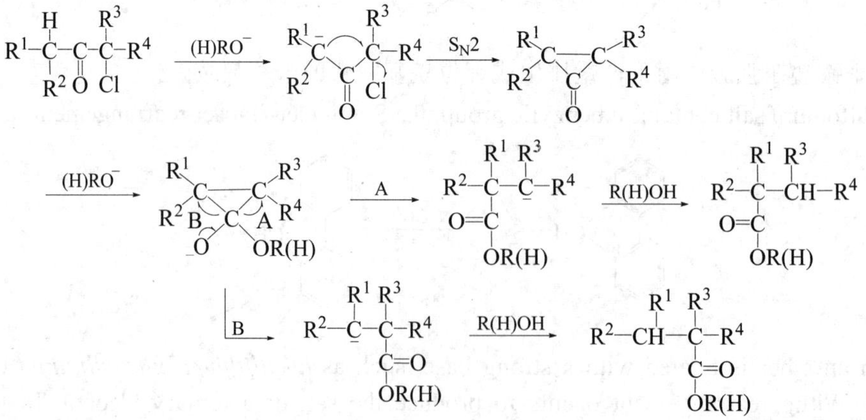

Cyclic α-halo ketones give ring contraction. An example is shown in the following diagram. An enol is initially formed, and then a cyclopropanone intermediate is generated. The cyclopropanone reacts with an alkoxide ion to produce a hemiacetal. Then the relief of angel strain drives the ring opening of the hemiacetal to yield a ring contraction product.

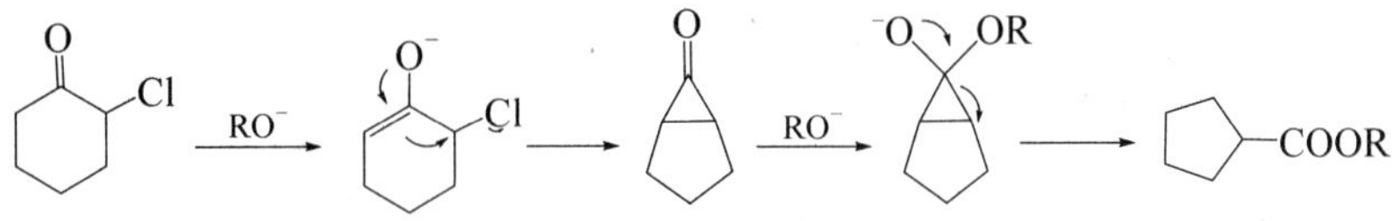

The following example shows how the α,α'-dibromo ketone undergoes the Favorskii rearrangement. The α hydrogen is removed by NaOEt to form a bromo substituted cyclopropanone. The rearrangement and the loss of a bromide anion to yield an α, β-unsaturated ester. If a α, α'-dihalo ketone bearing a α hydrogen on the non-halogenated side of the carbonyl reacts with an alkoxide ion, an α, β-unsaturated ester is also produced.

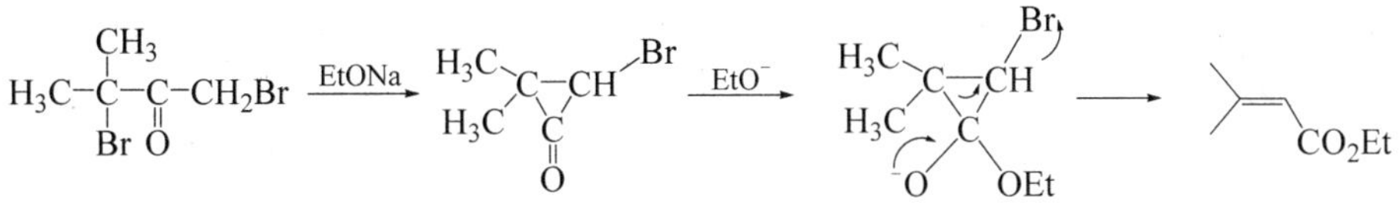

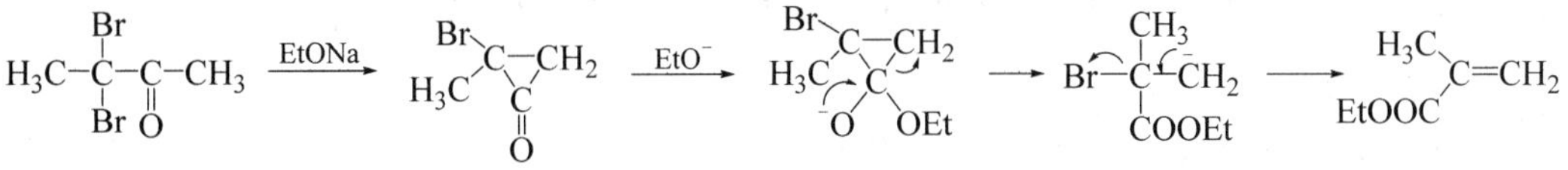

The unsymmetrical cyclopropanone opens to favor the less substituted α-carbon. It indicates that the stability order of the carbanion is 1°>2°>3°.

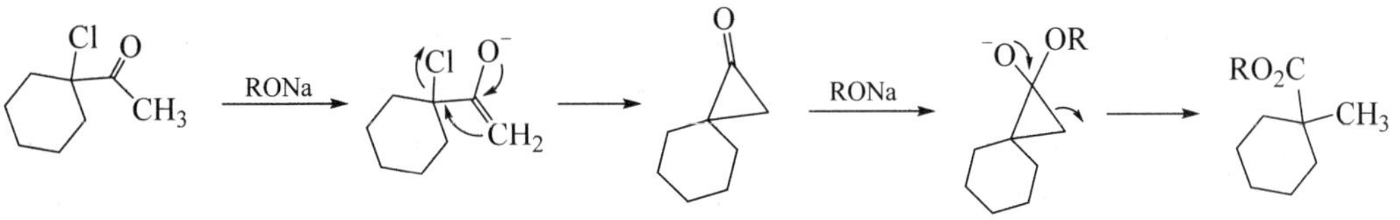

The rearrangement of the stereoisomeric substrates is proved to be stereospecific with configuration inversion at the chlorinated carbon.

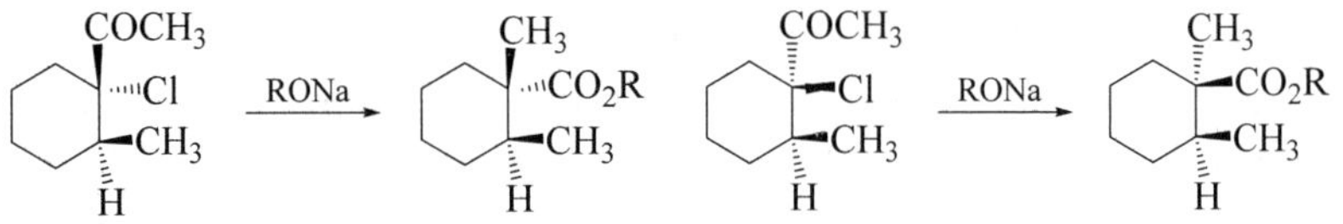

The α, β- dihalo ketone bearing a α hydrogen on the non-halogenated side of the carbonyl also gives a, β, γ-unsaturated ester.

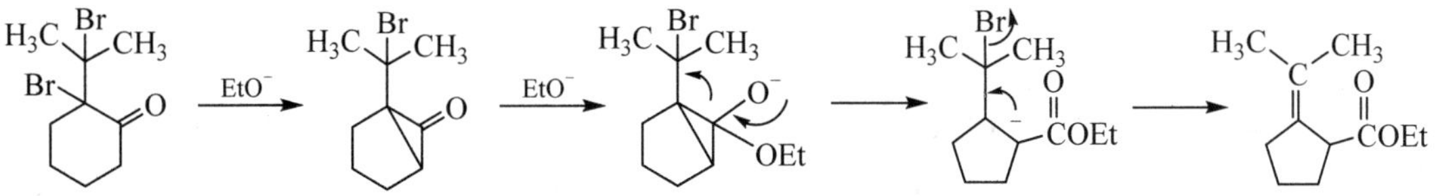

If the halo ketones do not have α-hydrogen, they also undergo the rearrangement to give the same type of product. The rearrangement is called quasi-Favorskii rearrangement. The addition of an alkoxide ion to the carbon atom of the carbonyl group generates a negative oxygen ion. The migration of R^1 and the loss of a chloride anion give the product. Demerol is prepared via the rearrangement.

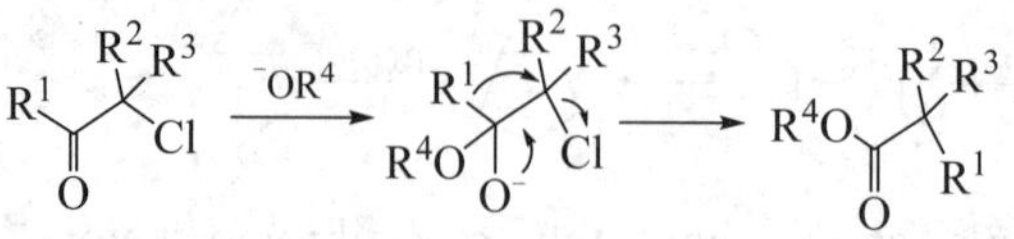

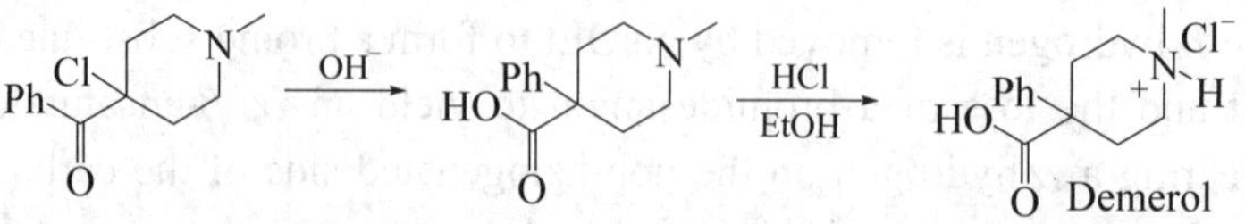

在醇钠、氢氧化钠、氨基钠等碱性催化剂存在下，α-卤代酮（氯代酮或溴代酮）失去卤原子，经过环丙酮中间体，重排成具有相同碳原子的羧酸酯、羧酸、酰胺的反应。如果生成的环丙酮中间体是不对称的，生成碳负离子稳定性高的一边开环概率更高。具有 α'-氢的 α,α-二卤代酮重排时，产物为 α,β-不饱和酯。α-卤代环酮重排得到环缩小的羧酸。

7.3 Rearrangements on aromatic rings（芳环重排）

7.3.1 The Fries rearrangement（Fries 重排）

In the Fries rearrangement, an acyl group of a phenyl ester migrates to the benzene ring, thus the phenyl ester is rearranged to a hydroxyl aryl ketone. When the benzene ring has a *meta*-directing group, it has an adverse effect on the Fries rearrangement. The reaction is *ortho* and *para* selective. One of the two products can be favored by changing the reaction conditions, such as temperature, solvent. The reaction is catalyzed by *Brønsted* or Lewis acids, such as HF, $AlCl_3$, BF_3, $TiCl_4$, or $FeCl_3$. The mechanism of the Fries rearrangement is drawn below. A phenyl ester reacts with a Lewis acid to generate a complex by the *coordination* of $AlCl_3$ to the phenolic oxygen atom. Then the complex dissociates to form a free acylium cation. The acylium cation *ortho* and *para* selectively reacts in a classical electrophilic aromatic substitution with the aromatic ring. The hydrogen is removed as hydrochloric acid where the chloride is derived from aluminium chloride. A low temperature favors *para* substitution, while a high temperature favors *ortho* substitution. The *ortho* product is also favored in non-polar solvents. If the solvent polarity increases, the ratio of the *para* product can increase.

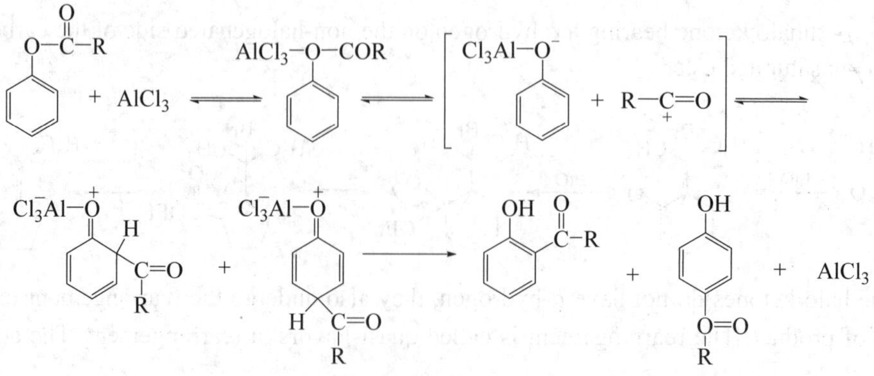

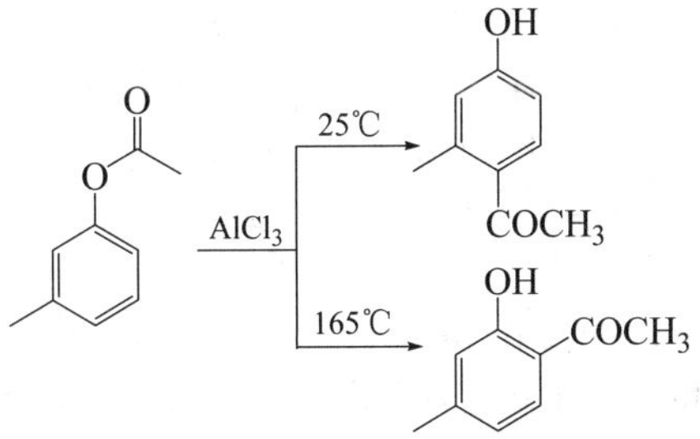

In addition to the thermal rearrangement discussed above, there is a photo-Fries rearrangement. The photo-Fries rearrangement involves a radical mechanism. In the presence of light, an acyl radical and a *para* hexadienone radical are generated. The *rearomatization* by the removal of the hydrogen gives the rearrangement product.

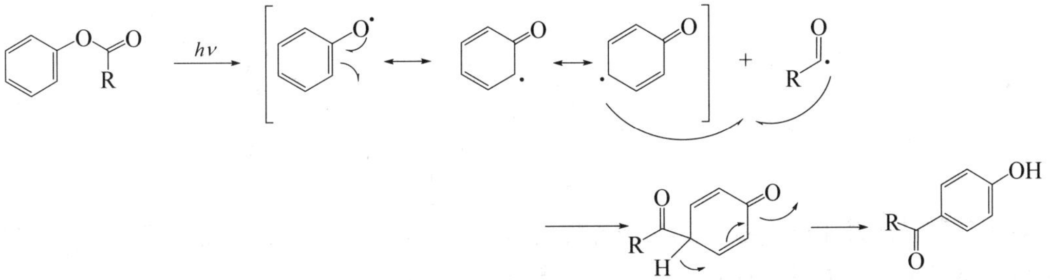

酰氧酚醚中的酰基重排到苯环上得到邻酰基苯酚及对酰基苯酚。Fries 重排是分子间重排，低温、极性溶剂条件下重排到对位，得到动力学稳定的产物，高温、非极性溶剂条件下重排到邻位，得到热力学稳定的产物。酚的芳环上带有间位定位基的酯不能发生此重排。

7.3.2 The Claisen rearrangement（Claisen 重排）

Allylic aryl ethers, when heated, rearrange to *o*-allylphenols. The reaction is called the Claisen rearrangement. In the Claisen rearrangement, an allylic group migrates to the *ortho* position. If both *ortho* positions are occupied, the allylic group migrates to the *para* position. When both the *para* and *ortho* positions are occupied, there is no rearrangement. Migration to the *meta* position has not been observed. The mechanism is concerted with pericyclic [3,3]-*sigmatropic* rearrangement.

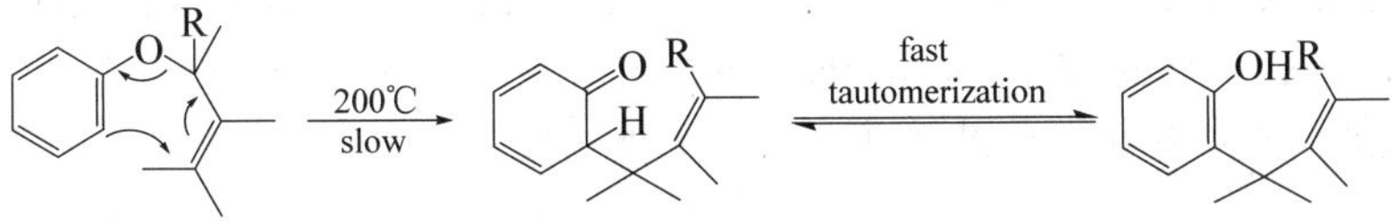

In the *ortho* migration, the allylic group always undergoes an allylic shift. That is, a substituent α to the oxygen is now γ to the ring, as is shown below. It has been demonstrated by ^{14}C labeling.

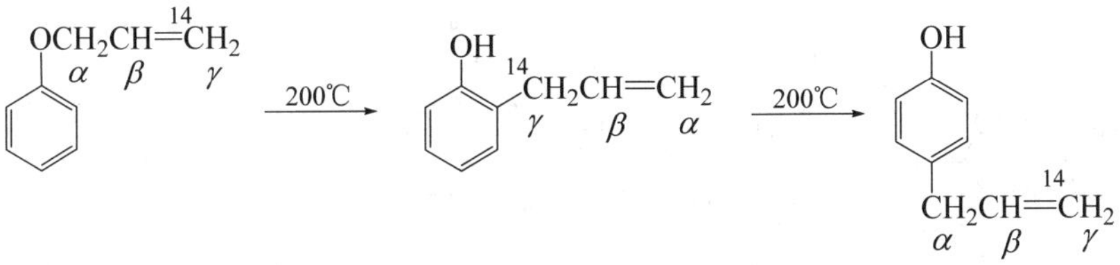

When both *ortho* positions are occupied, a second [3, 3]-sigmatropic rearrangement occurs, as is shown below. In the *para* - Claisen rearrangement, the allylic group is found exactly as it is in the original ether. That is, a substituent α to the oxygen in the original ether is now still α to the ring.

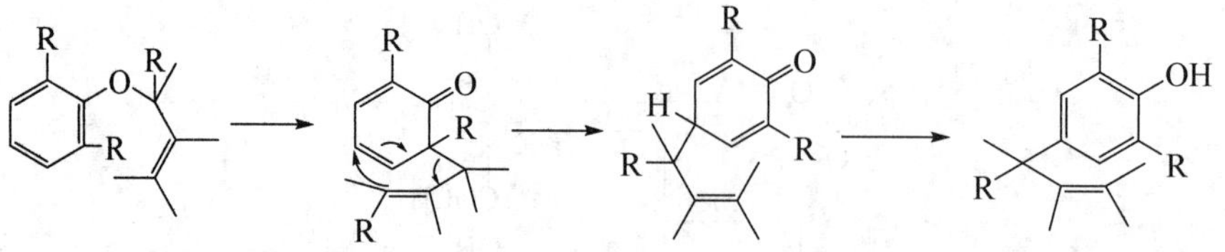

R O R R R O R R H R O R R R OH R R

It is noticed that the transition state of the Claisen rearrangement is in the chairlike form rather than the boatlike form, as is shown below. Allylic ethers of enols and the analogs of allylic aryl ethers also can undergo the Claisen rearrangement.

H Z H O ≠ H H O R H E HO

H E H O ≠ H H O H S E HO

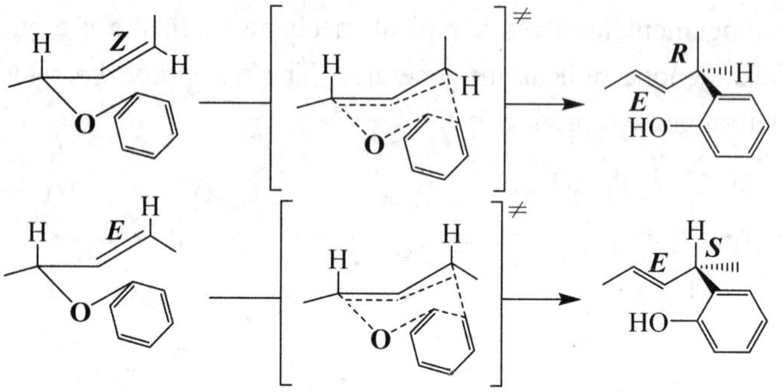

烯丙基芳基醚发生[3,3]-σ 迁移生成邻烯丙基苯酚。如果邻位被占，则继续迁移至对位得到对烯丙基苯酚。中间经过椅式的环状过渡态。

7.3.3 The Benzidine rearrangement（Benzidine 重排）

In the Benzidine rearrangement, hydrazobenzene is treated with acids to give about 70% of 4,4′-diaminobiphenyl and about 30% of 2,4′-diaminobiphenyl. The rearrangement is general for *N,N′*-diarylhydrazines. Usually, 4,4′-diaminobiaryl is the major product. A crossover experiment is carried out for the explanation of the formation of the major product. A mixture of 2,2′- dimethoxylhydrazobenzene and 2,2′-diethoxylhydrazobenzen produce a mixture of 3,3′-dimethoxyl-4,4′-diaminobiphenyl and 3,3′-diethoxyl-4,4′-diaminobiphenyl. No cross rearrangement products are formed. The experiment shows that the Benzidine Rearrangement is intramolecular rearrangement, instead of intermolecular rearrangement. The mechanism has been extensively studied. ^{15}N labeling at both nitrogens of hydrazobenzene and ^{14}C labeling at a *para* position show that both the new C—C bond formation and the N—N bond cleavage are in the rate-determining step. Thus the mechanism is concerted. The following [5,5′]-sigmatropic rearrangement gives 4,4′-diaminobiphenyl, as is shown below.

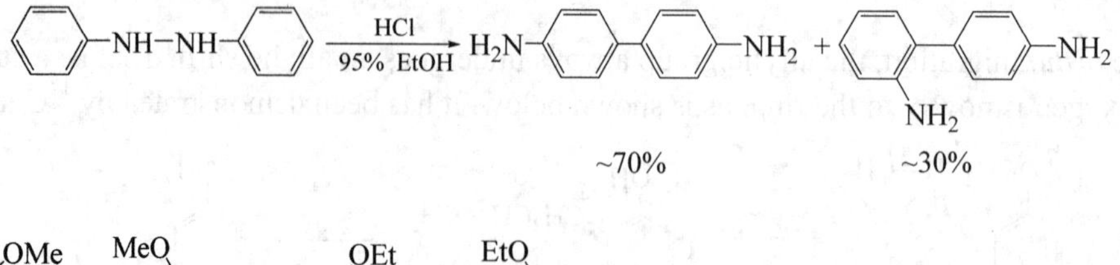

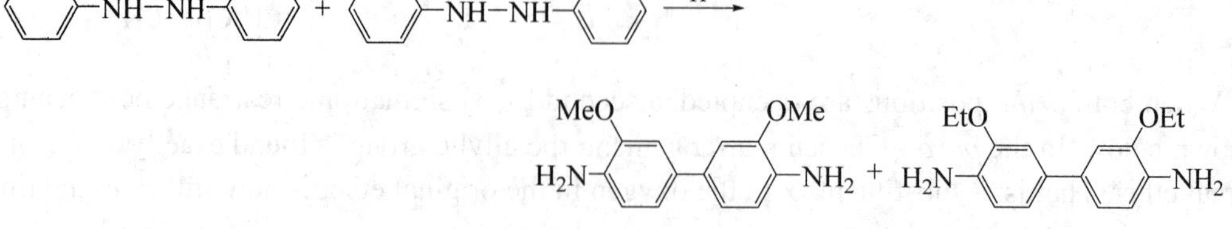

NH—NH —HCl, 95% EtOH→ H_2N— —NH_2 + —NH_2 (NH_2)

~70% ~30%

OMe MeO NH—NH + OEt EtO NH—NH —H^+→

MeO OMe H_2N— —NH_2 + EtO OEt H_2N— —NH_2

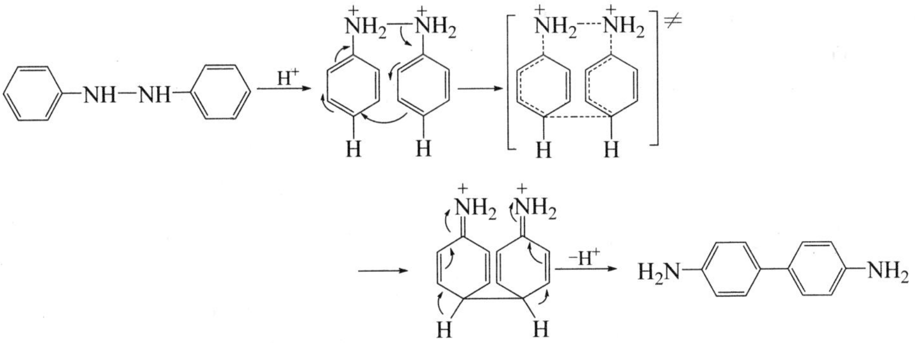

氢化偶氮苯在强酸催化下经[5,5]-σ 迁移生成 4,4′-二氨基联苯。

Problems

1. Write mechanisms for the following reactions.

(1)

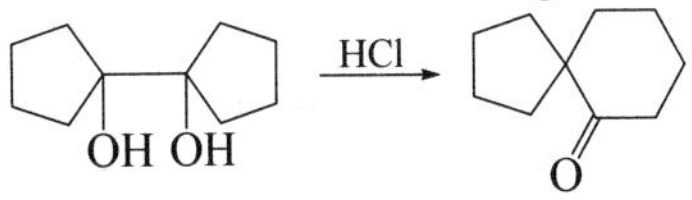

(2)

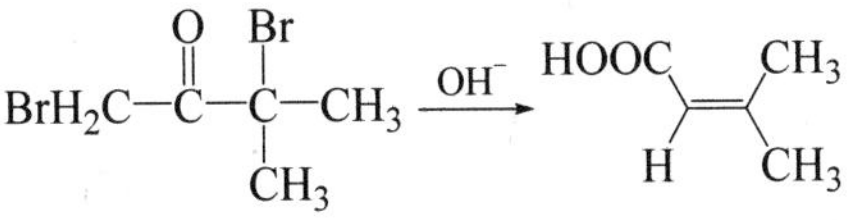

(3)

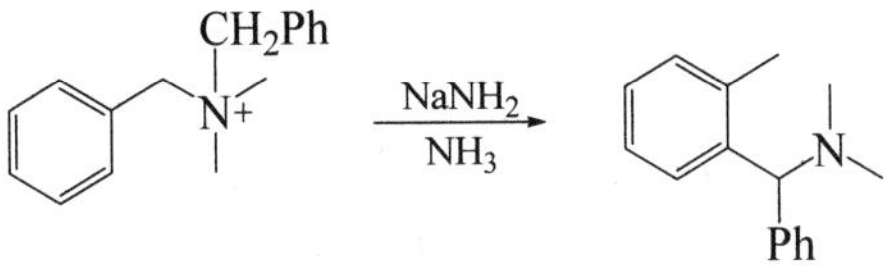

(4)

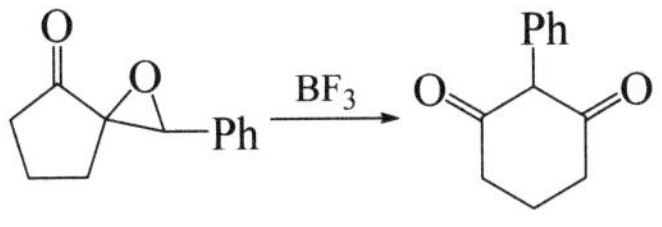

(5)

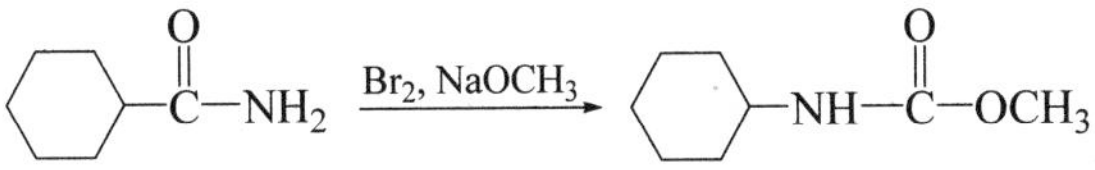

(6)

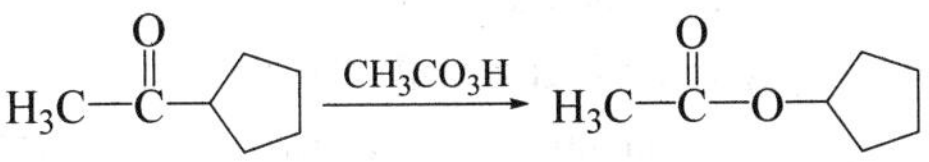

(7)

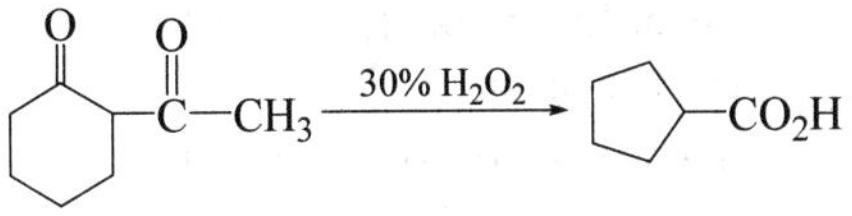

(8)

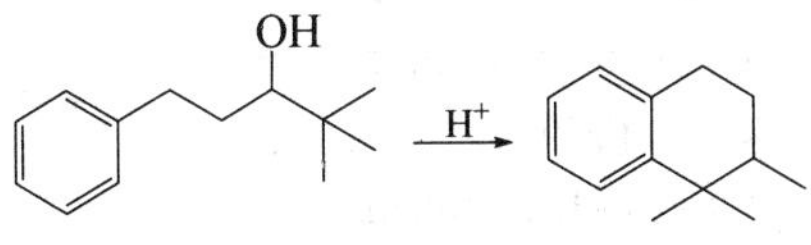

(9)

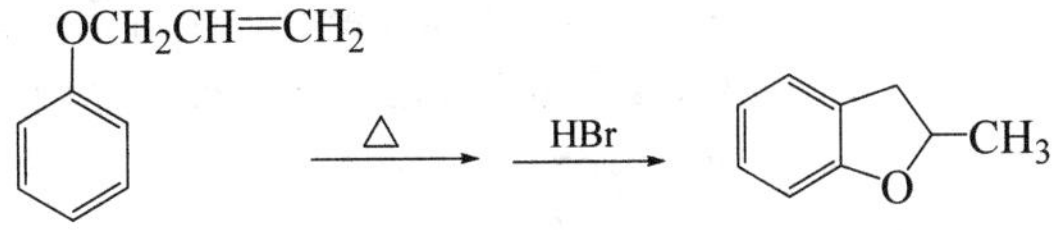

2. Write down the major products of the following reactions.

(1) 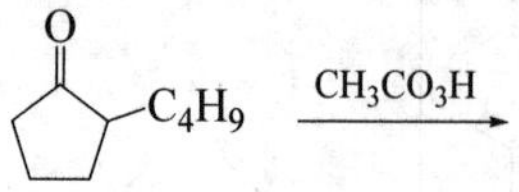 O, C_4H_9 $\xrightarrow{CH_3CO_3H}$

(2) $C—CH_3$, N—OH $\xrightarrow{PCl_5}$

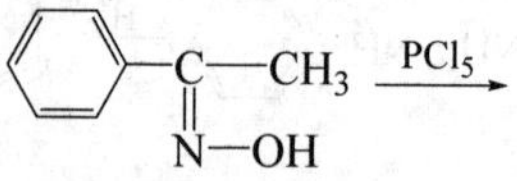

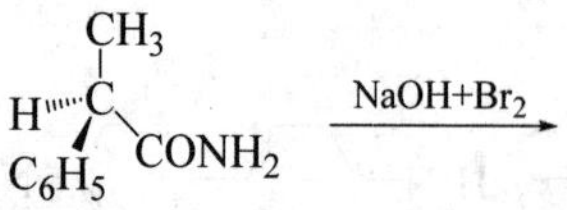

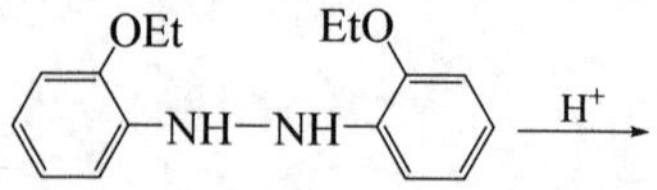

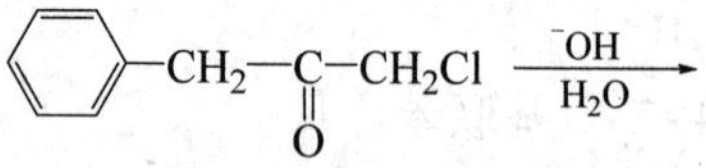

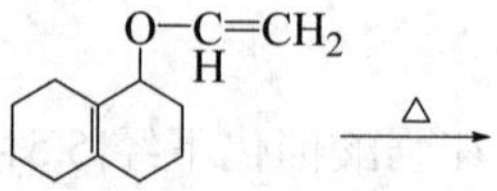

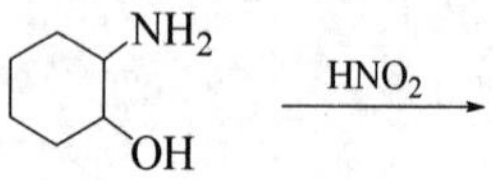

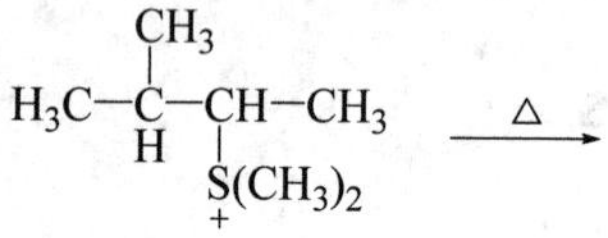

(3) CH_3, H, C, C_6H_5, $CONH_2$ $\xrightarrow{NaOH+Br_2}$

(4) OEt, EtO, —NH—NH— $\xrightarrow{H^+}$

(5) $—CH_2—C—CH_2Cl$, O $\xrightarrow[H_2O]{^-OH}$

(6) $O—C=CH_2$, H $\xrightarrow{\triangle}$

(7) NH_2, OH $\xrightarrow{HNO_2}$

(8) CH_3, $H_3C—C—CH—CH_3$, H, $\overset{+}{S}(CH_3)_2$ $\xrightarrow{\triangle}$

(9) O O $\xrightarrow{KOH}$

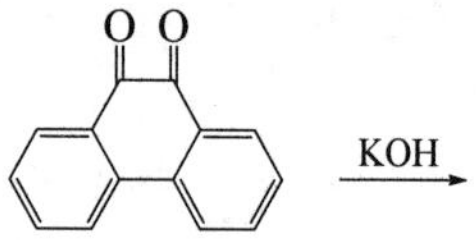

(10) 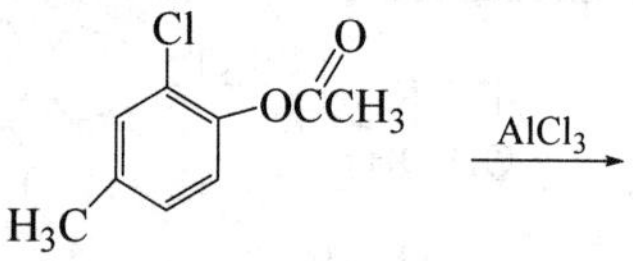 Cl, O, $OCCH_3$, H_3C $\xrightarrow{AlCl_3}$

(11) 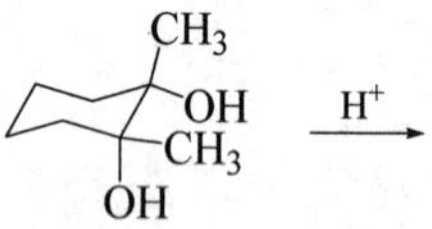 CH_3, OH, CH_3, OH $\xrightarrow{H^+}$

(12) 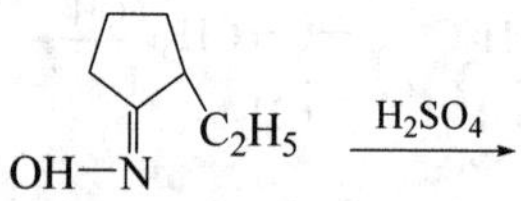 C_2H_5, OH—N $\xrightarrow{H_2SO_4}$

(13) 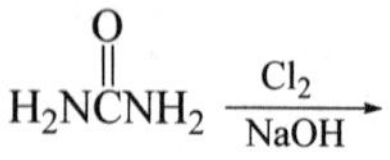 O, H_2NCNH_2 $\xrightarrow[NaOH]{Cl_2}$

(14) 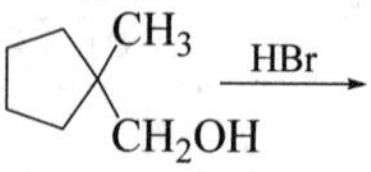 CH_3, CH_2OH $\xrightarrow{HBr}$

Vocabulary

rearrangement [ˌriːə'reɪndʒmənt] 重排
shift [ʃɪft] 移动
migration [maɪ'greɪʃn] 迁移
adjacent [ə'dʒeɪsnt] 邻近的
nucleophilic rearrangement 亲核重排
electrophilic rearrangement 亲电重排
free-radical rearrangement 自由基取代
sextet [seks'tet] 六隅体
nitrene ['nɪtriːn] 氮宾
carbocation [kɑːbə'keɪʃən] 碳正离子
acyl azide 酰基叠氮
strain [streɪn] 张力
tertiary ['tɜːʃərɪ] 叔的
intermediate [ˌɪntə'miːdiət] 中间体
torsion ['tɔːʃn] 扭转
substitution [ˌsʌbstɪ'tjuːʃn] 取代
alkylation [ˌælkɪ'leɪʃən] 烷基化
pinacol [pɪnə'kɒl] 频哪醇
unsymmetrical ['ʌnsɪ'metrɪkəl] 非对称的
epoxide [e'pɒksaɪd] 环氧化物
ring contraction 环缩小
deprotonation [deprəʊtɒ'neɪʃn] 去质子化
carboxylic acid 羧酸
acylation ['æsəleɪʃən] 酰化作用
ketene ['kiːtiːn] 烯酮
pyrolysis [paɪ'rɒlɪsɪs] 高温分解
isocyanate [aɪsəʊ'saɪəneɪt] 异氰酸酯
stereospecificity [stɪərɪəʊspesɪ'fɪsɪtɪ] 立体专一性
tautomerize [tɔː'tɒməraɪz] （使）发生互变（异构）现象

insertion [ɪn'sɜ:ʃn] 插入
hemiacetal [hemɪ'æsɪtæl] 半缩醛
quaternary ammonium salt 季铵盐
carbanion ['kɑ:bənaɪən] 碳负离子
nitrogen ylide 氮叶立德
sigmatropic [sɪgmə'trɒpɪk] σ移位的
alkyllithium ['ælkil 'lɪθi:əm] 烷基锂
phenyllithium ['fenəl 'lɪθi:əm] 苯基锂
ortho ['ɔ:θəʊ] 邻位的
Brønsted acid 布朗斯特酸
Lewis acid 路易斯酸
coordination [kəʊˌɔ:dn'eɪʃən] 配位
rearomatization [reiærəʊmətaɪ'zeɪʃn] 再芳香化

Chapter 8

Electrophilic and Nucleophilic Aromatic Substitutions
(芳香亲电取代与亲核取代)

The typical reaction of benzene and its derivatives is electrophilic substitution. Aromatic electrophilic substitution includes a variety of reactions such as *nitration*, sulphonation, *halogenation*, Friedel-crafts reactions etc.

8.1 Electrophilic aromatic substitution（芳香亲电取代）

A two-step mechanism has been proposed for electrophilic substitution reactions.

Step 1: Slow or rate-determining step

The reagent and catalyst undergo acid-base reaction to produce the attacking *electrophile* which then attacks the ring to form a positively charged *benzenonium intermediate* (also called a σ complex, an *arenium*) in the first slow step. The σ complex is not aromatic because the sp^3 hybrid carbon atom interrupts the p orbitals of the ring.

$$E—Nu + Cat \longrightarrow E^+ + Nu—Cat^-$$

σ complex

Step 2: fast step

A proton from the site of attack is removed from this intermediate, yielding a substituted product.

σ complex

The overall reaction is the substitution of an electrophile (E^+) for a proton (H^+) on the aromatic ring.

*Arene*s, especially benzene, have an exceptionally stable and unreactive aromatic system of delocalized electrons. Consequently, arenes only react with the most reactive of *electrophiles* and special catalysts are required to generate the highly electrophilic species, such as Br^+, $\overset{+}{N}O_2$, R^+, RCO^+ and D^+. Typical electrophilic aromatic substitutions of benzene are listed in the Table 8.1.

Table 8.1 Typical electrophilic aromatic substitutions of benzene

Reaction type	Typical equation	electrophile (E^+)
Halogenations	$C_6H_6 + Cl_2 \xrightarrow[\text{heat}]{\text{Fe or FeCl}_3} C_6H_5Cl + HCl$ Chlorobenzene	Cl^+ or Br^+
Nitration	$C_6H_6 + HNO_3 \xrightarrow[\text{heat}]{H_2SO_4} C_6H_5NO_2 + HCl$ Nitrobenzene	NO_2^+
Sulfonation	$C_6H_6 + H_2SO_4 \underset{\text{heat}}{\rightleftharpoons} C_6H_5SO_3H + HCl$ benzenesulfonic acid	SO_3H^+
Friedel-Crafts Alkylation	$C_6H_6 + RCl \xrightarrow[\text{heat}]{AlCl_3} C_6H_5R + HCl$ Arene	R^+
Friedel-Crafts Acylation	$C_6H_6 + RCOCl \xrightarrow[\text{heat}]{AlCl_3} C_6H_5COR + HCl$ Arene	RCO^+
Protonation *H-D exchange*	$C_6H_6 + D_2SO_4 \xrightarrow[\text{heat}]{D_2O} C_6D_6$ benzene-d_6	D_3O^+

8.2 Specific electrophilic aromatic substitution reactions(特殊的芳香亲电取代反应)

8.2.1 Halogenations（卤代反应）

Free halogens can attack activated benzene rings but Lewis acid catalyst is required. It is suggested that probably benzene first forms a π complex with the halogen molecule. For example, the Lewis acid $FeBr_3$ then polarizes the Br—Br bond and helps in the formation of a σ complex between benzene carbon and the electrophilic end of the polarized bromine by removing the initial bromide ion. Subsequent abstraction of hydrogen in the second step completes the reaction.

The reaction mechanism for the *bromination* of benzene is as follows:

Step 1: Generation of a stronger electrophile, Br^+.

$$Br_2 + FeBr_3 \longrightarrow Br^+ + FeBr_4^-$$

Step 2: The electrophile attacks the benzene ring to form a σ complex.

H Br+ slow H Br H Br H Br H Br

σ complex

Step 3: The proton is removed to give the substitution product.

H Br + $FeBr_4^-$ fast Br + HBr + $FeBr_3$

The order of reactivity of the halogen is $F_2 > Cl_2 > Br_2 > I_2$. Fluorine is too reactive for practical use. Under ordinary condition, ionization fails. In the presence of HNO_3, direct iodination has been achieved. The attacking electrophile I^+ is produced by HNO_3.

8.2.2 Nitration（硝化反应）

Nitration of benzene is effectuated with a mixture of concentrated nitric and sulphuric acids. In the absence of sulphuric acid the reaction is slow. It is suggested that H_2SO_4 acts as a strong acid protonates. HNO_3 acts to generate *nitronium* ion, $\overset{+}{NO_2}$.

The reaction mechanism for the nitration of benzene is as follows:

Step 1: Generation of the *nitronium* ion, $\overset{+}{NO_2}$.

$$HNO_3 + 2\,H_2SO_4 \rightleftharpoons \overset{+}{NO_2} + H_3O^+ + 2\,HSO_4^-$$

Step 2: Electrophile attack.

The nitronium ion then attacks the benzene ring to form a carbocation.

slow

nitronium ion

σ complex

Step 3: Loss of a proton.

A fast abstraction of hydrogen from the site of attack by the base (HSO_4^-) completes the reaction.

8.2.3 Sulfonation（磺化反应）

Sulfonation is a reversible reaction that produces *benzenesulfonic acid* by adding sulfur trioxide or fuming sulfuric acid. The reaction is reversed by adding hot aqueous acid to benzenesulfonic acid to produce benzene. The sulfur in sulfur trioxide is electrophilic because the oxygen atoms pull electrons away from it, as oxygen is very electronegative. The benzene attacks the sulfur (and subsequent proton transfers occur) to produce benzenesulfonic acid.

$-H^+$

$+\ \overset{+}{S}O_3H$ slow, $-H^+$ fast

Desulfonation follows the same mechanistic path as sulfonation, except in the opposite order. A proton adds to a ring carbon to form a σ complex, then loss of sulfur trioxide gives the desulfonated aromatic ring.

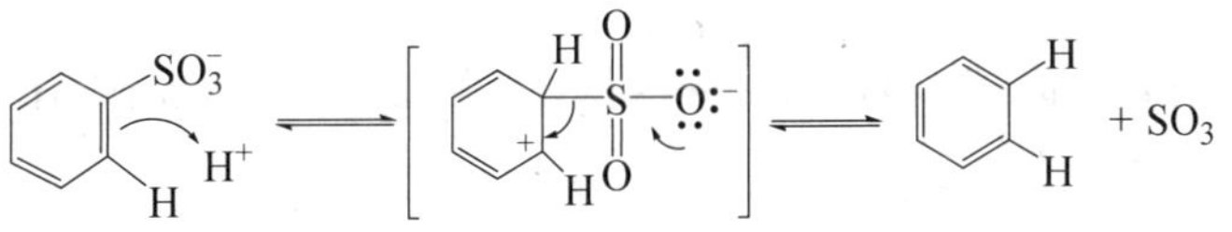

$$(SO_3 + H_2O \rightleftharpoons H_2SO_4)$$

8.2.4 Friedel-Crafts alkylation（Friedel-Crafts 烷基化反应）

This Lewis acid-catalyzed electrophilic aromatic substitution allows the synthesis of alkylated products via the reaction of arenes with alkyl halides. The alkylating reagent may also be aliphatic alcohols, alkenes, ethers, etc. in the presence of strong proton acids which generate the carbocation for the electrophilic attack. Since alkyl substituents activate the arene substrate, polyalkylation may occur. The general mechanism is shown below.

$$R{-}X + AlCl_3 \longrightarrow R^+ + X{-}\bar{A}lCl_3$$

$$ROH + H{-}X \xrightarrow{-X^-} R\overset{+}{O}H_2 \xrightleftharpoons{-H_2O} R^+$$

$$R{-}CH{=}CH_2 + H_2SO_4 \rightleftharpoons R\overset{+}{C}HCH_3 + HSO_4^-$$

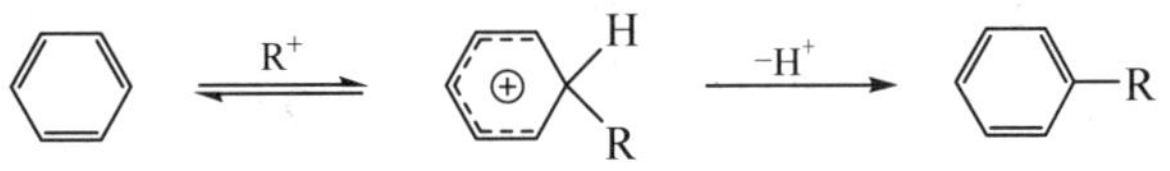

If the chloride is not on a tertiary carbon or secondary carbon, carbocation rearrangement reaction will occur. This reactivity is due to the relative stability of the tertiary and secondary carbocation over the primary carbocations. Friedel—Crafts reactions are possible with any carbocationic intermediate such as those derived from alkenes and a protic acid, Lewis acid, *enone*s, and *epoxide*s.

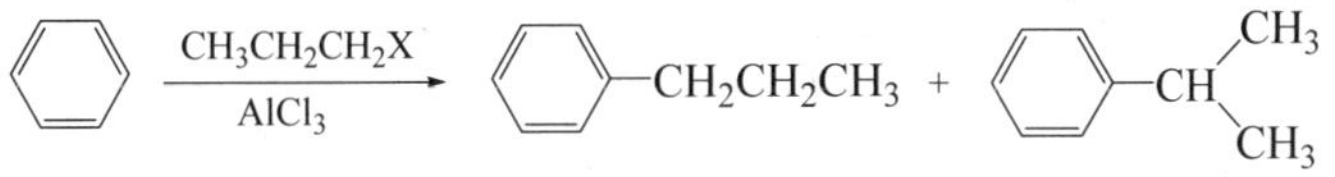

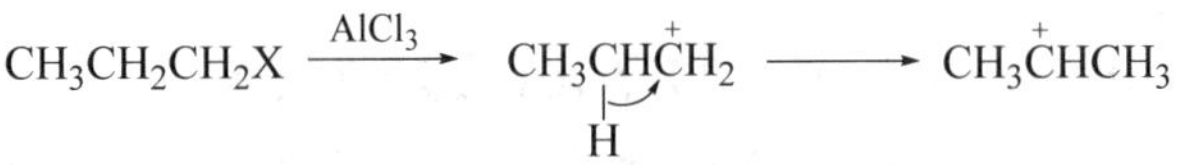

$$CH_3CH_2CH_2X \xrightarrow{AlCl_3} CH_3CH\overset{+}{C}H_2 \longrightarrow CH_3\overset{+}{C}HCH_3$$

8.2.5 Friedel-Crafts acylation （Friedel-Crafts 酰基化反应）

Friedel-Crafts acylation is the acylation of aromatic rings with an *acyl* chloride or acid anhydrides using a strong Lewis acid catalyst. Reaction conditions are similar to the Friedel-Crafts alkylation mentioned above. This reaction has several advantages over the alkylation reaction. Due to the electron-withdrawing effect of the carbonyl group, the ketone product is always less reactive than the original molecule, so multiple acylations do not occur. Also, there are no carbocation rearrangements, as the carbonium ion is stabilized by a resonance structure in which the positive charge is on the oxygen.

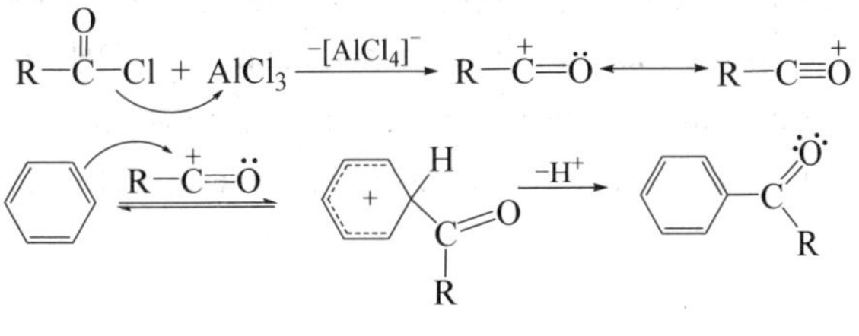

$$R{-}\overset{O}{\overset{\|}{C}}{-}Cl + AlCl_3 \xrightarrow{-[AlCl_4]^-} R{-}\overset{+}{C}{=}\ddot{O} \longleftrightarrow R{-}C{\equiv}\overset{+}{O}$$

$$\text{C}_6\text{H}_6 + CH_3CH_2CH_2COCl \xrightarrow{AlCl_3} C_6H_5C(=O)CH_2CH_2CH_3 \xrightarrow[HCl]{Zn\text{-}Hg} CH_3CH_2CH_2CH_2\text{—}C_6H_5$$

The first step consists of dissociation of a chloride ion to form an acyl cation (*acylium* ion). In some cases, the Lewis acid binds to the oxygen of the acyl chloride to form an adduct. The resulting acylium ion or a related adduct is subject to nucleophilic attack by the arene. Finally, chloride anion (or $AlCl_4^-$) deprotonates the ring (an arenium ion) to form HCl, and the $AlCl_3$ catalyst is regenerated. If desired, the resulting ketone can be subsequently reduced to the corresponding alkane substituent by either Wolff-Kishner reduction or Clemmensen reduction.

The Haworth reaction is a classic method for the synthesis of *tetralone*. In this reaction, benzene is reacted with *succinic anhydride*, the intermediate product is reduced and a second Friedel-Crafts acylation takes place with addition of acid.

$$\text{benzene} \xrightarrow[AlCl_3]{\text{succinic anhydride}} C_6H_5COCH_2CH_2CO_2H \xrightarrow{\text{reduction}} C_6H_5CH_2CH_2CH_2CO_2H \xrightarrow{H^+} \text{tetralone}$$

A reaction of phthalic anhydride with *N,N-diethylaminophenol* in the presence of zinc chloride gives the *fluorophore* rhodamine B:

$$2\ \text{HO—C}_6\text{H}_4\text{—NEt}_2 + \text{phthalic anhydride} \xrightarrow{H_2SO_4,\ heat} \text{rhodamine B (lactone form)}$$

8.2.6 Protonation and hydrogen exchange（质子化和氢交换反应）

Hydrogen exchange resulting from reversible protonation of an aromatic ring can be followed by the use of isotopic labels. Either *deuterium* or *tritium* can be used and the experiment can be designed to follow either the incorporation or the release of the isotope.

$$\text{C}_6\text{H}_6 + D_3O^+ \rightleftharpoons [\text{arenium ion (H, D)}] \rightleftharpoons \text{C}_6\text{H}_5\text{D} + HOD_2^+$$

8.2.7 Aromatic substitution by *diazonium* ions（通过形成重氮离子的芳香取代反应）

Aryl diazonium ions are weak electrophiles. These reagents react only with aromatic substrates having strong electron donating substitutents, such as —OH, $—NH_2$, —NHR, $—NR_2$, and the products are azo compounds with the coupling positions *para* to or on the same ring of fused aromatic rings. Aryl diazonium ions are usually generated by diazotization of aromatic amines.

$$PhN_2^+Cl^- + PhNMe_2 \xrightarrow{NaOAc} Ph{-}N{=}N{-}C_6H_4{-}NMe_2$$

$$PhN_2^+HSO_4^- + \text{HO}{-}C_{10}H_7 \xrightarrow{NaOH} Ph{-}N{=}N{-}C_{10}H_6{-}OH$$

8.2.8 Gattermann reaction（Gattermann 反应）

Formylation with $Zn(CN)_2$ and HCl is called the Gattermann reaction. This method can be successfully applied to phenols and their ethers and to many heterocyclic compounds. However it cannot be applied to aromatic amines. In the original version of this reaction the substrate was treated with HCl, HCN, $ZnCl_2$, but the use of $Zn(CN)_2$ and HCl makes the reaction more convenient to carry out and does not reduce yields. The mechanism of the Gattermann reaction has not been investigated very much, but there is an initial nitrogen containing product that is normally not isolated but is hydrolyzed to aldehyde.

$$C_6H_6 + Zn(CN)_2 \xrightarrow{HCl} C_6H_5{-}\overset{H}{C}{=}\overset{+}{N}H_2Cl^- \xrightarrow{H_2O} C_6H_5{-}CHO$$

$$C_6H_6 + ZnCl_2 \xrightarrow[HCl]{R(CN)_2} C_6H_5{-}CR{=}\overset{+}{N}H_2Cl^- \xrightarrow{H_2O} C_6H_5{-}\overset{O}{\overset{\|}{C}}R$$

The first stage consists of an attack on the substrate by species containing the nitrile and HCl to give an imine salt. In the second stage the salts are hydrolyzed to the products.

8.2.9 Gattermann-Koch Reaction（Gattermann-Koch 反应）

The formylation of benzene and alkylbenzene using carbonmonoxide and hydrogen chloride in the presence of aluminium chloride (catalyst) and a small amount of cuprous chloride (co-catalyst) under high pressure is known as Gattermann-Koch reaction.

$$C_6H_6 + CO + HCl \xrightarrow[Cu_2Cl_2]{AlCl_3} C_6H_5{-}CHO$$

$$H_3C{-}C_6H_5 + CO + HCl \xrightarrow[Cu_2Cl_2]{AlCl_3} H_3C{-}C_6H_4{-}CHO$$

In the case of alkylbenzene, the aldehyde group is introduced into the *para* position only. This method is used industrially to prepare arylaldehydes.

The Gattermann-Koch formylation is considered as a typical electrophilic aromatic substitution with high para *regioselectivity*. The most likely electrophile is the acylium ion $[HCO]^+$ in the ion pair $[HCO]^+[AlCl_4]^-$. Common factors such as electron density of the aromatic substrate, reactivity of electrophile, stability of reaction intermediates, and steric factors may influence the regioselectivity.

$$CO \xrightarrow{AlCl_3} \overset{+}{C}{=}\overset{-}{O}{-}AlCl_3 \xrightarrow[H^+]{Cl^-} Cl{-}\underset{H}{C}{=}\overset{+}{O}\cdots\overset{-}{AlCl_3} \longrightarrow [HC\equiv\overset{+}{O}]\ AlCl_4^- \xrightarrow{C_6H_6} C_6H_5CHO$$

8.2.10 Haloalkylation（卤烷基化反应）

When certain aromatic compounds are treated with formaldehyde and HCl, the CH_2Cl group is introduced into the ring in a reaction called chloromethylation. The reaction has also been carried out with other aldehydes and with HBr or HI. The reaction is successful for benzene, and alkyl-, alkoxy-, and halobenzenes. It is greatly hindered by *meta*-directing groups, which reduce yields or completely prevent the reactions. Amines and phenols are too reactive and usually give polymers unless deactivating groups are also present, but phenolic ethers and esters successfully undergo the reaction. Compounds of less reactivity can often be chloromethylated with chloromethyl methyl ether ($ClCH_2OMe$), or methoxyacetyl chloride ($MeOCH_2COCl$). Zinc chloride is the most common catalyst, but other Friedel-Crafts catalysts are also employed.

The initial step involves reaction of the aromatic compound with the aldehyde to form the hydroxyalkyl compound, and then HCl converts this to the chloroalkyl compound. The acceleration of the reaction by $ZnCl_2$ has been attributed to the raising of the acidity of the medium, causing an increase in the concentration of $HOCH_2^+$ ions.

$$C_6H_6 + HCHO + HCl \xrightarrow[70℃]{ZnCl_2} PhCH_2Cl + H_2O$$

$$MeOC_6H_5 + CH_3CHO + HCl \xrightarrow[70℃]{ZnCl_2} MeO{-}C_6H_4{-}CHClCH_3 \xrightarrow{NaOH} MeO{-}C_6H_4{-}CH{=}CH_2$$

8.2.11 Reimer-Tiemann reaction（Reimer-Tiemann 反应）

The formylation of activated aromatic ring compound with chloroform in alkaline solution is known as Reimer-Tiemann reaction. The method is useful only for phenols and certain heterocyclic compounds such as pyrroles and indoles. It leads preferentially to the formation of an *ortho* formylated product; when both the *ortho* positions are blocked, the incoming group occupies the *para* position.

$$C_6H_5O^- + CHCl_3 \xrightarrow{NaOH} \underset{\text{major}}{o\text{-}HOC_6H_4CHO} + \underset{\text{minor}}{p\text{-}HOC_6H_4CHO}$$

$$2\text{-}CH_3OC_6H_4OH + CHCl_3 \xrightarrow{NaOH} 4\text{-}HO\text{-}3\text{-}CH_3OC_6H_3CHO$$

The reaction is believed to proceed in the following manner. Firstly the dichlorocarbene (CCl_2:) an electron-deficient reactive species is formed by the reaction of chloroform with strong alkali. In the second step, it attacks the electron rich *ortho* position of the aromatic ring to form orthodichloromethylphenolate, which hydolysis to give hydroxyarylaldehyde.

$$CHCl_3 \xrightarrow{-H^+} {}^{-}CCl_3 \xrightarrow{-Cl^-} :CCl_2$$

8.2.12 Jacobsen reaction（Jacobsen 反应）

When polyalkyl or polyhalobenzenes are treated with sulphuric acid, the ring is sulphonated, but rearrangement also takes place. The reaction, known as Jacobsen reaction, is limited to benzene rings that have at least four substituents, which may be any combination of alkyl and halogen groups, where the alkyl groups may be ethyl or methyl and the halogen iodo, chloro, bromo. When isopropyl or *t*-butyl groups are on the ring, these groups are cleaved to give olefins. Since a sulphonic group can later be removed, the Jacobsen reaction can be used as a means of rearranging polyalkylbenzenes. The rearrangement always brings the alkyl or halo groups closer together than they were originally. Side products in this case are pentamethylbenzenesulphonic acid, 2, 4, 5-trimethyl benzenesulphonic acid, etc., indicating an intermolecular process, at least partially.

Common electrophiles and their generation:

Electrophiles	Generation	Electrophilic ability
NO_2^+	$HNO_3 + 2H_2SO_4 \rightleftharpoons NO_2^+ + H_3O^+ + 2HSO_4$	strong
$Br_2—MX_n$	$Br_2 + MX_n \rightleftharpoons Br_2—MX_n$	
$Br_2—\overset{+}{O}H_2$	$BrOH + H_3O^+ \rightleftharpoons Br_2—\overset{+}{O}H_2 + H_2O$	
$Cl_2—MX_n$	$Cl_2 + MX_n \rightleftharpoons Cl_2—MX_n$	
$Cl_2—\overset{+}{O}H_2$	$ClOH + H_3O^+ \rightleftharpoons Cl_2—\overset{+}{O}H_2 + H_2O$	
SO_3	$H_2S_2O_7 \rightleftharpoons H_2SO_4 + SO_3$	
RSO_2^+	$RSO_2Cl + AlCl_3 \rightleftharpoons RSO_2^+ + AlCl_4^-$	

R_3C^+	$R_3CX + AlCl_3 \rightleftharpoons R_3C^+ + AlCl_3X^-$	medium
	$R_3COH + H^+ \rightleftharpoons R_3C^+ + H_2O$	
	$R_2C=CR'_2 + H^+ \rightleftharpoons R_2C^+CHR'_2$	
$\overset{+}{C}(=O)R$	$RCOX + AlCl_3 \rightleftharpoons \overset{+}{C}(=O)R + AlCl_3X^-$	
H^+	$HX \rightleftharpoons H^+ + X^-$	
$R_2\overset{+}{C}OH$	$R_2CO + H^+ \rightleftharpoons R_2C=\overset{+}{O}H \rightleftharpoons R_2\overset{+}{C}OH$	
$R_2C=\overset{+}{O}-MX_n$	$R_2CO + MX_n \rightleftharpoons R_2C=\overset{+}{O}-MX_n$	
$HC\equiv\overset{+}{N}H$	$HCN + HX \rightleftharpoons HC\equiv\overset{+}{N}HX^-$	weak
NO^+	$HNO_2 + H^+ \rightleftharpoons NO^+ + N_2O$	
$Ar\overset{+}{N}\equiv N$	$ArNH_2 + HNO_2 + H^+ \rightleftharpoons Ar\overset{+}{N}\equiv N + 2H_2O$	

芳环上的亲电取代历程：①亲电试剂的产生；②π-络合物的形成；③σ-络合物的形成（决速步骤）；④失去 H^+，恢复芳环的结构。常见的反应有卤代、硝化、磺化、傅-克烷基化、傅-克酰基化、氢交换、重氮盐的偶联反应、卤烷基化等。

8.3 Directing effects of substituents（取代基的定位效应）

The reactivity of mono-substituted benzene and the orientation of incoming substituent depend on the nature of the substituent already present on the ring. A mono-substituted benzene on electrophilic substitution may give three possible di-substituted products (*ortho, meta* and *para* isomers).

The *substituent effects* on further substitutions of aromatic compounds need to be considered. These effects are a combination of *resonance and inductive effects*. Donation or withdrawal of electrons can occur via either a conjugative or an inductive effect. The activation or deactivation of the benzene ring toward electrophilic substitution may be correlated with the electron donating or electron withdrawing influence of the substituents, as measured by molecular dipole moments. In the following diagram we see that electron donating substituents activate the benzene ring toward electrophilic attack, and electron withdrawing substituents deactivate the ring (make it less reactive to electrophilic attack).

Resonance effects are those that occur through the π system and can be represented by resonance structures. These can be either electron donating (*e.g.* —OMe) where π electrons are pushed toward the arene or electron withdrawing (*e.g.* —C=O) where π electrons are drawn away from the arene. Inductive effects are those that occur through the σ system due to electronegativity type effects. These can be either electron donating (*e.g.* —Me) where σ electrons are pushed toward the arene or electron withdrawing (*e.g.* —CF_3, $^+NR_3$) where σ electrons are drawn away from the arene.

Activating Substitutents

C_6H_5—NH_2	C_6H_5—OH	C_6H_5—OCH_3	C_6H_5—CH_3
1.52	1.45	1.20	0.40

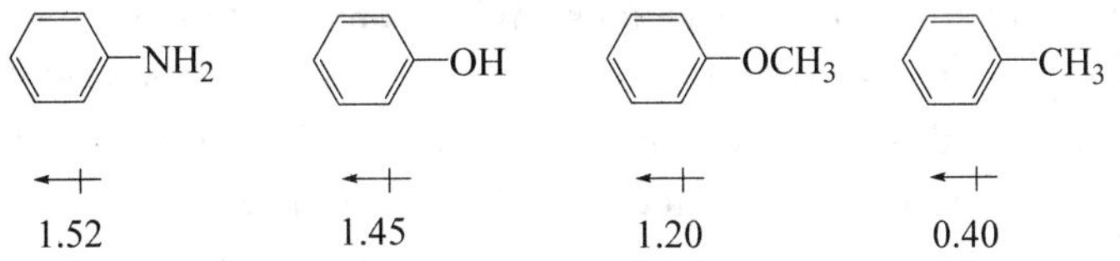

Deactivating Substitutents

C_6H_5—NO_2	C_6H_5—CN	C_6H_5—CO_2CH_3	C_6H_5—Cl
3.97	3.90	1.91	1.56

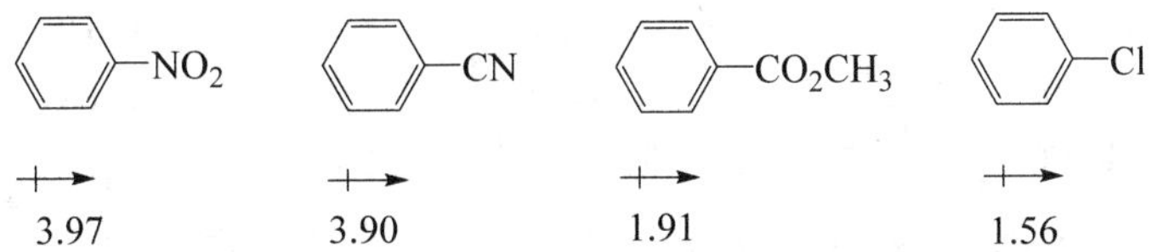

Electron donating groups with *lone pairs* (*e.g.* —OH, —OR, —NH_2, —NR_2, etc.) on the atoms adjacent to the π system activate the aromatic ring by increasing the electron density on the ring through a *resonance donating effect*. The resonance only allows electron density to be positioned at the *ortho*- and *para*- positions. Hence these sites are more nucleophilic, and the system tends to react with electrophiles at these *ortho*- and *para*- sites.

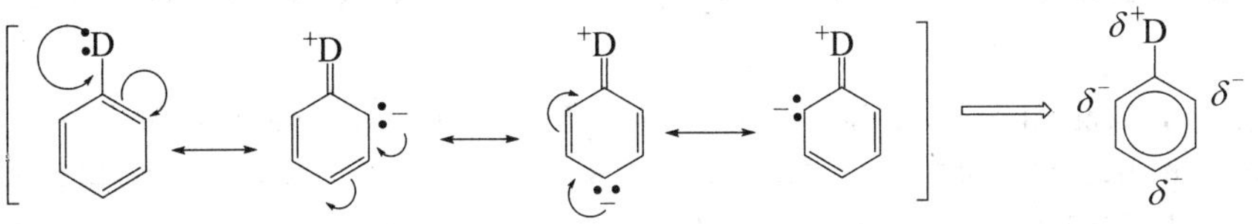

Electron withdrawing groups with π bonds to electronegative atoms (*e.g.* —C=O, —NO_2, —CN) adjacent to the π system deactivate the aromatic ring by decreasing the electron density on the ring through a *resonance withdrawing effect*. The resonance only decreases the electron density at the *ortho*- and *para*- positions. Hence these sites are less nucleophilic, and so the system tends to react with electrophiles at the *meta*- positions.

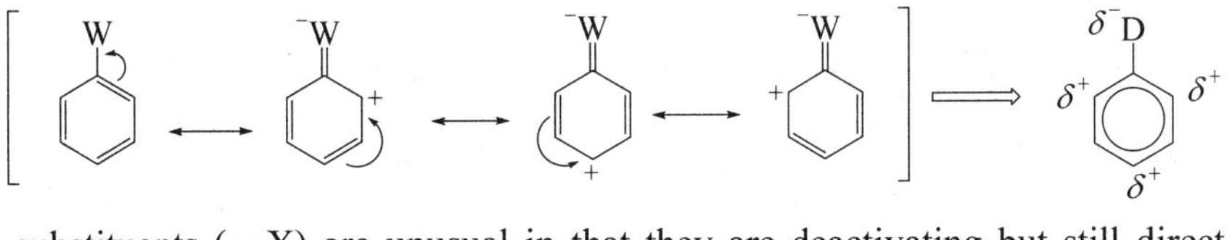

Halogen substituents (—X) are unusual in that they are deactivating but still direct *ortho*- / *para*-, because they are both inductive electron withdrawing (electronegativity) and resonance donating (lone pair donation). The inductive effect lowers the reactivity but the resonance effect controls the regiochemistry due to the stability of the intermediates.

Table 8.2 Orientation and reactivity effects of ring substituents

Electronic Effect	Examples	Activation Effect	Direction
Donation-conjugation	—OH, —OR, —OC_6H_5, —$OCOCH_3$ —NH_2, —NR_2, —$NHCOCH_3$, —C_6H_5	activating	*ortho*, *para*
Donation-inductive effect	—R	activating	*ortho*, *para* with some *meta*
donation-conjugation, withdrawal- inductive effect	—F, —Cl, —Br, —I	deactivating	*ortho* and (mostly) *para*
withdrawal-inductive effect	—CF_3, —$^+NR_3$, —$^+PR_3$, —$^+SR_2$	deactivating	*meta*
withdrawal-conjugation	—NO_2, —CN, —SO_3H, —SO_2R, —CO_2H —CO_2R, —$CONH_2$, —CHO, —COR	deactivating	*meta*

Substituents that activate the benzene ring toward electrophilic attack generally direct substitution to the *ortho* and *para* locations. With some exceptions, such as the halogens, deactivating substituents direct substitution to the *ortho* and *para* locations (Table 8.2).

Substituents with *–I* and *–C* effects e.g., NO_2, COR, CO_2R, CN are deactivating and *meta* directing, however, their deactivating influence is less than the positively charged $–N^+Me_3$ group. Substituents with *+I* and *+C* effects like alkyl groups are activating and *ortho*, *para* directing. The $–O^-$ substituent e.g., in a phenoxide ion ArO^-, is also of *+I* and *+C* type and these effects are much stronger compared with those in an alkyl group.

第一类定位基(邻、对位定位基)，包括致活基—OH、—OR、$—OC_6H_5$、$—OCOCH_3$、$—NH_2$、—NHR、$—NR_2$、$—NHCOCH_3$、$—C_6H_5$、—R，致钝基—F、—Cl、—Br、—I。

第二类定位基（间位定位基），包括吸电子诱导基团$—CF_3$、$—^+NR_3$、$—^+PR_3$、$—^+SR_2$，吸电子共轭基团$—NO_2$、—CN、$—SO_3H$、$—SO_2R$、$—CO_2H$、$—CO_2R$、$—CONH_2$、—CHO、—COR，它们均为致钝基团。

8.4 Effects of multiple substituents on electrophilic aromatic substitution（多取代基对芳香亲电取代反应的影响）

The regioselectivity for further substitution of disubstituted benzenes can usually be predicted by looking at the cumulative effects of the substituents. Aromatic substitution reactions of disubstituted benzenes usually lead to a mixture of products because of competing orientation influences of two substituents to afford a mixture of products. If both substituents are activating groups, the position of the incoming electrophile is primarily governed by the orientation influence of the stronger of the two activating groups. If one substituent is activating while the other is deactivating, the position of the entering group will be dictated mainly by the activating group. The position between the two *meta* substituents is least attacked by the incoming electrophile due to the steric hindrance.

8.5 Nucleophilic aromatic substitution（芳香亲核取代反应）

8.5.1 Nucleophilic aromatic substitution by the addition-elimination mechanism（加成-消除机理的芳香亲核取代）

The addition-elimination mechanism uses one of the vacant π* orbitals for bonding interaction with the *nucleophile*. This permits addition of the nucleophile to the aromatic ring without displacing any of the existing substitutents. This apparent nucleophilic substitution reaction is surprising, since aryl halides are generally incapable of reacting by either a S_N1 or S_N2 pathway.

$$C_6H_5\text{—}Cl + NaOH \xrightarrow{350℃} C_6H_5\text{—}OH + C_6H_5\text{—}O\text{—}C_6H_5 + NaCl$$

The presence of electron-withdrawing groups (such as nitro) *ortho-* and *para-* to the chlorine substantially enhance the rate of substitution, as shown below. This two-step mechanism is characterized by initial addition of the nucleophile (hydroxide ion or water) to the aromatic ring, followed by loss of a halide anion from the negatively charged intermediate.

$$p\text{-}ClC_6H_4NO_2 \xrightarrow[150℃]{1\,mol/L\ NaOH} p\text{-}HOC_6H_4NO_2 + NaCl$$

$$2,4\text{-}(NO_2)_2C_6H_3Cl \xrightarrow[150℃]{10^{-3}mol/L\ NaOH} 2,4\text{-}(NO_2)_2C_6H_3OH + NaCl$$

$$2,4,6\text{-}(NO_2)_3C_6H_2Cl \xrightarrow[40℃]{H_2O} 2,4,6\text{-}(NO_2)_3C_6H_2OH + NaCl$$

Sep 1 $\quad C_6H_5X + Y^- \xrightarrow{slow} [C_6H_5XY]^-$ (resonance-stabilized anion)

Sep 2 $\quad [C_6H_5XY]^- \xrightarrow{fast} C_6H_5Y + X^-$

Such addition-elimination processes generally occur at sp^2 or sp hybridized carbon atoms, in contrast to aliphatic S_N1 and S_N2 reactions. When applied to aromatic halides, this mechanism is called S_NAr. Some distinguishing features of the three common nucleophilic substitution mechanisms are summarized in Table 8.3.

Table 8.3 Comparison of S_N1, S_N2 and S_NAr

Mechanism	Number of steps	Bond formation timing	Carbon hybridization
S_N1	two	after bond breaking	usually sp^3
S_N2	one	simultaneous with bond breaking	usually sp^3
S_NAr	two	prior to bond breaking	usually sp^2

The role of the leaving group in determining the reaction rate is the somewhat different from S_N2 and S_N1 substitution at alkyl groups. In those cases, bond strength is the dominant factor, so the order of reactivity of the halogens is $I^- > Br^- > Cl^- > F^-$. In nucleophilic aromatic substitution, the formation of the addition intermediate is the rate-determining step, so the ease of C—X bond breaking does not affect the rate. The order of reactivity is often $F^- > Cl^- > Br^- > I^-$, which is the result of the polar effect of the halogen. The stronger bond dipoles associated with the more electronegative halogens favor the addition step and thus increase the overall rates of reaction.

Alkoxy groups are very poor leaving groups in S_N2 reactions but can act as leaving groups in aromatic substitution. The rate-determining step is the addition reaction, the alkoxide can be eliminated in energetically favorable *rearomation*. *Sulfonyl* and nitro groups can also be displaced.

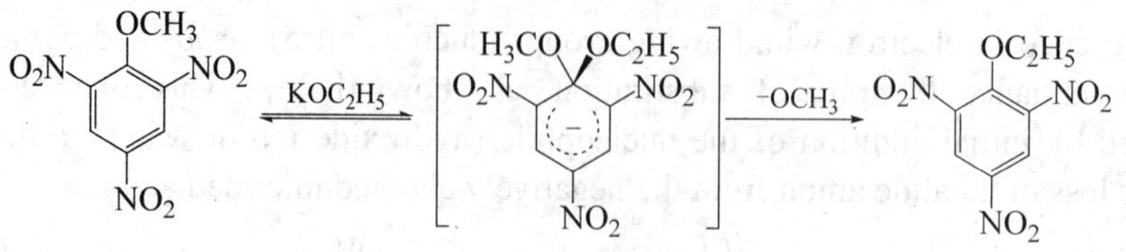

OCH_3 O_2N NO_2 NO_2 $\xrightarrow{KOC_2H_5}$ [H_3CO OC_2H_5 O_2N NO_2 NO_2] $\xrightarrow{-OCH_3}$ OC_2H_5 O_2N NO_2 NO_2

Meisenheimer complexes

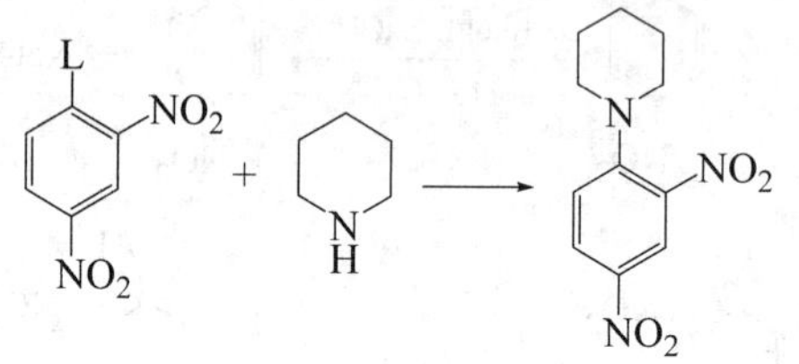

L NO_2 NO_2 + N H ⟶ N NO_2 NO_2

L=Cl, Br, I, SO_2Ph, *p*-$O_2NC_6H_5O^-$

One of the most historically significant examples of aromatic nucleophilic substitution is the reaction of amines with 2,4-dinitrofluorebenze by Sanger to develop a method for identification of the *N*-terminal amino acid in proteins. This process opened the way for structural characterization of proteins and other biopolymers.

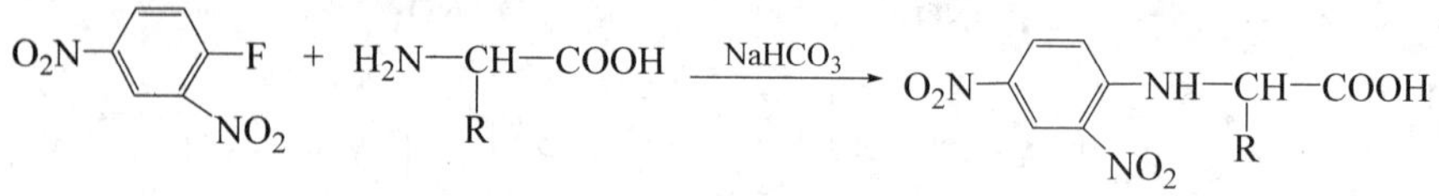

O_2N—⟨⟩—F (NO_2) + H_2N—CH(R)—COOH $\xrightarrow{NaHCO_3}$ O_2N—⟨⟩—NH—CH(R)—COOH (NO_2)

2-Halopyridines and other π-deficient nitrogen *heterocycle*s are excellent reactants for nucleophilic aromatic substitution.

N Cl + $NaOC_2H_5$ ⟶ N OC_2H_5

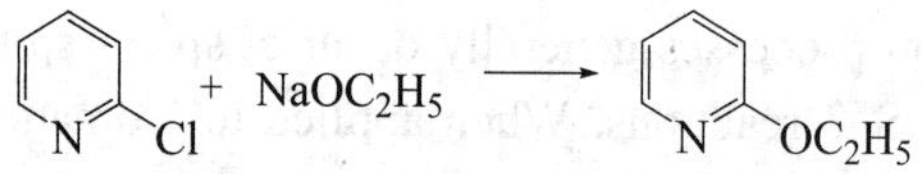

加成-消除历程：吸电子基有利于降低芳环的电子云密度，特别是邻、对位电子云密度降低得更为显著，从而有利于亲核试剂对这些位置发生亲核加成，生成离域的碳负离子中间体(即 Meisenheimer 络合物)，再失去离去基团，恢复苯环结构。这一过程又称为 S_NAr 历程。离去基团的离去能力 $I^- > Br^- > Cl^- > F^-$。

8.5.2 Nucleophilic aromatic substitution by the elimination-addition mechanism（消除-加成机理的芳香亲核取代）

There is a good evidence that the synthesis of phenol from chlorobenzene does not proceed by the addition-elimination mechanism (S_NAr) described above. For example, treatment of *para-chlorotoluene* with sodium hydroxide solution at temperatures above 350℃ give an equimolar mixture of *meta-* and *para-*cresols (hydroxytoluenes). Chloro and bromobenzene reacted with the very strong base sodium amide ($NaNH_2$ at low temperature (−33℃ in liquid ammonia) to give good yields of aniline (aminobenzene). However, *ortho-chloroanisole* give exclusively *meta-methoxyaniline* under the same conditions. These reactions are described by the following equations.

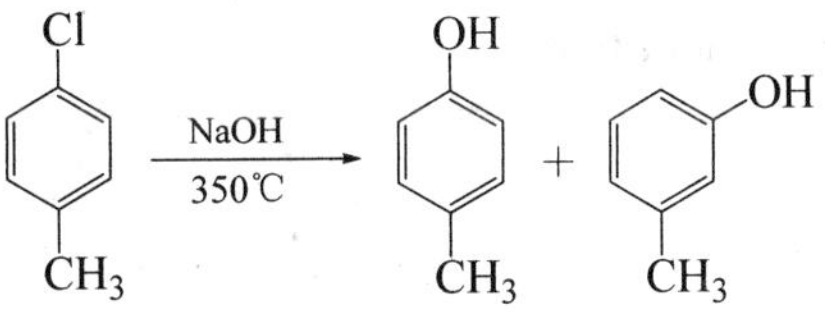

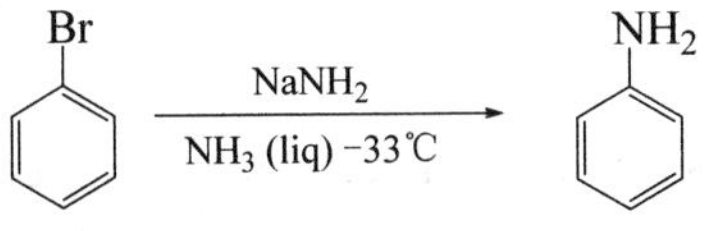

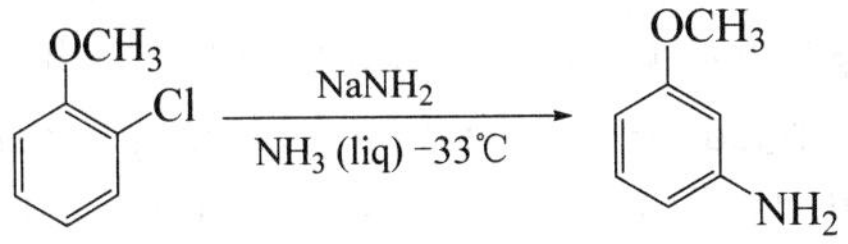

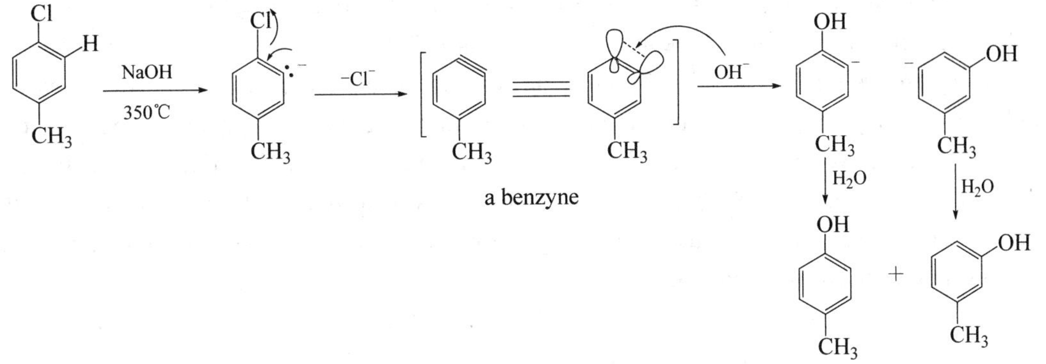

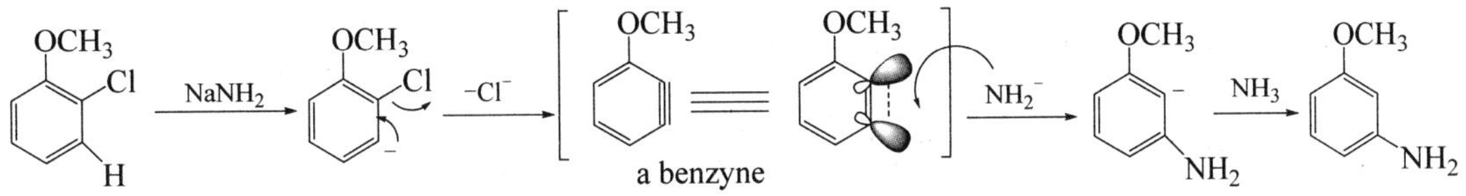

The explanation for this curious repositioning of the substituent group lies in a different two-step mechanism we can refer to as an elimination-addition process. The intermediate in this mechanism is an unstable *benzyne* species. An orbital picture of benzyne shows the normal benzene arrangement of p orbitals perpendicular to the plane of the ring. The other bond of the triple bond results from two adjacent sp^2 hybridized carbons overlapping in π fashion in the plane of the ring. Obviously, these orbitals do not overlap very well because they are not parallel. While it is convenient to use a structure with a triple bond to represent benzyne, we must recognize that one bond of this triple bond is not a typical π bond and is highly reactive. In contrast to the parallel overlap of p-orbitals in a stable alkyne triple bond, the p-orbitals of a benzyne are tilted *ca.*120° apart, so the reactivity of this incipient triple bond to addition reactions is greatly enhanced. In the absence of steric hindrance (top example) equal amounts of *meta-* and *para-*cresols are obtained. The entering nucleophile need not always enter at the carbon to which the leaving group was bound. The steric bulk of the methoxy group and the ability of its ether oxygen to stabilize an adjacent anion result in a substantial bias in the addition of amide anion or ammonia. The mechanism was summarized below:

Step 1: Deprotonation

Step 2: Elimination of the leaving group

Step 3: Addition of the Nucleophile

Step 4: Protonation

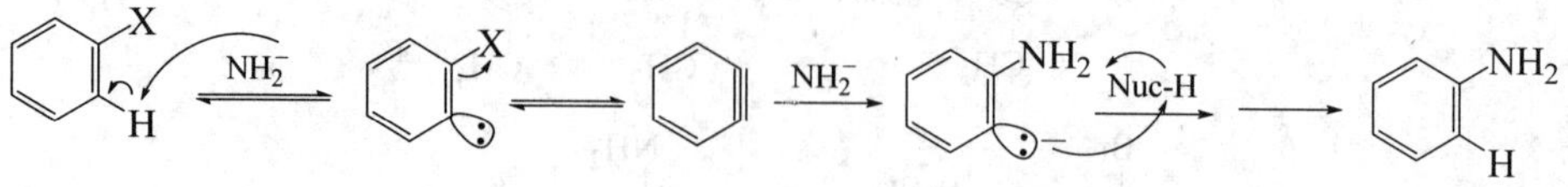

The elimination-addition mechanism is facilitated by structural effects that favor removal of a hydrogen from the ring by a strong base. Relative reactivity also depends on the halide. The order $Br^- > I^- > Cl^- > F^-$ has been established in the reaction of aryl halides with KNH_2 in liquid ammonia. The order has been interpreted as representing a balance between two effects. The polar order favoring proton removal would be $F^- > Cl^- > Br^- > I^-$, but this is largely overwhelmed by the order of leaving group ability $I^- > Br^- > Cl^- > F^-$, which reflects bond strengths.

The regiochemistry of the nucleophilic addition is influenced by ring substitutents. Electron withdrawing groups tend to favor addition of nucleophile at the more distant end of the "triple bond", since this permits maximum stabilization of the developing negative charge. Electron donating groups have the opposite effect. The direct effects probably arise mainly through interaction of the substituent with the electron pair that is localized on the *ortho* carbon by the addition step. Selectivity is usually not high, formation of both possible products from monosubstituted benzynes is common.

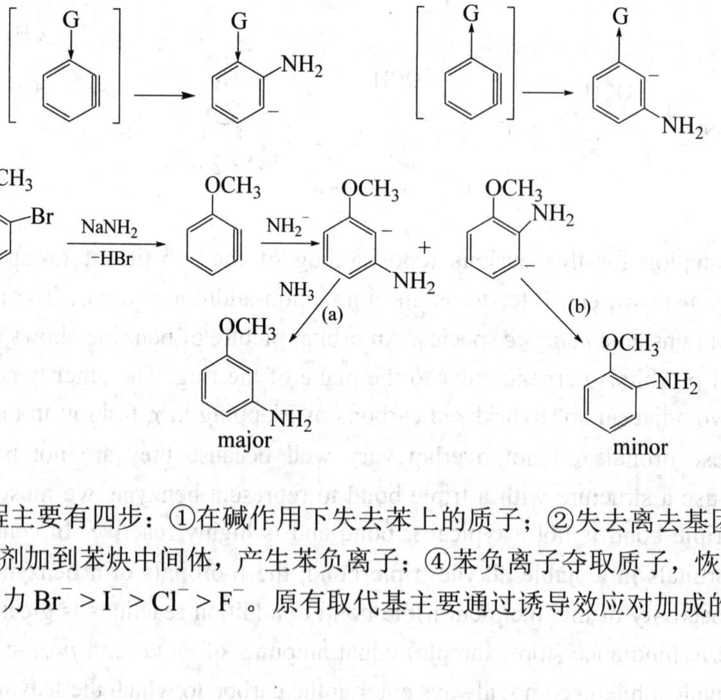

消除-加成历程主要有四步：①在碱作用下失去苯上的质子；②失去离去基团，产生苯炔中间体；③亲核试剂加到苯炔中间体，产生苯负离子；④苯负离子夺取质子，恢复苯环结构。离去基团的离去能力 $Br^- > I^- > Cl^- > F^-$。原有取代基主要通过诱导效应对加成的区域选择性产生影响。

Problems

1. Analyze the results of the studies of intramolecular electrophilic substitution that are described below. Write mechanisms for each of the cyclizations and comment on the relation between ring size and the outcome of cyclization.

(1) 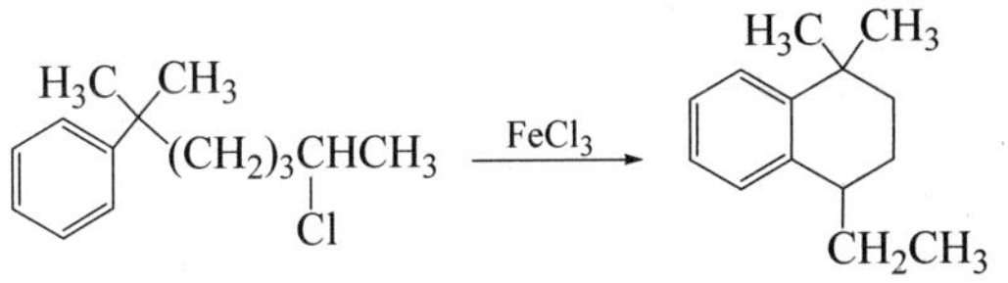 H_3C CH_3 $(CH_2)_3CHCH_3$ Cl $\xrightarrow{FeCl_3}$ H_3C CH_3 CH_2CH_3

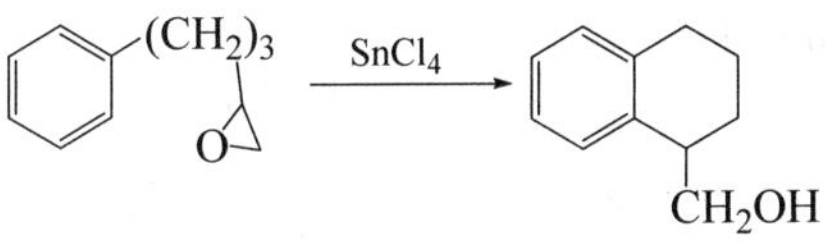

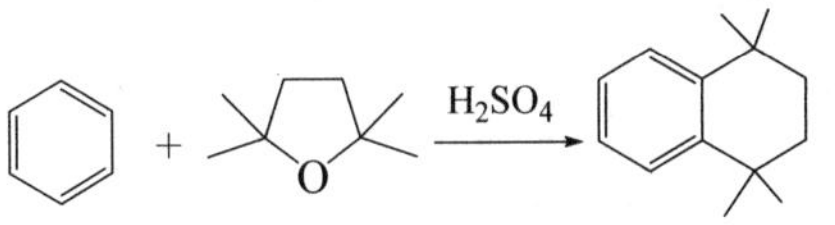

(2) $(CH_2)_3$ O $\xrightarrow{SnCl_4}$ CH_2OH

(3) + O $\xrightarrow{H_2SO_4}$

(4) $CH_2CH_2COCH_2COCH_3$ Cl $\xrightarrow[NH_3(liq)]{KNH_2}$ O $COCH_3$

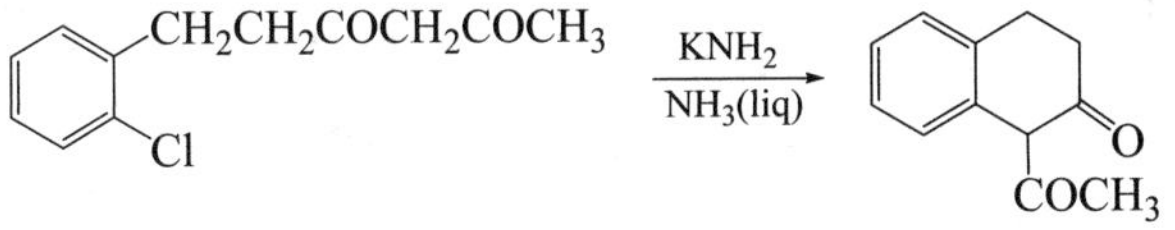

(5) 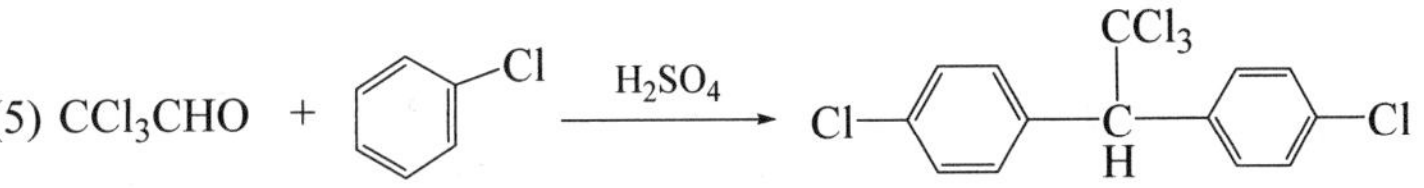 CCl_3CHO + Cl $\xrightarrow{H_2SO_4}$ Cl–C(H)(CCl_3)–Cl

2. Explain the outcome of the following reactions by a mechanism showing how the product could be formed.

(1) 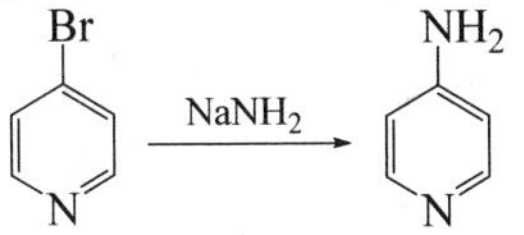 Br N $\xrightarrow{NaNH_2}$ NH_2 N

(2) Br N OC_2H_5 $\xrightarrow{NaNH_2}$ NH_2 N OC_2H_5 major + NH_2 N OC_2H_5 minor

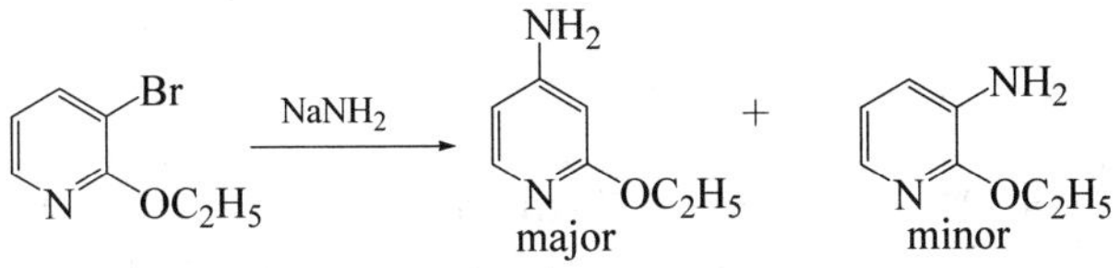

3. Complete the following reactions.

(1) 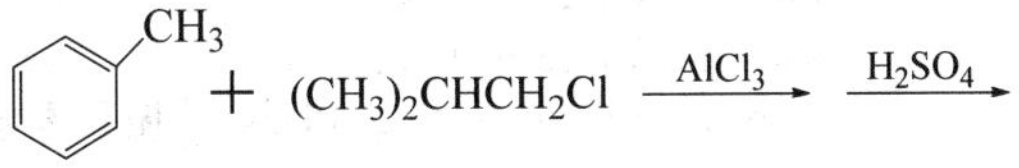 CH_3 + $(CH_3)_2CHCH_2Cl$ $\xrightarrow{AlCl_3}$ $\xrightarrow{H_2SO_4}$

(2) $\xrightarrow[\geqslant 165℃]{H_2SO_4}$ $\xrightarrow[H_2SO_4]{HNO_3}$

(3) $\xrightarrow[H_2SO_4]{}$

(4) + COCl O_2N $\xrightarrow{AlCl_3}$ $\xrightarrow[H_2SO_4]{HNO_3}$

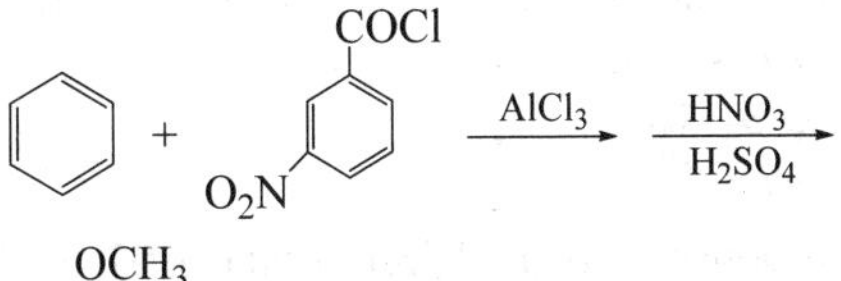

(5) 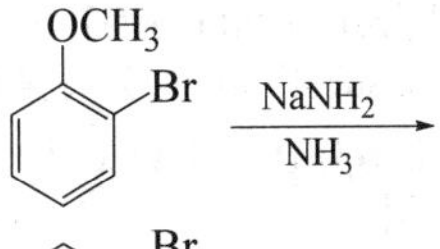 OCH_3 Br $\xrightarrow[NH_3]{NaNH_2}$

(6) 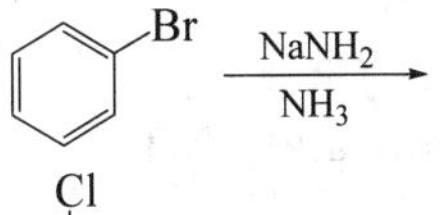 Br $\xrightarrow[NH_3]{NaNH_2}$

(7) 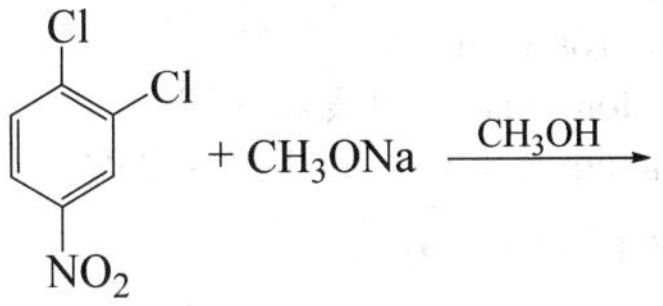 Cl Cl NO_2 + CH_3ONa $\xrightarrow{CH_3OH}$

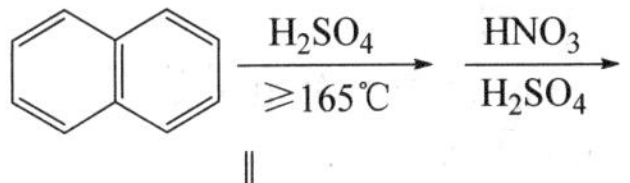

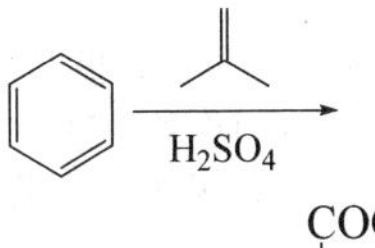

(8) 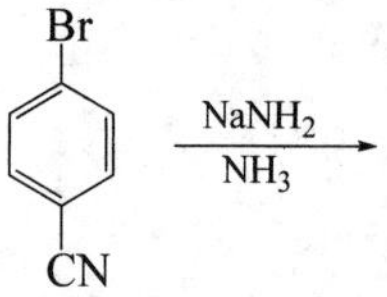 Br–C_6H_4–CN $\xrightarrow[NH_3]{NaNH_2}$

4. Synthesis the following products from the starting materials.

(1) 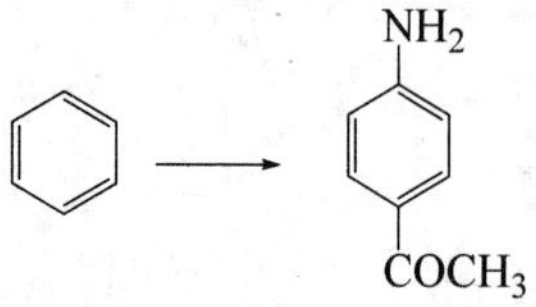 benzene ⟶ NH_2, $COCH_3$

(2) Cl ⟶ NH_2, O_2N, NO_2

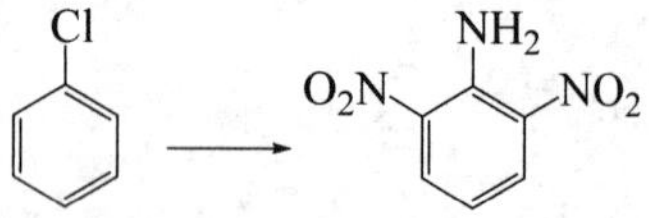

(3) 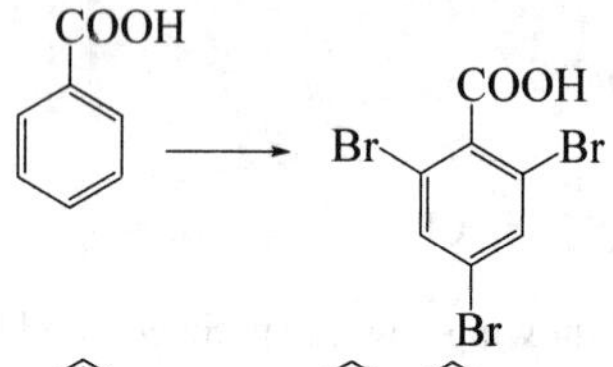 COOH ⟶ COOH, Br, Br, Br

(4) 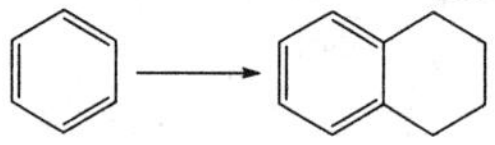 benzene ⟶

Vocabulary

electrophilic aromatic substitution 芳香亲电取代反应
benzenonium intermediate 苯正离子中间体
arenium 芳正离子
electrophile [ɪ'lektrəˌfaɪl] 亲电试剂
halogenation [hælədʒə'neɪʃən] 卤代反应
nitration [naɪ'treɪʃən] 硝化反应
sulfonation [sʌlfə'neɪʃən] 磺化反应
Friedel-Crafts alkylation 傅-克烷基化反应
Friedel-Crafts acylation 傅-克酰基化反应
H-D exchange 氢-重氢交换反应
bromination [brəʊmɪ'neɪʃn] 溴代反应
nitronium [nɪ'troʊnɪəm] 硝基正离子
benzenesulfonic acid 苯磺酸
desulfonation [desəlfɒ'neɪtaɪən] 去(脱)磺化反应
enone 不饱和羰基化合物
epoxide [e'pɒksaɪd] 环氧化合物
acyl ['æsəl] 酰基
acylium 酰基阳离子
tetralone ['tetrəloʊn] 萘满酮
succinic anhydride 琥珀酰酐
N,N-diethylaminophenol *N,N*-二乙氨基苯酚
fluorophore ['flʊərəfɔː] 荧光团
deuterium [djuː'tɪriəm] 氘
tritium ['trɪtiəm] 氚
diazonium [ˌdaɪə'zoʊnɪəm] 重氮
directing effects 定位效应
substituent effects 取代基效应
resonance and inductive effects 共振和诱导效应
arene [ə'riːn] 芳烃
electron donating groups 给电子基
lone pairs 孤电子对
resonance donating effect 给电子共振效应
electron withdrawing groups 吸电子基
resonance withdrawing effect 吸电子共振效应
ortho- 邻-
para- 对-
meta- 间-
regioselectivity [riːdʒiːəʊsɪlek'tɪvɪtɪ] 区域选择性
nucleophilic aromatic substitution 芳香亲核取代
nucleophile ['njuːkliːəˌfaɪl] 亲核试剂
rearomation 再芳烃化
sulfonyl ['sʌlfəˌnɪl] 磺酰基
heterocycle ['hetəroʊsaɪkl] 杂环
para-chlorotoluene 对-氯甲苯
ortho-chloroanisole 邻氯-苯甲醚
meta-methoxyaniline 间-甲氧基苯胺
benzyne ['benzaɪn] 苯炔

Chapter 9

Oxidation and Reduction Reactions (氧化还原反应)

Oxidation occurs when a carbon or hydrogen that is connected to a carbon atom in a structure is replaced by more *electronegative* atoms such as oxygen, nitrogen, or halogen. In the oxidation reaction, C—H bond or C—C bond is broken, and C—O, C—N or C—X bond is formed. *Reduction* is a reaction that results in replacement of electronegative atoms by hydrogen or carbon. In the reduction reaction, C—O, C—N or C—X bond is broken and C—H bond or C—C bond is formed.

There are various kinds of oxidations. Their mechanisms are also various. Among them, oxidations of C═C bond, alcohol, aldehyde, ketone are more important. In the lab, *potassium permanganate* ($KMnO_4$), *chromic acid* (H_2CrO_4), peracid, *ozone* (O_3), peroxide, etc. are often used as *oxidants*. Reduction reactions including catalytic hydrogenation, metal-ammonia reduction and hydride reduction will be discussed in this chapter.

底物分子得到电负性大于碳的氧(氮、氯等)的反应或分子中失去氢的反应称为氧化反应。分子失去电负性大于碳的氧（氮、氯等）的反应或得到氢的反应称为还原反应。

9.1 Oxidation of alkenes and alkynes（烯烃和炔烃的氧化）

9.1.1 Oxidation of alkenes（烯烃的氧化）

(1) Hydroxylation（羟基化）

When an alkene is treated with aqueous potassium permanganate (pH > 8) at lower temperature or *osmium tetroxide* in pyridine solution, dihydroxylated products (glycols) are obtained. Both reactions appear to proceed by the same mechanism, as is shown below. A cyclic osmate intermediate or a cyclic manganate intermediate is formed, respectively. The metallocyclic intermediate may be isolated in the osmium reaction. It can be an evidence for the mechanism. In basic solution, the purple permanganate anion is reduced to the green manganate ion, providing a nice color test for the double bond functional group. After hydrolysis, *cis*-1,2-cyclohexanediol or *cis*-1,2-cyclopentanediol is obtained, respectively. From the mechanism we would expect *syn-stereoselectivity* in the bonding to oxygen, and regioselectivity is not an issue. It is not practicable to use quantitative OsO_4, for it is very expensive. If H_2O_2 is used as an oxidant, catalytic amount of OsO_4 is enough for the hydroxylation reaction of alkenes.

H O O Os O O → H O O Os O O H $\xrightarrow[NaHSO_3]{H_2O}$ H OH OH H

a cyclic osmate intermediate

cis-1,2-cyclohexanediol

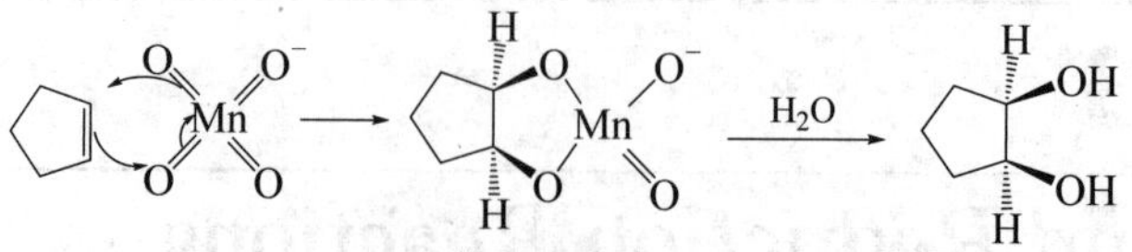

a cyclic manganate intermediate

cis-1,2-cyclopentanediol

烯烃被碱性（中性）、稀、冷 $KMnO_4$ 或 OsO_4 氧化为邻二醇，中间经过环状锰酸酯或锇酸酯，立体化学为顺式加成。

(2) *Epoxidation*（环氧化）

Some oxidation reactions of alkenes give cyclic ethers. In the reactions, both carbons of a double bond are bonded to the same oxygen atom. These products are called epoxides or *oxiranes*. The reactions of alkenes with peracids, RCO_3H, are an important method for preparing epoxides. Epoxidations by peracids always have *syn*-stereoselectivity, and seldom give structural rearrangement. Two examples are illustrated below. Oxidation of *cis*-2-butene with RCO_3H gives *cis*-2,3-dimethyloxirane, while oxidation of *trans*-2-butenes affords *trans*-2,3-dimethyloxirane. In the peracid, an oxygen atom is electron negative, thus it can be regarded as a nucleophile; the other oxygen atom is electron positive, therefore, it can be regarded as an electrophile. The electron positive oxygen is added to the C═C bond. The epoxide product is formed. A carboxylic acid is formed as a byproduct. From the mechanism, we can see that the epoxidation reaction is a kind of electrophilic addition reaction. Therefore, the electron density is higher; the rate of the reaction speed is higher.

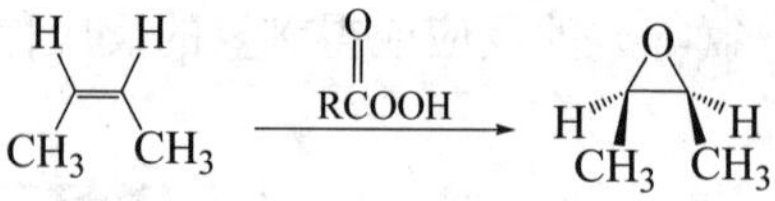

cis-2-butene → *cis*-2, 3-dimethyloxirane

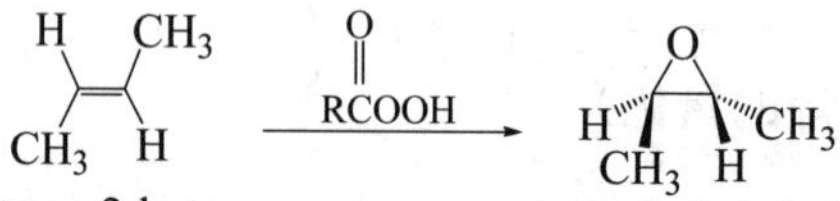

trans-2-butene → *trans*-2, 3-dimethyloxirane

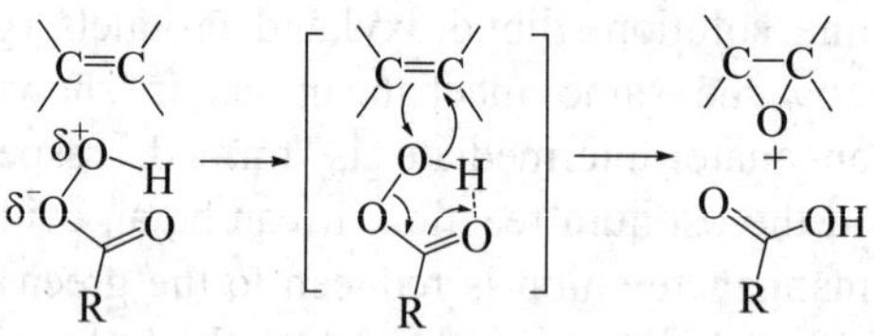

Epoxides may be further cleaved by aqueous acid to give glycols that are often diastereomeric. Proton *transfer* from the acid catalyst to the oxygen of the epoxide generates the conjugate acid of the epoxide. Then a nucleophile, such as water, attacks the three-membered ring from the opposite side of the oxygen atom, which results in the cleavage of the three-membered ring. The result is *anti*-hydroxylation of the double bond, in contrast to the *syn*-stereoselectivity of the hydroxylation of alkenes.

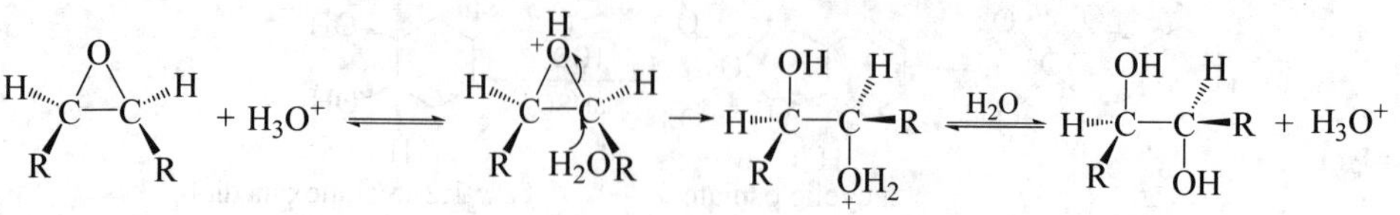

C═C 用有机过氧酸氧化得到环氧化物，具有立体专一性，再在酸或碱作用下开环得到外消旋的反式二醇。

(3) Oxidative cleavage of double bonds （双键的氧化断裂）

When an alkene is treated with ozone at low temperature, an *ozonide* is formed. The ozonide is often decomposed with zinc and acetic acid or catalytic hydrogenation to give two equivalents of aldehyde, or two equivalents of ketone, or one equivalent of each. The products of the reaction are dependent on the groups attached to the alkene. The mechanism is drawn below. Ozone rapidly adds to carbon-carbon double bonds. A molozonide is formed by initial *syn*-addition of ozone on the C═C bond. The formed molozoide is extremely unstable; it undergoes rearrangement via an O—O bond cleavage and a C—C bond cleavage and the subsequent two C—O bonds formation to generate an ozonide intermediate. The decomposition of the final ozonide to carbonyl products by either a reductive or oxidative workup are also shown below.

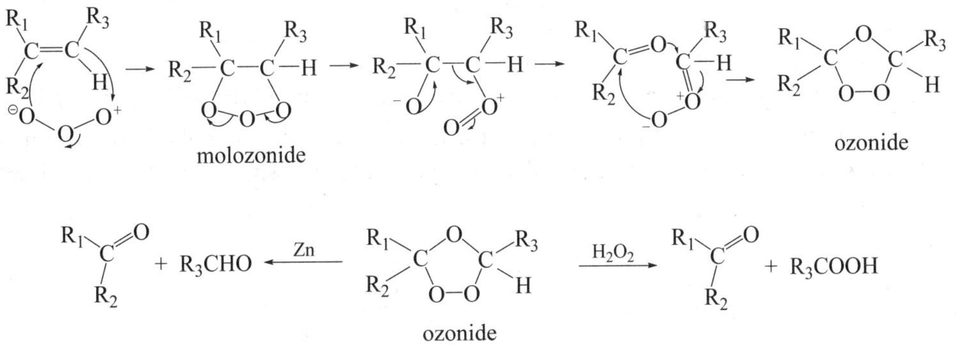

molozonide

ozonide

Zn: ozonide → $R_1R_2C{=}O$ + R_3CHO

H_2O_2: ozonide → $R_1R_2C{=}O$ + R_3COOH

ozonide

C═C 用 O_3 氧化，经还原水解得醛或酮。经氧化水解得羧酸或酮。

If an alkene reacts with neutral or acidic potassium permanganate ($KMnO_4$) or acid dichromate, the C═C bond is also broken to give ketone, carboxylic acid or carbon dioxide. The products of the reaction are also dependent on the groups attached to the alkene.

$KMnO_4$, HO^-, △; $KMnO_4$, H^+ → + CO_2; $KMnO_4$, HO^-, △ → + CO_2

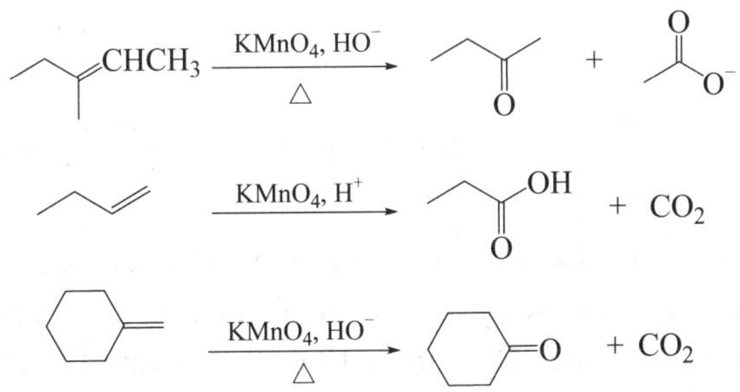

酸性 $KMnO_4$ 把烯烃氧化为 C═C 断裂的羰基化合物或羧酸或酮酸混合物。

9.1.2 Oxidation of alkynes（炔烃的氧化）

Oxidative cleavage of alkynes is generally more difficult than alkenes, but the treatment of an alkyne with ozone or potassium permanganate can give two equivalents of carboxylic acids or one equivalent of carboxylic acid and one equivalent of carbon dioxide. The products are dependent on the structure of the alkyne. Some examples are illustrated below.

$$CH_3CH_2CH_2-\equiv-CH_2CH_3 \xrightarrow{O_3} \xrightarrow{H_2O} CH_3CH_2CH_2COOH + CH_3CH_2COOH$$

$$CH_3CH_2CH_2-\equiv-CH_2CH_3 \xrightarrow[H^+, 25℃]{KMnO_4} \xrightarrow{H^+} CH_3CH_2CH_2COOH + CH_3CH_2COOH$$

炔烃被酸性 $KMnO_4$ 氧化为羧酸，在中性条件下可得邻二酮。

9.2 Oxidation of alcohols（醇的氧化）

9.2.1 Oxidation of monohydroxyl alcohols（单羟基醇的氧化）

(1) Dehydrogenation with copper（用铜脱氢氧化）

There are two methods for oxidation of alcohols. One is dehydrogenation reaction; the other is oxidation with an oxidant. When simple 1° or 2° -alcohols in the gaseous state is exposed to a hot copper surface, two hydrogen atoms of the alcohol are lost; the alcohol can be converted to an aldehyde or a ketone.

伯醇在 Cu 催化下脱氢得醛；仲醇可得酮。

(2) Oxidation with chromic acid and chromic acid derivatives(用铬酸和铬酸衍生物氧化）

Chromic acid is the most common oxidant for the oxidation of alcohols. Some examples are illustrated below. Oxidations of simple 2°-alcohols give ketones, while oxidations of simple 1°-alcohols give carboxylic acids.

$$CH_3CH_2CH(OH)CH_3 \xrightarrow[H_2SO_4]{CrO_3} CH_3CH_2C(=O)CH_3$$

$$\text{cyclohexanol} \xrightarrow[H_2SO_4]{Na_2Cr_2O_7} \text{cyclohexanone}$$

$$\text{C}_5\text{H}_9-CH(OH)CH_2CH_3 \xrightarrow{H_2CrO_4} \text{C}_5\text{H}_9-C(=O)CH_2CH_3$$

$$CH_3CH_2CH_2CH_2OH \xrightarrow{H_2CrO_4} [CH_3CH_2CH_2CHO] \xrightarrow[\text{oxidation}]{\text{further}} CH_3CH_2CH_2COOH$$

The currently accepted mechanism of oxidation with chromic acid is essential that proposed by Westheimer. The first step shows the formation of a chromate ester. Then the chromate ester with a base undergoes E2 elimination to generate the carbonyl compound, and Cr (Ⅳ) species is generated. The base in the second step may be water. From the first and the second step, we can see that α-hydrogen of the alcohol is necessary for the oxidation reaction. Thus 3° - alcohols can not be oxidized. In some cases no external base is involved. The proton is directly transferred to one of the H_2CrO_3 oxygens. Cr (Ⅳ) species also can oxidize the alcohol to a carbonyl compound, and Cr (Ⅳ) species is reduced to Cr (Ⅲ) species. A free-radical may involve in the reaction. If one R group is hydrogen, the alcohol is primary alcohol. It gives an aldehyde, and this step can not be stopped, as further oxidation occurs to give a carboxylic acid. Since chromate reagents are a dark orange-red color (chromium Ⅵ) and chromium Ⅲ compounds are normally green, the progress of these oxidations is easily observed.

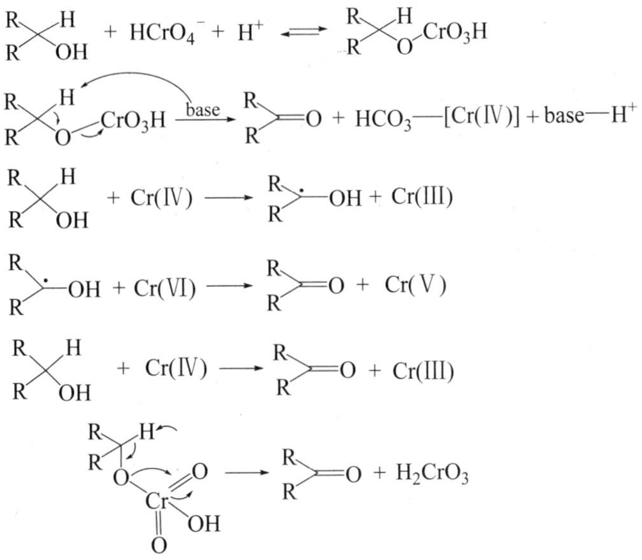

$$R_2CH(OH) + HCrO_4^- + H^+ \rightleftharpoons R_2CH{-}O{-}CrO_3H$$

$$R_2CH{-}O{-}CrO_3H \xrightarrow{\text{base}} R_2C{=}O + HCO_3{-}[Cr(IV)] + base{-}H^+$$

$$R_2CH(OH) + Cr(IV) \longrightarrow R_2\dot{C}{-}OH + Cr(III)$$

$$R_2\dot{C}{-}OH + Cr(VI) \longrightarrow R_2C{=}O + Cr(V)$$

$$R_2CH(OH) + Cr(IV) \longrightarrow R_2C{=}O + Cr(III)$$

$$R_2CH{-}O{-}Cr(=O)_2(OH) \longrightarrow R_2C{=}O + H_2CrO_3$$

伯醇被 $KMnO_4$ 或酸性 $K_2Cr_2O_7$ 氧化一般得酸；仲醇可得酮，在激烈的条件下得碳-碳键断裂的产物。叔醇不易被氧化，但在酸性条件下经烯烃氧化得到碳-碳键断裂的产物。

Some chromic acid derivatives, such as, pyridinium chlorochromate, $C_5H_5\overset{+}{N}H(CrO_3)Cl^-$, commonly named by PCC, or Sarrett reagent, CrO_3-Py, can oxidize a primary alcohol to an aldehyde. Some examples are illustrated below.

$$CH_3CH_2CH_2CH_2OH \xrightarrow[CH_2Cl_2, 25℃]{PCC} CH_3CH_2CH_2CHO$$

$$CH_3(CH_2)_4C{\equiv}CCH_2OH \xrightarrow[CH_2Cl_2, 25℃]{\text{Sarrett}} CH_3(CH_2)_4C{\equiv}CCHO$$

Jones reagent, a solution of CrO_3 in aqueous sulfuric acid, can selectively oxidize unsaturated secondary alcohols to ketones, while carbon-carbon double bonds remain unchanged. An example is shown below. If a primary alcohol is oxidized by Jones reagent, a carboxylic acid is produced.

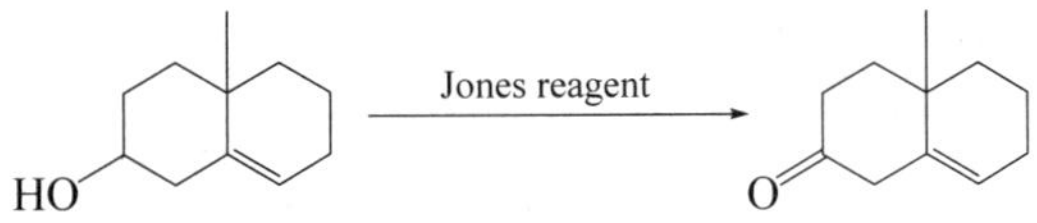

HO $\xrightarrow{\text{Jones reagent}}$ O

CrO_3 + 吡啶 + CH_2Cl_2 为 Collin 试剂; CrO_3 +吡啶+ HCl + CH_2Cl_2/DMF 为 PCC 试剂; CrO_3/硫酸为 Jones 试剂，它们可选择性地氧化烯醇为烯醛（酮），不影响 C═C。

(3) Oxidation with DMSO(用二甲亚砜氧化)

Swern reagent, DMSO-$(COCl)_2$-Et_3N, also can oxidize alcohols. A primary alcohol is converted to an aldehyde by Swern reagent, while 2°-alcohols give ketones, as are shown below.

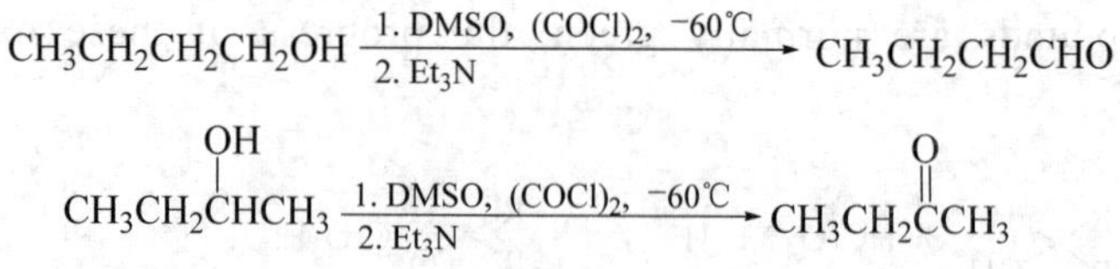

$$CH_3CH_2CH_2CH_2OH \xrightarrow[2.\ Et_3N]{1.\ DMSO,\ (COCl)_2,\ -60℃} CH_3CH_2CH_2CHO$$

$$CH_3CH_2CH(OH)CH_3 \xrightarrow[2.\ Et_3N]{1.\ DMSO,\ (COCl)_2,\ -60℃} CH_3CH_2C(=O)CH_3$$

The mechanism of the Swern oxidation is drawn below. *Oxalyl chloride* reacts with DMSO at low temperature. Chlorodimethylsulfonium chloride is formed with the release of carbon dioxide and carbon monoxide. Chlorodimethylsulfonium chloride is the reactive species that oxidizes the alcohol. A base captures α-hydrogen of the alcohol or the hydrogen of a methyl group from DMSO, which results in the final elimination. An aldehyde or a ketone is produced along with the byproduct formation of $(CH_3)_2S$.

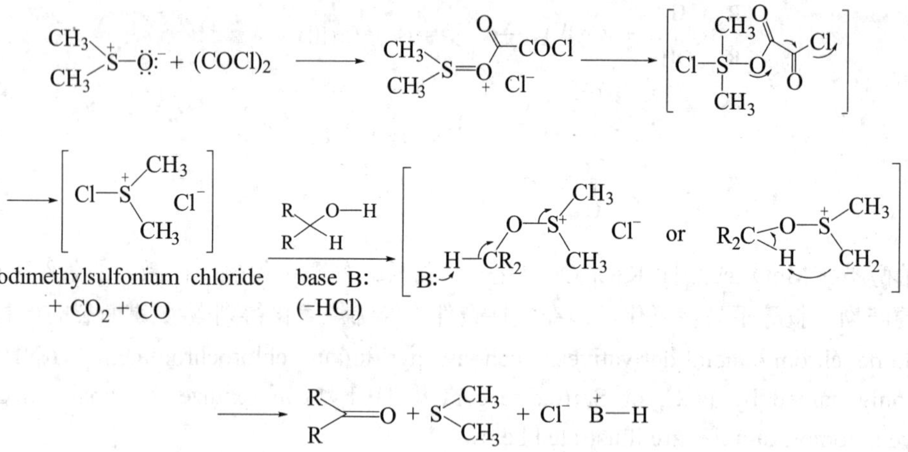

$$(CH_3)_2\overset{+}{S}-\ddot{O}^- + (COCl)_2 \longrightarrow (CH_3)_2S{=}\overset{+}{O}-C(=O)COCl\ Cl^- \longrightarrow [Cl-S(CH_3)_2-O-C(=O)-C(=O)-Cl]$$

$$\longrightarrow [Cl-\overset{+}{S}(CH_3)_2\ Cl^-] \xrightarrow[\text{base B:}\ (-HCl)]{R_2CH-OH} [B:\ H-CR_2-O-\overset{+}{S}(CH_3)_2\ Cl^-\ \text{or}\ R_2C(H)-O-\overset{+}{S}(CH_3)-\overset{-}{C}H_2]$$

chlorodimethylsulfonium chloride + CO_2 + CO

$$\longrightarrow R_2C{=}O + S(CH_3)_2 + Cl^-\ \overset{+}{B}-H$$

DMSO-$(COCl)_2$-Et_3N 为 Swern 试剂，氧化伯醇为醛，氧化仲醇为酮。

(4) Oxidation with MnO_2 or $BaMnO_4$（用二氧化锰或锰酸钡氧化）

Allylic alcohols and benzylic alcohols can be oxidized selectively by an activated MnO_2 in suitable organic solvents. Primary alcohols give aldehydes, while secondary alcohols afford ketones, and C═C double bonds remain in the oxidation reactions. Some examples are illustrated below. The third example shows that barium permanganate ($BaMnO_4$) has similar effect as MnO_2 in the oxidation reaction.

$$Bu_3Sn-CH{=}CH-CH{=}C(CH_3)-CH_2OH \xrightarrow[CH_2Cl_2]{MnO_2} Bu_3Sn-CH{=}CH-CH{=}C(CH_3)-CHO\ \ 89\%$$

$$R-CH{=}C(CH_3)-CH(CH_3)OH \xrightarrow[acetone]{MnO_2} R-CH{=}C(CH_3)-C(=O)CH_3\ \ 75\%$$

(R = the boxed allenic carotenoid moiety bearing HO, H_3C, CH_3 groups)

$$SEM-O-CH_2-CH(CH_3)-(\text{tetrahydropyran})-C(=CH_2)-CH_2OH \xrightarrow[CH_2Cl_2]{BaMnO_4} \cdots-C(=CH_2)-CHO\ \ 98\%$$

SEM=$(CH_3)_3SiCH_2CH_2OCH_2$−

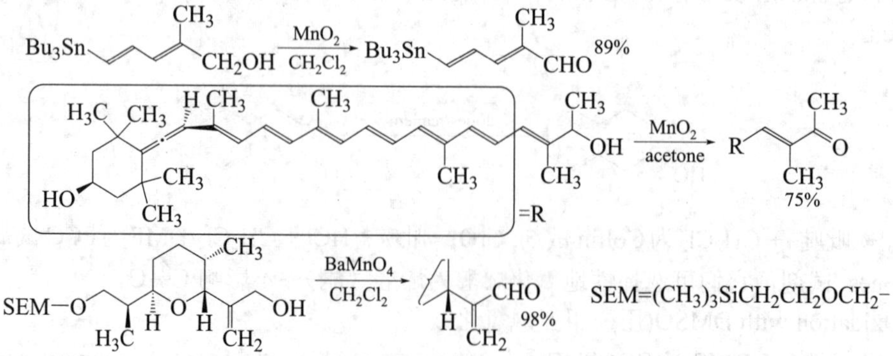

MnO_2 或 $BaMnO_4$ 有较高的选择性，它们可选择性地氧化烯醇为烯醛（酮），不影响 C═C。

9.2.2 Oxidation of the *vicinal* glycols（邻二醇的氧化）

When the vicinal glycols are treated with lead tetraacetate [$Pb(OAc)_4$] or periodic acid (HIO_4), they are easily cleaved under mild conditions. The products are two equivalents of aldehyde, two equivalents of ketone, or one equivalent of each. The structures of the products are dependent on the groups attached to the two carbons which bear hydroxyl groups. A general equation for these oxidations is shown below. A cyclic-intermediate of ester containing I or Pb is formed, respectively. *cis*-glycols react much more rapidly than *trans*-glycols. The fact that added acetic acid retards the oxidation reaction by $Pb(OAc)_4$ also supports the cyclic-intermediate mechanism. The two reagents are complementary, for HIO_4 can be used in water, and $Pb(OAc)_4$ can be used in organic solvents.

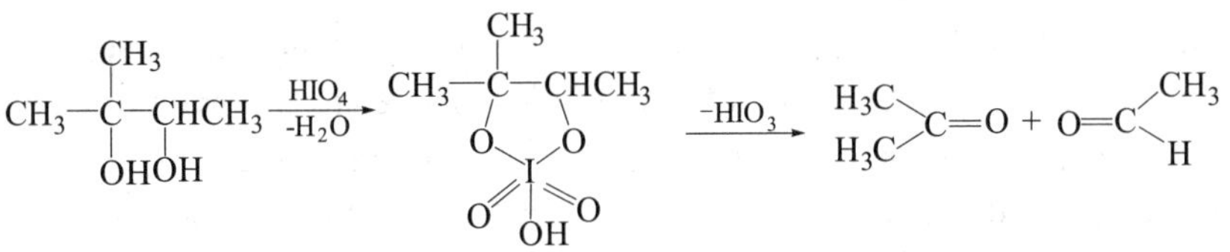

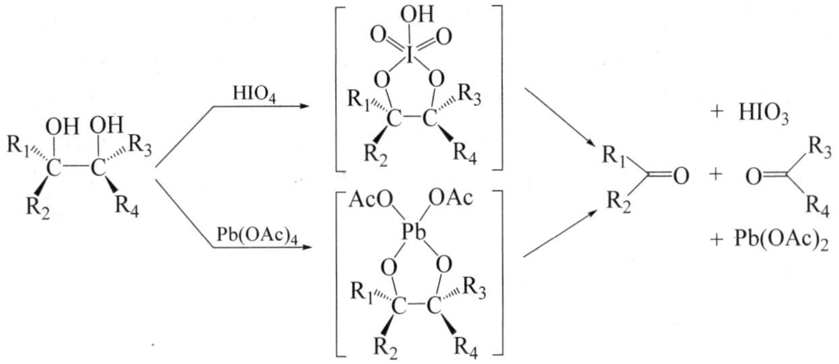

$$(CH_3)_2C(OH)-CH(OH)CH_3 \xrightarrow[-H_2O]{HIO_4} \text{cyclic periodate ester} \xrightarrow{-HIO_3} (H_3C)_2C=O + O=C(CH_3)H$$

$$R_1R_2C(OH)-C(OH)R_3R_4 \xrightarrow{HIO_4 \text{ or } Pb(OAc)_4} [\text{cyclic I or Pb ester}] \longrightarrow R_1R_2C=O + O=CR_3R_4 + HIO_3 \text{ (or } Pb(OAc)_2)$$

Other compounds containing oxygens or nitrogens on adjacent carbons, such as *β*-amino alcohols, *α*-hydroxyl aldehydes and ketones, *α*-diketones, also can be oxidized by HIO_4 or $Pb(OAc)_4$. These compounds also undergo the C—C bond cleavage. These reactions are shown below.

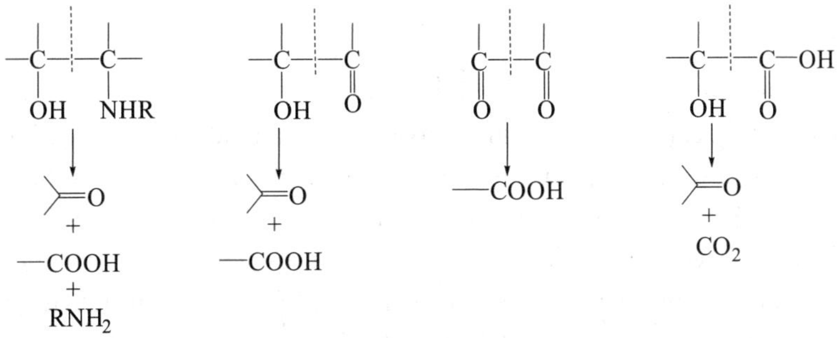

$$-\overset{|}{\underset{OH}{C}}-\overset{|}{\underset{NHR}{C}}- \longrightarrow \;>C=O + -COOH + RNH_2$$

$$-\overset{|}{\underset{OH}{C}}-\underset{O}{\overset{|}{C}} \longrightarrow \;>C=O + -COOH$$

$$\underset{O}{\overset{|}{C}}-\underset{O}{\overset{|}{C}} \longrightarrow -COOH$$

$$-\overset{|}{\underset{OH}{C}}-\underset{O}{C}-OH \longrightarrow \;>C=O + CO_2$$

The reaction of *α*-amino-*β*-hydroxyl propanoic acid with HIO_4 results in two C—C bonds cleavage, and the products are formaldehyde, methanoic acid, carbon dioxide, and ammonia.

$$H_2C(OH)-CH(NH_2)-C(=O)-OH \xrightarrow{HIO_4} HCHO + HCOOH + CO_2 + NH_3$$

邻二醇可被 HIO_4 或 $Pb(OAc)_4$ 氧化，首先生成环状中间体，发生 C—C 键断裂，生成醛或酮，每断开一组 C—C 键，消耗一分子 HIO_4。1,3-二醇或羟基间隔更远的二醇不反应。$Pb(OAc)_4$ 还可使不能形成环状中间体的邻二醇氧化。但顺式二醇氧化速度更快。*β*-氨基醇、α-羟基醛或酮、邻二酮也可被 HIO_4 或 $Pb(OAc)_4$ 氧化。例如 *α*-氨基-*β*-羟基丙酸能被 HIO_4 或 $Pb(OAc)_4$ 氧化为甲醛、甲酸、CO_2 及 NH_3。

9.3 Oxidation of aldehydes and ketones（醛和酮的氧化）

9.3.1 Oxidation of aldehydes（醛的氧化）

(1) Oxidation with permanganate（用高锰酸盐氧化）

The oxidation of an aldehyde to a carboxylic acid is very common. The most popular oxidizing agents are permanganate in acid, neutral, or basic solution. There are two main mechanisms. One is a free-radical mechanism; the other is an ionic one. In the free-radical process, the aldehydic hydrogen is abstracted, and acyl radical is formed. Then the acyl radical reacts with ·OH, which is from the oxidizing agent, to give a carboxylic acid. In the ionic process, the initial step is the addition of a species ZO^- to the carbonyl bond when the oxidation reaction is carried out in alkaline solution. An oxygen anion intermediate is formed. The aldehydic hydrogen is then lost as a proton to a base. At the same time, Z leaves with its electron pair. A salt of a carboxylic acid is yield. If the oxidation is carried out in neutral solution, the addition of ZOH to the carbonyl bond occurs. The aldehydic hydrogen is lost again, and Z leaves with its electron pair at the same time. A carboxylic acid is produced.

$$RCHO \longrightarrow R\dot{C}O \xrightarrow{\cdot OH} RCOOH$$

$$RCHO + ZO^- \longrightarrow RCH(O^-)(OZ) \xrightarrow{B^-} RCOO^- + B{-}H + Z^-$$

$$RCHO + ZOH \longrightarrow RCH(OH)(OZ) \xrightarrow{B^-} RCOOH + B{-}H + Z^-$$

(2) Oxidation with acid dichromate（用铬酸氧化）

Chromic acid is frequently employed in the oxidation of an aldehyde to a carboxylic acid. The mechanism is drawn below. Step 1 shows the formation of a chromate ester via the addition step of OH group to the carbonyl carbon atom. Then the aldehydic hydrogen is abstracted with a base to give a carboxylic acid (Step 2). At the same time, Cr (Ⅵ) is reduced to Cr (Ⅳ). Steps 6 and 7 undergo a similar oxidation by Cr (Ⅴ), which is generated by an electron-transfer process, and Cr (Ⅴ) is reduced to Cr (Ⅲ). Both steps 3 and 4 show the formation of an acyl free-radical. However, in step 3, it is Cr (Ⅵ) that abstracts the aldehydic hydrogen, which results in the formation of the acyl free-radical; in step 4, it is Cr (Ⅳ) that abstracts the aldehydic hydrogen to generate a new free-radical. The following loss of water produces the acyl free-radical. The acyl free-radical reacts with OH group, which is from H_2CrO_4, to give a carboxylic acid.

Step 1 $RCHO + H_2CrO_4 \rightleftharpoons R_2C(OH)(O—CrO_3H)$

Step 2 $RCH(OH)(O—CrO_3H) \xrightarrow{B^-} RCOOH + Cr(IV) + B—H$

Step 3 $RCHO + Cr(VI) \longrightarrow R\dot{C}O$

Step 4 $RCH(OH)_2 + Cr(IV) \longrightarrow R\dot{C}(OH)_2 \xrightarrow{-H_2O} R\dot{C}O + Cr(III)$

Step 5 $R\dot{C}O + H_2CrO_4 \longrightarrow RCOOH + Cr(V)$

Step 6 $RCHO + Cr(V) \rightleftharpoons R_2C(OH)(O—Cr(V))$

Step 7 $R_2C(OH)(O—Cr(V)) \xrightarrow{B^-} RCOOH + Cr(III) + B—H$

醛被酸性 $KMnO_4$ 或酸性 $K_2Cr_2O_7$ 氧化为羧酸。

(3) Oxidation with other reagents（用其它试剂氧化）

A weak oxidant, such as Ag^+ (Tollens reagent), or Cu^{2+} (Fehling reagent or Benedict reagent) can easily oxidize an aldehyde to a salt of a carboxylic acid. The oxidation reaction can be a useful test for the aldehyde, for there is a silver mirror in the Tollens reaction, while there appear brick-red precipitate in the Fehling and Benedict reaction. The following is the Tollens oxidation reaction.

$$RCHO + 2AgOH \longrightarrow RCOO^- + 2Ag + H_2O$$

醛被弱氧化剂 Tollens、Fehling、Benedict 试剂氧化为羧酸盐。

9.3.2 Oxidation of ketones（酮的氧化）

Saturated ketones are generally inert to be oxidized under the oxidation conditions that convert aldehydes to carboxylic acids. Therefore, neither Tollens reagent nor Fehling reagent or Benedict reagent can oxidize saturated ketones. However, the ketones can be oxidized under vigorous acid-catalyzed oxidation conditions, such as with nitric or chromic acids, which result in carbon-carbon bond cleavage at the carbonyl group.

In the presence of peroxyacids, ketones, especially cyclic ketones undergo the Baeyer-Villiger oxidation to give “oxygen insertion” products. In the reaction, a C—C bond is cleaved and a C—O bond is formed. The mechanism and examples see chapter 7 (7.1.3).

酮可以被过氧酸或过氧化氢氧化为羧酸酯，醛可以被过氧酸或过氧化氢氧化为羧酸，称为 Baeyer-Villiger 重排。

9.4 Oxidation of aromatic side chains and the aromatic rings(芳香侧链和芳环的氧化)

Many oxidizing agents, such as permanganate, chromic acid and nitric acid, can oxidize alkyl chain or aromatic rings to give COOH groups. The most often applied alkyl chain is methyl group, although longer chains also can be cleaved to give COOH groups. However, if the alkyl chain is a tertiary alkyl group, it cannot be oxidized, and the aromatic ring is cleaved. If functional groups are in the α position on the side chain, the ease of oxidation can be greatly increased. Functional groups also can be present in other positions on the side chain. If a phenyl group is in the α position on the side chain, the oxidation reaction stops at the diaryl ketone step. The reactivity order in the oxidation of aromatic side chains is $CHR_2 > CH_2R > CH_3$.

The mechanism is not clear, but the formation of Ar_2CH free-radical or cation may be the rate-determining step.

$$C_6H_5CH_2CH_3 \xrightarrow[\text{or } K_2CrO_7]{KMnO_4} C_6H_5COOH$$

$$Ar_2CH_2 + CrO_3 \longrightarrow Ar_2CO$$

$$Ar_2CH_2 \longrightarrow Ar_2CH\cdot$$

$$Ar_2CH_2 \longrightarrow Ar_2CH^+$$

Strong enough oxidizing agents, such as ruthenium tetroxide and $NaIO_4$, V_2O_5 and air, can oxidize aromatic rings. Lead tetraacetate or nickel peroxide can oxidize *o*-diamine, as is shown below. In the reaction, the aromatic ring is cleaved, and a *cis*, *cis*-dicyano compound is produced.

$$o\text{-}C_6H_4(NH_2)_2 \xrightarrow{Pb(OAc)_4} NC\text{-}CH{=}CH\text{-}CH{=}CH\text{-}CN$$

酸性$KMnO_4$、酸性$K_2Cr_2O_7$或浓HNO_3能氧化具有*α*-H的侧链为羧酸，一般情况下芳环不被氧化，环上有强给电子基OH和NH_2等时，芳环可被氧化，稠环芳烃的电子云密度大的环易被氧化。

9.5 Reduction reactions(还原反应)

9.5.1 Catalytic hydrogenation(催化氢化，催化加氢)

The reduction by addition of two hydrogen atoms can be carried out in the presence of a metal catalyst, so it is called catalytic hydrogenation. Alkenes, alkynes, imines, nitriles, aldehydes and ketones can easily undergo catalytic hydrogenation, but carboxylic acids, carboxylic acid esters, carboxylic acid amides cannot. The ease of reduction of various functional groups is listed in Table 9.1. The order is approximate, but the table gives us a fairly good indication. From the table, we can see that it is very difficult for benzene to be reduced by catalytic hydrogenation.

Table 9.1 The ease of reduction of various functional groups toward catalytic hydrogenation.

Functional groups	Products	Ease of Reduction
RCOCl	RCHO, RCH_2OH	Easiest
RNO_2	RNH_2	↓
RC≡CR′	RCH═CHR′(*cis*) RCH_2CH_2R'	
RCHO	RCH_2OH	
RCH═CHR′	RCH_2CH_2R'	
RCOR′	RCH(OH)R′	
$PhCH_2OR$	$PhCH_3$+ROH	
naphthalene	tetralin	Most difficult
RCO_2R'	RCH_2OH+R′OH	
RCONHR′	RCH_2NHR'	
benzene	cyclohexane	

Note: The groups are listed in approximate order of ease of reduction.

(1) Hydrogenation of C═C and C≡C bonds（碳碳双键和三键的氢化）

Platinum, palladium and nickel are the most widely used hydrogenation catalysts. In the presence of the catalyst, the addition of 1mol H_2 to a C═C results in reduction of an alkene, while the addition of 2mol of H_2 to a C≡C results in reduction of an alkyne. Both the reduction reactions give saturated compounds. However, the latter reduction reaction is actually via the formation of an alkene intermediate by hydrogen addition to the triple bond, and the alkene intermediate can not be isolated, for the catalyst is very effective. Lindlar's catalyst, which is prepared by deactivating a palladium catalyst by treating it with lead acetate and quinoline, is less effective, thus it permits an alkyne to be converted to an alkene, without further reduction to an alkane. In the reaction, the addition of hydrogen is *steroselectively syn*, thus a *cis* alkene is produced. However, metal-ammonia reduction can realize steroselectively *anti* addition, and a *trans* alkene is generated. Some examples are shown below.

$$CH_3CH_2CH{=}CH_2 \xrightarrow{H_2\,/\,Pt,\,Pd,\,or\,Ni} CH_3CH_2CH_2CH_3$$

$$CH_3CH_2CH_2C{\equiv}CH \xrightarrow{H_2\,/\,Pt,\,Pd,\,or\,Ni} CH_3CH_2CH_2CH_2CH_3$$

$$CH_3C{\equiv}CCH_3 \xrightarrow{H_2\,/\,Lindlar's\,cat.} (H_3C)(H)C{=}C(H)(CH_3)\ (cis)$$

The mechanism of hydrogenation of C═C may be drawn below. Firstly, hydrogen is adsorbed onto the surface of the metal catalyst. Then the H—H bond is weakened by the metal catalyst. The π bond binds to the catalyst. Two hydrogen atoms shift from the metal surface to the carbons of the double bond, thus the double bond is saturated. The saturated hydrocarbon leaves the metal surface, for it is more weekly adsorbed.

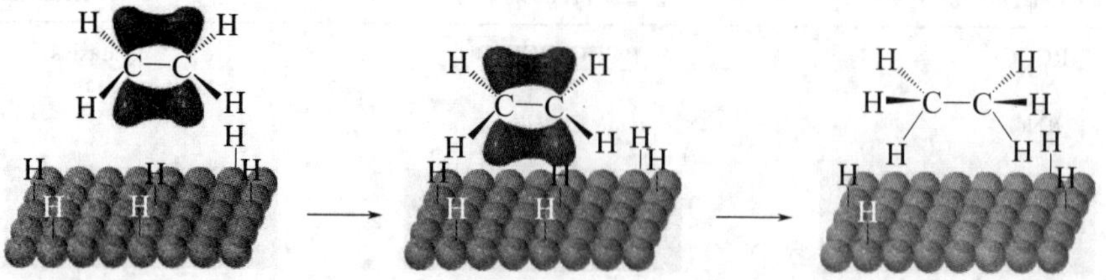

From the mechanism of reduction reaction of an alkene, we can see that the addition of hydrogen is steroselectively *syn*. Two examples are drawn below. The first example shows *Fischer projections* of the products. The second example shows the *syn*-addition from the less *steric hindrance* side.

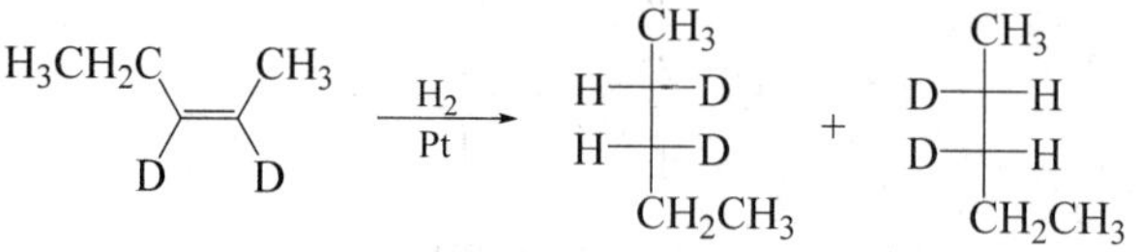

cis- $\xrightarrow[Pt]{H_2}$ *erythro* enantiomers

H_2, PtO_2, EtOH, ~100MPa, 25℃ … not

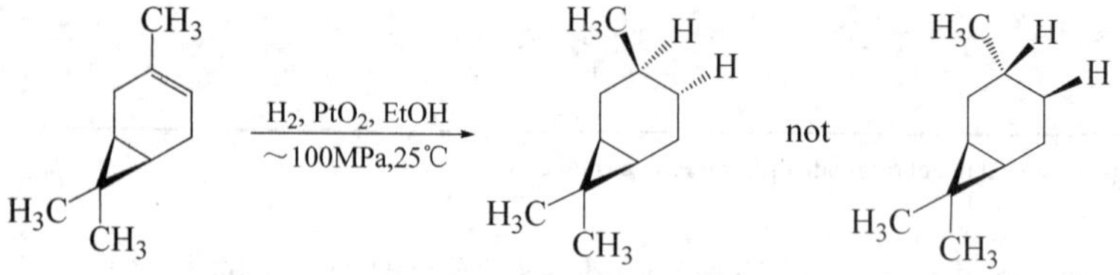

Pt、Pd、Rh、Ni 等及其氧化物作为催化剂，反应在催化剂表面上进行，烯烃加氢得顺式加成产物，从位阻小的方位加成。Willkinson 催化剂：$(Ph_3P)_3RhCl$ 对烯、炔催化加氢时，$—NO_2$、—Cl、—C═O、—N═N、—CN 等不被还原，立体化学为顺式加成。

The mechanism of reduction reaction of an alkyne is similar to the alkene. Both the alkene and the alkyne must be adsorbed on the surface of the metal catalyst before the addition of hydrogen to a multiple bond. However, the alkyne adsorbs more strongly to the catalytic surface than do the alkene. The alkyne preferentially occupies reactive sites on the catalyst. Subsequent transfer of hydrogen to the adsorbed alkyne proceeds slowly, compared to the corresponding transfer to an adsorbed alkene molecule. But the formed alkene is very rapidly hydrogenated to an alkane. The mechanism explains the facts that alkenes undergo the catalytic hydrogenation more rapidly than do alkynes. The Lindlar catalyst can adsorb alkynes and reduces alkynes to alkenes, but the formed alkenes cannot be adsorbed sufficiently on the catalyst, thus the further reduction reactions do not occur.

Lindlar 催化剂：　$Pd / BaCO_3 / Pb(OAc)_2$ 催化炔烃为顺式烯烃。

(2) Hydrogenation of C═O, C═N and C≡N bonds （碳氧双键、碳氮双键和碳氮三键的氢化）

Palladium on carbon can catalyze addition of 1mol H_2 to a C═O in an aldehyde, and the aldehyde gives a primary alcohol, while Raney Ni-catalyzed addition of 1mol H_2 to a C═O in a ketone affords a secondary alcohol. Both the two reductions usually have no chemoselectivity. The examples are illustrated below.

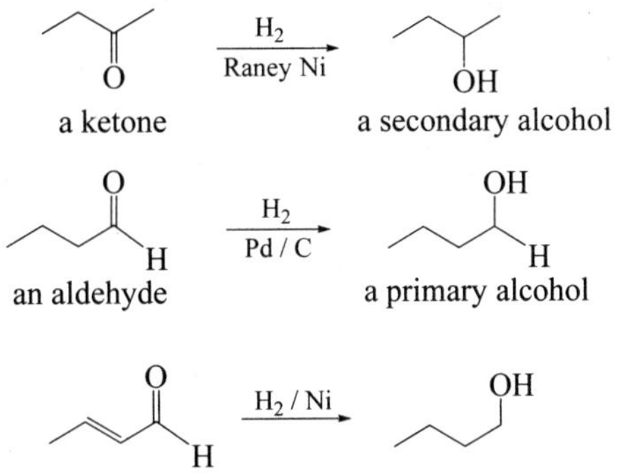

$\xrightarrow[\text{Raney Ni}]{H_2}$

a ketone a secondary alcohol

$\xrightarrow[\text{Pd / C}]{H_2}$

an aldehyde a primary alcohol

$\xrightarrow{H_2 / Ni}$

醛、酮分别被催化氢化为伯醇、仲醇，反应不具有选择性，C═C 与 C≡C 能同时被氢化。

When an acyl chloride undergoes catalytic hydrogenation, the addition of hydrogen to a C═O also occurs, and a primary alcohol is produced via the formation of an aldehyde intermediate. Further reduction of the intermediate occurs immediately after its formation. If the catalyst is partially deactivated, the catalytic hydrogenation of the acyl chloride can stop at the step of the formation of aldehyde. The reaction is called Rosenmund reduction.

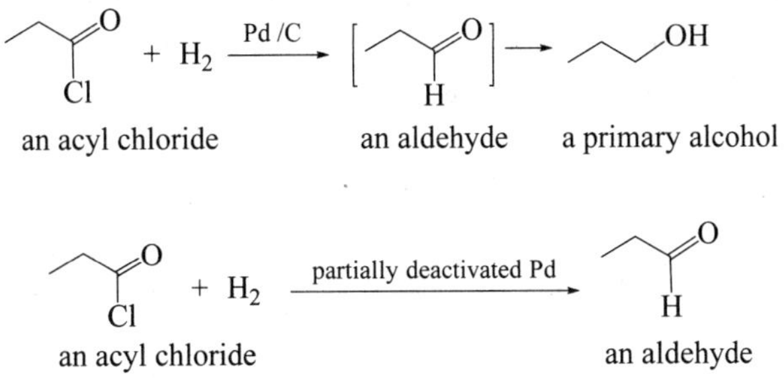

$+ H_2 \xrightarrow{Pd/C}$

an acyl chloride an aldehyde a primary alcohol

$+ H_2 \xrightarrow{\text{partially deactivated Pd}}$

an acyl chloride an aldehyde

Catalytic hydrogenation of imines gives secondary amines, while catalytic hydrogenation of nitriles produces primary amines.

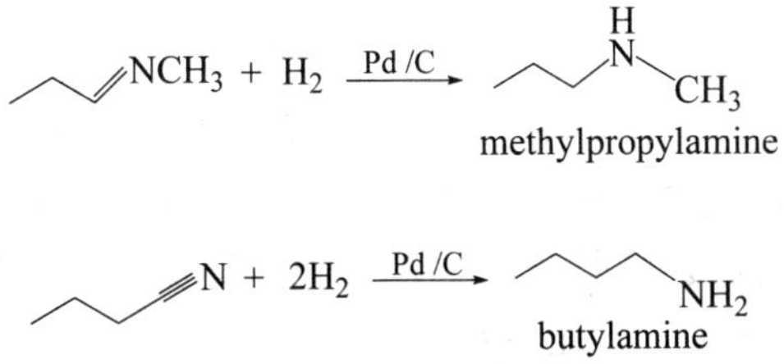

$NCH_3 + H_2 \xrightarrow{Pd/C}$

methylpropylamine

$N + 2H_2 \xrightarrow{Pd/C}$ NH_2

butylamine

催化氢化亚胺得到仲胺，催化氢化腈得到伯胺，催化氢化硝基化合物得到伯胺。

含有不饱和键的化合物如烯烃、炔烃、亚胺、腈、硝基化合物、醛、酮容易发生催化氢化，加氢生成饱和或不饱和度降低的化合物，而羧酸、酯、酰胺不易被催化氢化。Rosenmund 还原：Pd/$BaSO_4$/喹啉/硫化物，还原酰卤为醛，而此时羰基不被还原。

用 Pd 或 Pt 作为催化剂，用于脱卤、脱苄氢解：当苄基、取代苄基和杂原子相连时可氢解为甲苯或取代甲苯，利用苄基作为保护基。脱硫氢解，一般用 Raney Ni 作为催化剂。

9.5.2 Metal-ammonia reduction（金属-氨还原）

Metal-ammonia reduction is a reduction reaction by addition of an election, which is from an *alkali metal*, and a proton, which is from ammonia.

活泼金属-质子给予体（H_2O、ROH、RCOOH、NH_3 等）在一定条件下可作为还原剂。

(1) Reduction of alkynes（炔烃还原）

If an alkyne is treated with sodium or lithium in *liquid ammonia*, the C≡C undergoes a *trans* addition of hydrogen, and a *trans* alkene is produced. When a C═C and a C≡C are in a same molecule, the C≡C may be reduced, but the C═C remains. The examples are shown below.

$$\text{2-butyne} \xrightarrow[\text{NH}_3\text{ (liq.)}]{\text{Na or Li}} \textit{trans}\text{-2-butene}$$

$$\xrightarrow[\text{NH}_3\text{ (liq.)}]{\text{Na or Li}}$$

The mechanism of the reduction reaction can be drawn below. An alkali metal in liquid ammonia loses an electron to form the metal cation. Then the electron adds to an alkyne, and a radical anion is formed. The radical anion is *protonated* by ammonia to give a vinylic radical. Another electron adds to the vinylic radical again to form a vinylic anion. Then the vinylic anion is protonated by ammonia again to give a *trans* alkene. Thus two electrons from the metal and two protons from ammonia are involved in the reduction reaction.

$$\text{M} \xrightarrow{\text{NH}_3} \text{M}^+(\text{NH}_3)_x + \text{e}^-(\text{NH}_3)_x$$

$$\text{H}_3\text{C}{-}\text{C}{\equiv}\text{C}{-}\text{CH}_3 + \text{e}^- \longrightarrow \text{H}_3\text{C}{-}\dot{\text{C}}{=}\ddot{\text{C}}^-{-}\text{CH}_3 \xrightarrow[-^-\text{NH}_2]{\text{H}{-}\text{NH}_2}$$

a radical anion

$$\text{H}_3\text{C}{-}\dot{\text{C}}{=}\text{C(H)CH}_3 \xrightarrow{\text{e}^-} \text{H}_3\text{C}{-}\ddot{\text{C}}^-{=}\text{C(H)CH}_3 \xrightarrow[-^-\text{NH}_2]{\text{H}{-}\text{NH}_2} \text{H}_3\text{C(H)C}{=}\text{C(H)CH}_3$$

a vinylic radical　　a vinylic anion　　a *trans* alkene

非末端炔被还原时，碱金属溶于液氨产生金属离子及溶剂化的电子，该溶剂化的电子加到炔烃三键上，形成一个阴离子自由基，阴离子自由基从液氨中夺取质子形成乙烯基自由基，再加上一个溶剂化的电子形成乙烯基负离子，负离子再质子化得到反式烯烃。分子中既有C═C 与 C≡C 时，只还原 C≡C，不还原 C═C。

末端炔能与金属形成带负电荷的炔化物，排斥电子，不被还原。

(2) Reduction of benzene and its derivatives（苯及其衍生物的还原）

When benzene or its derivatives are treated with an alkali metal, such as sodium or lithium or potassium, in liquid ammonia, *non-conjugated* 1,4-cyclohexadiene or 1,4-cyclohexadiene derivatives are produced. The reduction reaction is called the Birch reduction.

$$\xrightarrow[\text{NH}_3\text{ (liq.), EtOH}]{\text{Na or Li}}$$

The mechanism of the Birch reduction is given below. The initial electron addition produces a radical-anion. The radical-anion may have many resonance contributors. Two of them are shown below. The following protonation of the radical-anion by the week acid ammonia gives a radical. The radical is *delocalized*. Then the radical accepts a second electron to give an anion. The anion also has some resonance contributors, which are shown below. The generated anion is delocalized over three carbon atoms, and is protonated on the central carbon. Thus a non-conjugated diene is produced, and the isolated double bonds in the product do not react under the reaction conditions. In a word, the benzene ring undergoes the Birch reduction by a stepwise addition of two electrons and two protonations after each electron addition.

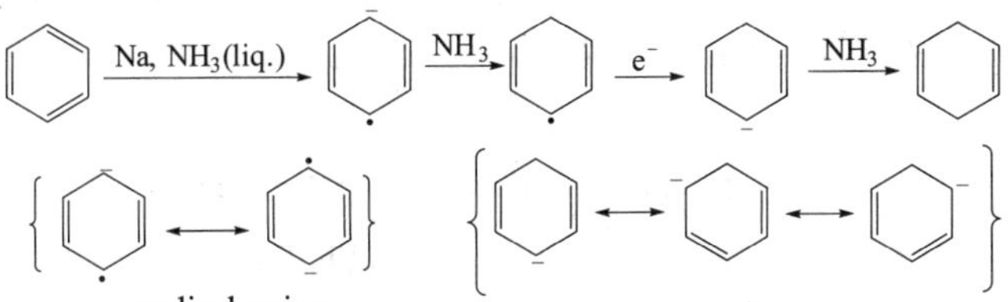

Na, NH_3(liq.) — NH_3 — e^- — NH_3

radical-anion　　anion

If the benzene ring has some substituents, they may influence the regioselcetivity of the Birch reduction, for the product is determined by the site of the first protonation, and the second protonation is nearly always opposite to the first protonation. *Electron-donating substituents*, such as methyl group, favor protonation at an unoccupied site *ortho* to the substituent, while *electron-withdrawing substituents*, such as carboxyl group, favor *para* protonation. It may be contributed to the theory that electron withdrawing substituents can stabilize the carbanions, thus the reduction takes place on carbon atoms bearing electron withdrawing substituents.

CH_3 — Na or Li / NH_3 (liq.) , EtOH — CH_3

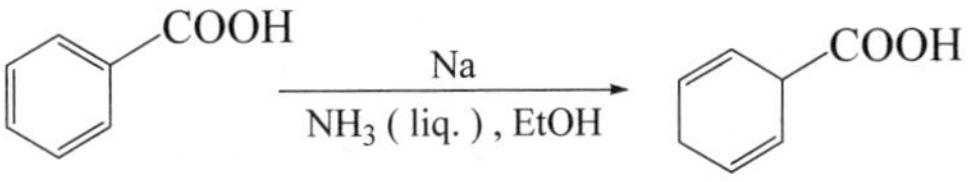

COOH — Na / NH_3 (liq.) , EtOH — COOH

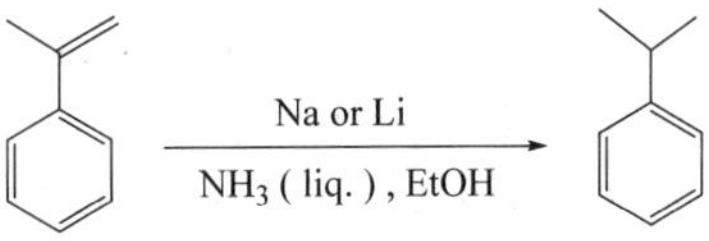

Na or Li / NH_3 (liq.) , EtOH

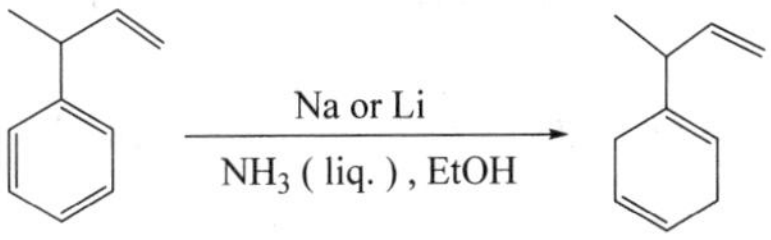

Na or Li / NH_3 (liq.) , EtOH

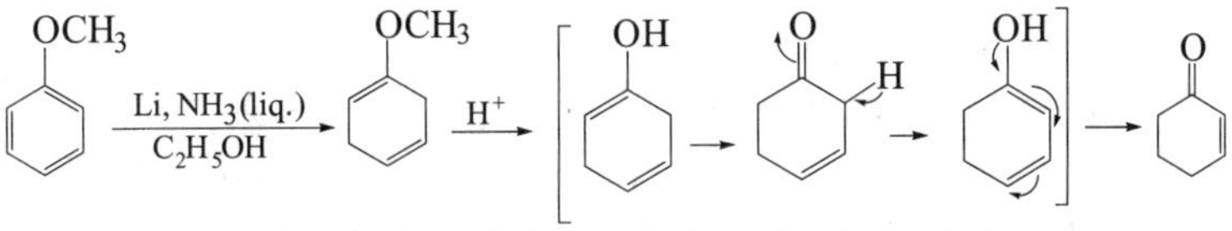

OCH_3 — Li, NH_3(liq.) / C_2H_5OH — OCH_3 — H^+ — [OH — O, H — OH] — O

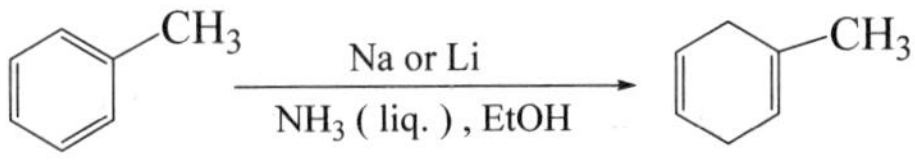

苯及其衍生物被碱金属 Na、K、Li 在液氨中还原为非共轭的环己二烯及衍生物。过程为碱金属溶于液氨产生金属离子及溶剂化的电子，该溶剂化的电子加到苯环双键上，形成一个阴离子自由基，阴离子自由基从液氨中夺取质子形成可共振的自由基，再加上一个溶剂化的电子形成烯丙基负离子，负离子再质子化得到非共轭的环己二烯及衍生物。

芳环上连吸电子基加速反应，生成取代基不在双键碳原子上的产物。

芳环上连供电子基使反应速率降低，生成取代基在双键碳原子上的产物。

萘环中一个苯环上有烷基或羟基等取代基时，还原反应选择性地发生在另一个环上。

碱金属 Na、K、Li 在液氨中只还原 α,β-不饱和酮的 C═C 为 C—C，不还原 C═O。

苄基、烯丙基等与杂原子之间的键在液氨中被碱金属裂解，最终被氢解，构型不变。

(3) Bimolecular reduction of ketones（酮的双分子还原）

When a ketone is treated with sodium in liquid ammonia or with magnesium in benzene, a pinacol is formed. Two examples are given below.

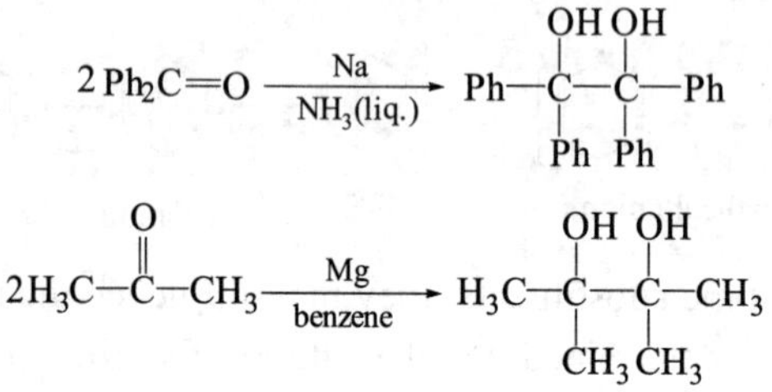

$$2\,Ph_2C{=}O \xrightarrow[NH_3(liq.)]{Na} Ph{-}C(OH)(Ph){-}C(OH)(Ph){-}Ph$$

$$2\,H_3C{-}C(=O){-}CH_3 \xrightarrow[benzene]{Mg} H_3C{-}C(OH)(CH_3){-}C(OH)(CH_3){-}CH_3$$

The mechanism of the bimolecular reduction is shown below. The initial electron addition to a C═O produces an anion radical. Then the anion radical undergoes a coupling reaction. After hydrolysis, a pinacol is yielded.

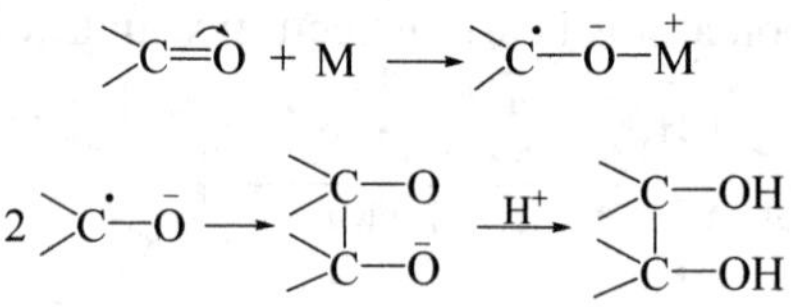

$$>C{=}O + M \longrightarrow >\dot{C}{-}\bar{O}{-}\overset{+}{M}$$

$$2\,>\dot{C}{-}\bar{O} \longrightarrow \begin{matrix}>C{-}\bar{O}\\ |\\ >C{-}\bar{O}\end{matrix} \xrightarrow{H^+} \begin{matrix}>C{-}OH\\ |\\ >C{-}OH\end{matrix}$$

Na(Hg)和 Mg(Hg)等活泼金属向 C═O 提供一个电子，产生中间体负离子自由基，阴离子自由基再经过自由基二聚，产生双氧负离子，进一步夺氢得到频那醇 pinacol。

(4) Reduction of esters（酯的还原）

The treatment of an ester with sodium in ethanol gives an alcohol. An example and a mechanism are shown below. The initial electron addition to a C═O produces an anion radical. Then the anion radical reacts with H^+ to generate a radical. The transfer of another electron by the metal to the radical gives an anion. Then the anion reacts with H^+ to produce an alcohol.

$$C_2H_5OCO(CH_2)_8COOC_2H_5 \xrightarrow{Na,\ C_2H_5OH} HOCH_2(CH_2)_8CH_2OH$$

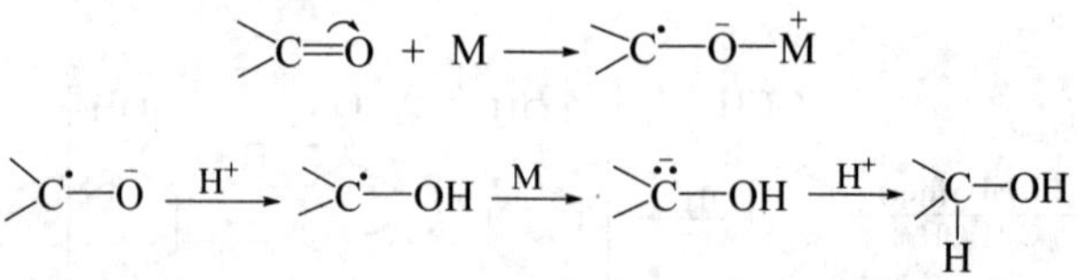

$$>C{=}O + M \longrightarrow >\dot{C}{-}\bar{O}{-}\overset{+}{M}$$

$$>\dot{C}{-}\bar{O} \xrightarrow{H^+} >\dot{C}{-}OH \xrightarrow{M} >\ddot{\bar{C}}{-}OH \xrightarrow{H^+} >CH{-}OH$$

If an ester reacts with sodium in liquid ammonia or in dimethylbenzene, a α-hydroxylketone is produced. An example and a mechanism are shown below. An anion radical is initially formed by an alkali metal again. However, there is no strong donator of proton in the reaction. Therefore, the anion radical prefers to undergo dimmerization, and α-diketone is formed. The diketone reacts with the metal to yield an α-enediol. Isomerization of the enediol gives a α-hydroxylketone.

$$C_2H_5OCO(CH_2)_8COOC_2H_5 \xrightarrow{\text{Na / xylene}} \xrightarrow{CH_3COOH} (CH_2)_8 \text{ (ring: C=O, CHOH)}$$

$$\text{R(R'O)C=O} + M \longrightarrow \text{R(R'O)}\dot{C}-\bar{O}-\overset{+}{M} \xrightarrow{\text{dimerization}} \text{R'O(R)C(}O^-\text{)}-\text{C(}O^-\text{)(R)OR'}$$

$$\text{RC(=O)–C(=O)R} \xrightarrow{M} \text{RC(}O^-\text{)=C(}O^-\text{)R} \longrightarrow \text{RC(OH)=C(OH)R} \longleftrightarrow \text{RC(=O)–CH(OH)R}$$

酯被碱金属Na、K、Li-液氨或Na-二甲苯还原时，由于体系中没有强质子供体，负离子自由基二聚并失去烷氧基得到邻二酮，邻二酮继续得到电子，α-烯二醇负离子再质子化为α-烯二醇，然后异构化为α-羟基酮。

9.5.3 Hydride reduction（氢化物还原）

Hydride reduction is reduction by addition of a hydride ion. The reduction product is formed by the subsequent protonation.

For the reduction reactions, it is often necessary to reduce one group in a molecule without affecting another reducible group. In some cases, the strategy of protecting groups can realize selective reduction. However, the more promising and direct way is to find a reducing agent that will do this. The metallic hydrides are one of the most common broad-spectrum selective reducing agents.

Among the metallic hydrides, $LiAlH_4$ and $NaBH_4$ are the most commonly used. These reagents have two main advantages over other reducing agents. They usually do not reduce C═C or C≡C. The reaction is broad and general. All four hydrogens in $LiAlH_4$ can be used for reduction. Compared with $NaBH_4$, $LiAlH_4$ is a stronger reducing reagent, for Al—H bonds are more polar than B—H bonds. Common solvents for the reduction by $LiAlH_4$ are THF or ether. Water and alcohol must be removed before the reduction, since they react readily with $LiAlH_4$.

$LiAlH_4$活性强，选择性低。只能在乙醚或THF使用。

Some compounds easily undergo the hydride reduction by $LiAlH_4$, while others difficultly undergo the reaction. The ease of the reduction of various functional groups is listed in Table 9.2.

Table 9.2 The ease of the reduction of various functional groups with $LiAlH_4$ in ether

Substrate	Product	Ease of Reduction
RCHO	RCH_2OH	Easiest
RCOR	RCHOHR	↓
RCOCl	RCH_2OH	↓
Lactone	Diol	↓
Epoxide (RHC–O–CHR)	RCH_2CHOHR	↓
RCOOR′	RCH_2OH+R′OH	↓
RCOOH	RCH_2OH	↓
RCOO-	RCH_2OH	↓
$RCONR'_2$	$RCH_2NR'_2$	↓
RC≡N	RCH_2NH_2	↓
RNO_2	RNH_2	Most difficult
$ArNO_2$	ArN═NAr	

Compared with $LiAlH_4$, $NaBH_4$ is more selective. The other advantage is that $NaBH_4$ can be used in water or alcoholic solvents. Thus compounds which are not soluble in ethers can be reduced. In general, $NaBH_4$ reduces carbonyl compounds in the order: aldehydes > α,β-unsaturated aldehydes > ketones > α,β-unsaturated ketones. However, $NaBH_4$ cannot reduce a carboxylic acid, a carboxylic acid ester and a carboxylic acid amide.

$NaBH_4$还原性较弱，选择性高，可在水、醇溶液中使用。能还原醛、酮，不还原羧酸及其衍生物，也不脱卤，不还原C═C、C≡C和硝基等。

(1) Hydride reduction of aldehydes and ketones （醛酮的氢化物还原）

An aldehyde is reduced to a primary alcohol by a metallic hydride, and a ketone gives a secondary alcohol.

1. $NaBH_4$ 2. H^+, H_2O

an aldehyde → a primary alcohol

1. $LiAlH_4$ 2. H+, H_2O

a ketone → a secondary alcohol

The mechanism is shown below. The initial addition of a hydride ion to a C═O leads to the formation of an anion. At the same time, AlH_3 is released. Then the formed anion reacts with AlH_3 to produce another anion. After hydrolysis, an alcohol is yielded.

$$[H_3Al-H]^{\ominus} + >C=O \longrightarrow AlH_3 + >C(H)-O^- \longrightarrow [>C(H)-O-AlH_3]^- \xrightarrow[H_2O]{H^+} >C(H)-O-H$$

When the reduction of cyclohexanone is carried out, the stereochemistry may be studied. If the R is hydrogen, **b** is the major product, for the product is more stable. When R is not hydrogen, **a** is the major product, for a hydride may attack at the carbonyl carbon atom from the less hindrance side, thus the activity energy of the reduction reaction is low. The reduction speed is high, and **a** is the major product.

1. $LiAlH_4$ or $NaBH_4$ 2. H_2O

R = H, **b** is the major product

R≠H, **a** is the major product

(2) Hydride reduction of a carboxylic acid and its derivatives by $LiAlH_4$（用氢化铝锂还原羧酸及其衍生物）

A carboxylic acid is reduced to a primary alcohol by $LiAlH_4$, while a carboxylic acid ester affords a primary alcohol and another alcohol. An amide can be reduced to a primary amine, a secondary amine or a tertiary amine. The product is dependent on the type of the amide. The examples are shown below.

a carboxylic acid $\xrightarrow[\text{2. } H^+, H_2O]{\text{1. } LiAlH_4}$ a primary alcohol

an ester $\xrightarrow[\text{2. } H^+, H_2O]{\text{1. } LiAlH_4}$ a primary alcohol $+ CH_3OH$

$\xrightarrow[\text{2. } H_2O]{\text{1. } LiAlH_4}$ a primary amine

$\xrightarrow[\text{2. } H_2O]{\text{1. } LiAlH_4}$ a secondary amine

$\xrightarrow[\text{2. } H_2O]{\text{1. } LiAlH_4}$ a tertiary amine

Reduction mechanism of an ester or acid by $LiAlH_4$

$\xrightarrow{H_3O^+}$ RCH_2OH + aluminum salts

Reduction mechanism of an amide by $LiAlH_4$

$\xrightarrow{H_2O}$ $RCH_2NR'_2$ + aluminum salts

The mechanism of the reduction of an ester or an amide is shown above. As in the reductions of aldehydes and ketones, the first step in each case is the *irreversible* addition of hydride to the electrophilic carbonyl carbon. The carbonyl oxygen coordinates to a Lewis acidic metal (Al). The coordination undoubtedly enhances that electrophilic character of the carbonyl carbon. The reduction course of esters is different from the reduction course of amides. It may be contributed to the nature of the different hetero atom substituents on the carbonyl group. Amide anions are poorer leaving groups than alkoxide anions. On the other hand, oxygen forms especially strong bonds to aluminum. The addition of hydride to the carbonyl carbon atom generates a tetrahedral intermediate. The intermediate has a polar oxygen-aluminum bond. Neither the hydrogen nor the alkyl group (R) is a possible leaving group, so one of the two remaining substituents must be lost, if this tetrahedral species is to undergo elimination to reform a hetero atom double bond. For the ester, by eliminating an aluminum alkoxide (R'O—Al), an aldehyde is formed. The formed aldehyde undergoes further reduction and the following hydrolysis, a 1°-alcohol is produced. In the case of the amide, the more basic nitrogen may act to eject a metal oxide species (e.g. Al—O$^-$), and the resulting iminium double bond would then be reduced to an amine.

In the case of a carboxylic acid, the mechanism of its reduction reaction may be drawn as below. The carboxylic acid reacts with $LiAlH_4$ (a base) to form a lithium salt of the acid, and AlH_3 is released. Then AlH_3 approaches the formed salt, and a transfer of a hydrogen anion from Al to the carbonyl carbon atom generates a tetrahedral intermediate. The elimination of $LiOAlH_2$ gives an aldehyde, which undergoes further reduction by another $LiAlH_4$ molecule to produce an alcohol.

$$RCOOH + LiAlH_4 \longrightarrow RCOOLi + H_2 + AlH_3$$

$$R{-}\overset{O}{\overset{\|}{C}}{-}OLi \xrightarrow{AlH_3} R{-}\underset{H}{\overset{OAlH_2}{C}}{-}O^-Li^+ \xrightarrow{-LiOAlH_2} R{-}\underset{H}{C}{=}O \xrightarrow{\text{further reduction}} RCH_2OH$$

If several hydrogen atoms in $LiAlH_4$ are replaced by larger groups, such as alkoxy, its reduction activity becomes lower, thus an acyl chloride or an ester can be reduced to an aldehyde at low temperature. The formed aldehyde can be isolated.

$$CH_3CH_2CH_2COOCH_3 \xrightarrow[H_2O]{[(CH_3)_2CHCH_2]_2AlH,\,-78℃} CH_3CH_2CH_2\overset{O}{\overset{\|}{C}}H$$

$$CH_3CH_2CH_2CH_2COCl \xrightarrow[H_2O]{LiAl[OC(CH_3)_3]_3H,\,-78℃} CH_3CH_2CH_2CH_2\overset{O}{\overset{\|}{C}}H$$

$LiAlH_4$ 能将羰基化合物、羧酸及衍生物还原为醇（除酰胺外）。属于亲核加成反应，不还原 C═C 与 C≡C。大位阻氢负离子还原剂 $LiAl[OC(CH_3)_3]_3$ 可使酰氯还原停留在醛的阶段，$[(CH_3)_2CHCH_2]_2AlH$ 使酯还原停留在醛的阶段。

$LiAlH_4$ 中氢负离子为亲核试剂还原环氧化合物，发生 S_N2 反应，选择性为 H^- 进攻取代基较少的碳原子，得到仲醇或叔醇。

$LiAlH_4$ 还原酰胺、腈、肟和脂肪族硝基化合物得胺。

9.6 Other reduction reactions（其它还原反应）

9.6.1 Reduction with *diborane*（用双硼烷还原）

Diborane does not react easily with alkene double bonds, but when it dissolves in ether or THF, it is *dissociated* into a solvent coordinated monomer, $R_2O{-}BH_3$, and the solvated monomer adds rapidly to the alkene double bonds under mild condition. The addition of diborane to the alkyne triple bonds also can be carried out. This remains a primary application of this reagent, but it also can be used rapid and complete reduction of carboxylic acids, amides and nitriles. The ease of reduction of various functional groups with boranes is listed in Table 9.3. From the table, we can see that the ease of reduction of various functional groups with boranes is different from the ease of reduction of various functional groups with $LiAlH_4$, for example, the reduction of a carboxyl acid ester with boranes is much more difficult than its reduction with $LiAlH_4$, and an acyl chloride is inert for the reduction with boranes. Diborane also provides one of the best methods for reducing carboxylic acids to 1°-alcohols.

Table 9.3 The ease of reduction of various functional groups with borane

Substrate	Product	Ease of reduction
RCOOH	RCH_2OH	Easiest
RCH=CHR	$(RCH_2CHR)_3B$	↓
RCOR	RCHOHR	
RCN	RCH_2NH_2	
H, O, H, R, R (epoxide)	RCH_2CHOHR	
RCOOR′	RCH_2OH+R′OH	Most difficult
RCOCl		Inert

(1) Reduction of alkenes（烯烃的还原）

Borane is electron deficient. When diborane reacts with an alkene in an ether or THF, the solvated monomer $R_2O—BH_3$ is generated in situ, and BH_3, a Lewis acid, can bond to the π electrons of a double bond by displacement of the ether moiety. The binding might generate a dipolar intermediate. The intermediate has a carbon cation and a negatively-charged borane. A hydride shift from borane to the carbocation center generates a neutral product. It is noticed that all three hydrogen atoms in borane are potentially reactive, so the formed alkyl borane product from the first addition may serve as the hydroboration reagent for two additional alkenes molecules. Therefore, a trialkylborane is produced. After oxidation, the trialkylborane converts to an alcohol, as is shown below.

$$(CH_3)_2C{=}CH_2 \xrightarrow[\text{ether}]{BH_3} \left[(CH_3)_2\overset{+}{C}—CH_2—\overset{-}{B}H_3\right] \longrightarrow (CH_3)_2CH—CH_2—BH_2$$

$\{$ H—BH_2 … Transition state for a One-step Addition $\}$

$$(CH_3)_2C{=}CH_2 \xrightarrow[\text{ether}]{B_2H_6} (CH_3)_2CH—CH_2—BH_2 \xrightarrow{(CH_3)_2C=CH_2} ((CH_3)_2CH—CH_2)_3—B \xrightarrow{H_2O_2} (CH_3)_2CH—CH_2—OH + B(OH)_3$$

The mechanism of oxidation of a trialkylborane to an alcohol involves a migration of R group from the boron atom to the oxygen atom, as is shown below. The reaction of trialkylborane with H_2O_2 in the presence of OH^- gives an anion intermediate. Then the intermediate undergoes the migration to generate monoalkoxylborane, which undergoes the second and the third migration to give trialkoxylborane. The trialkoxylborane is hydrolyzed to produce an alcohol.

$$R_2B—R + H_2O_2 \xrightarrow{OH^-} R_2\overset{-}{B}(—R)(—O—OH) \longrightarrow R_2B—O—R + OH^-$$

$$\xrightarrow{H_2O_2} RB(O—R)_2 \xrightarrow{H_2O_2} B(OR)_3 \xrightarrow{H_2O} 3ROH + B(OH)_3$$

(2) Reduction of carboxylic acids and nitriles（羧酸和腈的还原）

Carboxylic acids can be easily reduced to alkoxyboranes, which are hydrolyzed to an alcohol.

The reduction is initialed by the addition of electron-deficient borane to the carbonyl oxygen. Then a hydrogen transfers from borane to the carbonyl carbon to generate an alkoxyborane. However, an acyl chloride can not undergo the reduction, for its carbonyl oxygen has low basicity, and the electron-deficient borane cannot add to the carbonyl oxygen. Nitriles can be reduced by diborane to 1°-amines.

B_2H_6含缺电子中心，与C═C经四元环过渡态发生加成得到三烷基硼，再水解得到醇。B_2H_6还可还原羧酸为醇，还原腈、酰胺为伯胺，不能还原酰卤和酯。

9.6.2 Reduction with aluminum isopropoxide in isopropyl alcohol（用异丙醇中的异丙醇铝还原）

An aldehyde or a ketone reacts with aluminum isopropoxide in isopropyl alcohol to give a primary alcohol or a secondary alcohol. This is called the Meerwein-Ponndorf-Verley reduction. The reaction is reversible, and the reverse reaction is known as the Oppenauer oxidation. If the acetone is removed by distillation, the equilibrium is shifted. The reaction can be carried out under very mild conditions and is highly specific for aldehydes and ketones. The reaction usually involves a cyclic transition state. The hydrogen attacks at the carbonyl carbon, and at the same time one isopropyl is removed as acetone. Thus many functional groups can be present without themselves being reduced.

$$RCOR' + Al[OCH(CH_3)_2]_3 \longrightarrow \left[\text{cyclic transition state: } (H_3C)_2HCO,\ OCH(CH_3)_2 \text{ on Al; } H_3C,\ CH_3,\ H,\ R,\ R'\right] \longrightarrow RR'CHOAl[O{-}CH(CH_3)_2]_2 + CH_3COCH_3 \longrightarrow RR'CHOH$$

$$\text{6-}O_2N\text{-2-tetralone} \xrightarrow{Al[O-CH(CH_3)_2]_3} \text{6-}O_2N\text{-2-tetralol (OH)} + CH_3COCH_3$$

The overall reduction of common functional groups with H^- was listed in Table 9.4.

Table 9.4 The overall reduction of common functional groups with H^-

Reducing Reagants	Reductable groups					
	$R_2C{=}\overset{+}{N}HR$	RCOX	RCOH	RCOR	RCOOR	$RCONH_2$	RCOOM
$LiAlH_4$	$-NH_2$	RCH_2OH	RCH_2OH	R_2CHOH	RCH_2OH	RCH_2NH_2	RCH_2OH
$LiAlH[OC(CH_3)_3]_3$	—	RCHO	RCH_2OH	R_2CHOH	RCH_2OH	RCHO	—
$NaBH_4$	$-NH_2$	—	RCH_2OH	R_2CHOH	—	—	—
$NaBH_3CN$	$-NH_2$	—	RCH_2OH	—	—	—	—
B_2H_6	—	—	RCH_2OH	R_2CHOH	—	RCH_2NH_2	RCH_2OH
AlH_3	—	RCH_2OH	RCH_2OH	R_2CHOH	RCH_2OH	—	—
$[(CH_3)_2CHCH_2]_2AlH$	—	—	RCH_2OH	R_2CHOH	RCHO	RCHO	RCH_2OH

异丙醇铝还原醛、酮为醇，该反应为Meerwein-Ponndorf-Verley反应，反应经六元环过渡态完成。因此如果体系中含其它官能团，体系也不受影响。反应条件温和。其逆反应为

Oppenauer 氧化法，产物为酮（如果体系中有 C═C，双键不受影响）。

9.6.3 Wolff-Kishner reduction and Clemmensen reduction（Wolff-Kishner-黄鸣龙还原）

An aldehyde or a ketone reacts with excess *hydrazine* in the presence of a strong base to produce the corresponding hydrocarbon. The reaction is called Wolff-Kishner reduction. A high-boiling hydroxylic solvent, such as diethylene glycol, is commonly used to achieve the high temperature. The strongly basic conditions preclude the reaction application to base sensitive compounds. The examples are shown below. The second example shows that acetal group cannot be reduced under the reaction conditions.

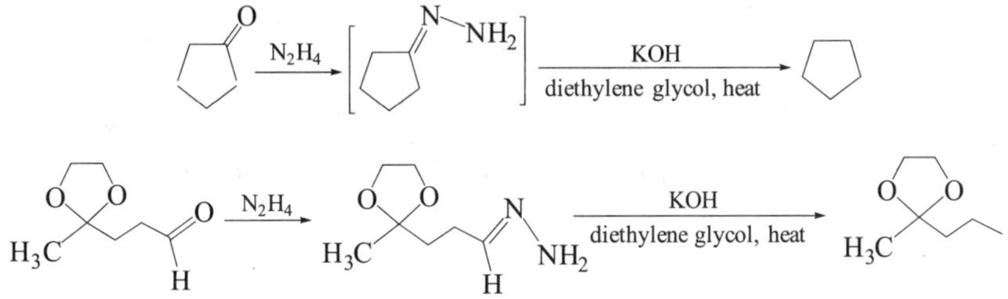

N_2H_4 ; NH_2 ; KOH ; diethylene glycol, heat

N_2H_4 ; H_3C ; NH_2 ; KOH ; diethylene glycol, heat

The mechanism is given below. The initial addition of hydrazine to the carbonyl compounds gives a *hydrazone*. Then *tautomerization* of the formed hydrazone to an azo isomer occurs. Hydrogen in the azo isomer is removed by a strong base. Nitrogen gas is released under high temperature, and a carbon anion is formed. The carbon anion captures hydrogen to give a hydrocarbon.

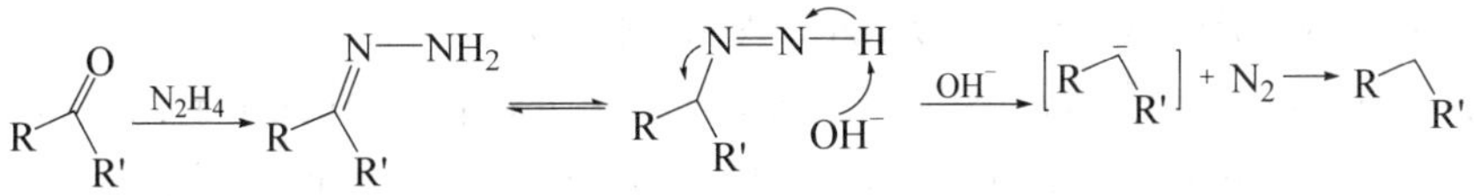

N_2H_4 ; $N—NH_2$; N=N—H ; OH^- ; $+ N_2$

羰基在碱性条件下被水合肼还原为亚甲基。

If an aldehyde or a ketone is base sensitive, Clemmensen reduction may be used to change the carbonyl to methylene. It is noticed that acid sensitive carbonyl compounds cannot be used in the reaction. The mechanism is drawn below. The mercury does not participate in the reaction. It serves only to provide a clean metal surface.

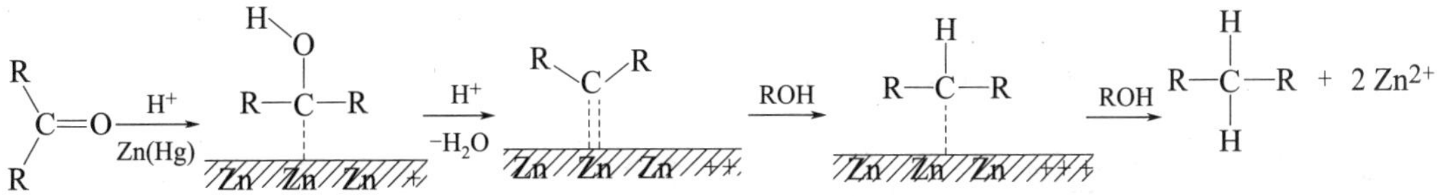

H^+, Zn(Hg) ; H^+, $-H_2O$; ROH ; ROH ; $+ 2\ Zn^{2+}$

Clemmensen 还原：羰基在酸性条件下被 Zn(Hg)还原为亚甲基。共轭 C═C 也被还原，但孤立双键不被还原。

9.7 Disproportionation reactions（歧化反应）

9.7.1 Cannizzaro reaction（Cannizzaro 反应）

When a non-enolizable aldehyde is heated in strong aqueous base, one aldehyde molecule is reduced to an alcohol, while another molecule is oxidized to a salt of a carboxylic acid. The reaction is called Cannizzaro reaction, which is a kind of disproportionation reaction. It is noticed that the reaction is restricted with non-enolizable aldehydes. The addition of OH^- with an aldehyde gives an oxygen anion. The hydride is transferred from the anion to the aldehyde, and a salt of carboxylic acid and an alcohol are produced.

$$2HCHO \xrightarrow{NaOH} HCOONa + CH_3OH$$

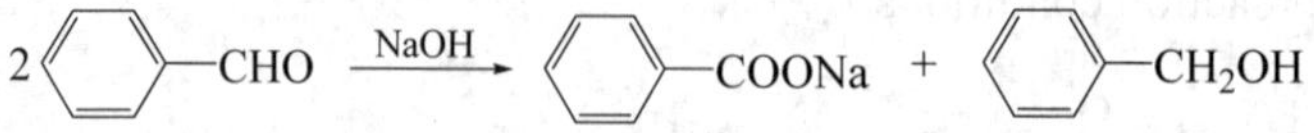

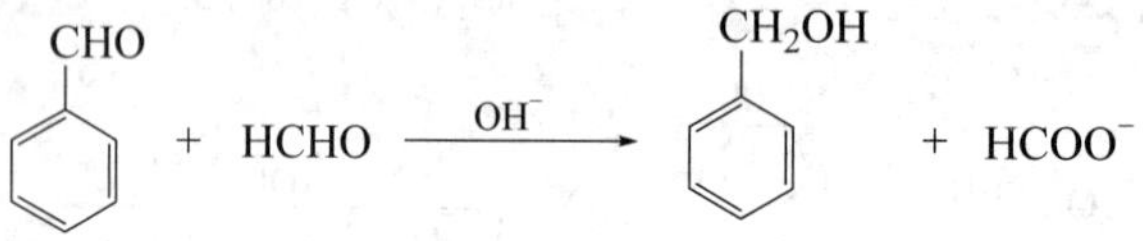

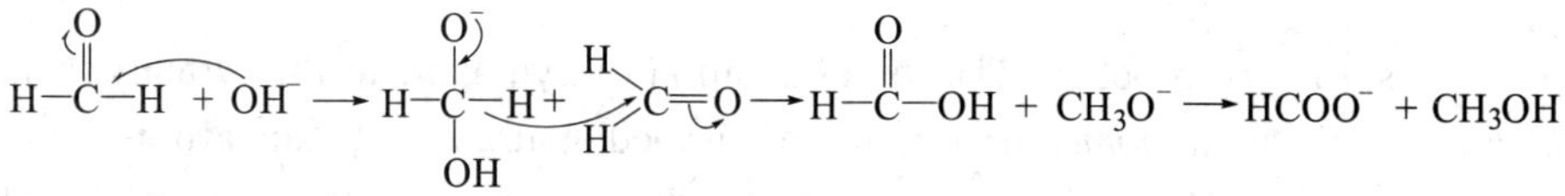

$$2\ C_6H_5-CHO \xrightarrow{NaOH} C_6H_5-COONa + C_6H_5-CH_2OH$$

$$C_6H_5CHO + HCHO \xrightarrow{OH^-} C_6H_5CH_2OH + HCOO^-$$

$$H-\overset{O}{\overset{\|}{C}}-H + OH^- \longrightarrow H-\overset{O^-}{\underset{OH}{C}}-H + \overset{H}{\underset{H}{>}}C{=}O \longrightarrow H-\overset{O}{\overset{\|}{C}}-OH + CH_3O^- \longrightarrow HCOO^- + CH_3OH$$

Cannizzaro 反应：不含 α-H 的醛在浓碱作用下发生自氧化-还原反应，其中一分子醛被氧化为酸，另一份子醛被还原为醇。

9.7.2 *Benzilic acid* rearrangement（Benzilic Acid 重排）

When a α-diketone is treated with base, a carbonyl group is reduced to a hydroxyl group, while the other is oxidized to a salt of a carboxylic acid. In the reaction, 1, 2-phenyl shift occurs. Thus the benzilic acid rearrangement is also a kind of disproportionation reaction. The example and the mechanism have been discussed in chapter 7.1.1.

Benzilic Acid 重排：α-二酮经碱处理，二芳基乙二酮重排为二芳基乙醇酸的反应。

Problems

1. Write mechanisms for the following reactions.

(1) 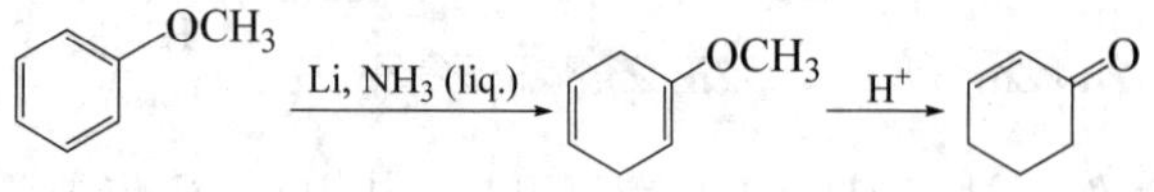 OCH_3 $\xrightarrow{Li,\ NH_3\ (liq.)}$ OCH_3 $\xrightarrow{H^+}$ O

(2) 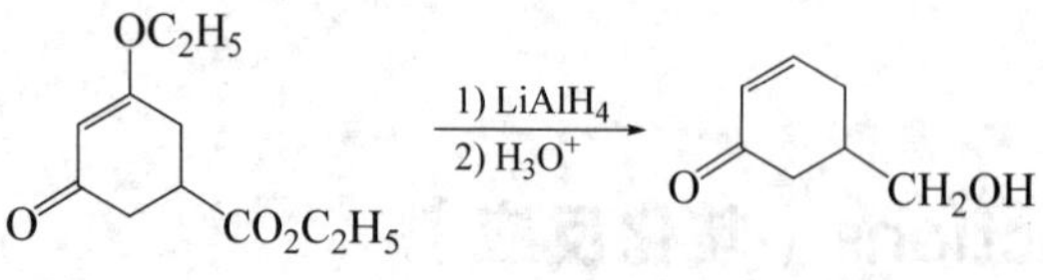 OC_2H_5, O, $CO_2C_2H_5$ $\xrightarrow[2)\ H_3O^+]{1)\ LiAlH_4}$ O, CH_2OH

2. Write down the reagents of the following reactions

(1) 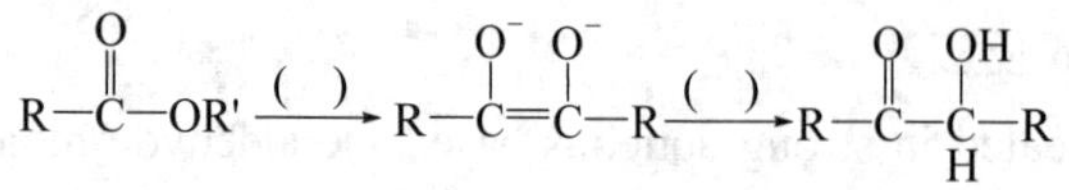 $R-\overset{O}{\overset{\|}{C}}-OR' \xrightarrow{(\ \ )} R-\overset{O^-}{C}=\overset{O^-}{C}-R \xrightarrow{(\ \ )} R-\overset{O}{\overset{\|}{C}}-\overset{OH}{\underset{H}{C}}-R$

(2) 1-naphthol $\xrightarrow{(\ )}$ 5,8-dihydro-1-naphthol (OH)

(3) CO_2Me-substituted N-CH_2Ph pyrrolidinone $\xrightarrow{(\ )}$ CO_2Me-substituted N-CH_2Ph pyrrolidine

(4) 3-ethoxy-2-cyclohexenone (O, OC_2H_5) $\xrightarrow{(\ )}$ 3-ethoxy-2-cyclohexenol (OH, OC_2H_5) $\xrightarrow{(\ )}$ 2-cyclohexenone (O)

3. Synthesize the following product from given starting Materials.

(1) $HC\equiv CH \longrightarrow$ (Fischer projection: C_2H_5; H—OH; H—OH; C_2H_5)

(2) cyclohexanol (OH) $\longrightarrow$ cyclopentene—CHO

(3) 4-methylphenol (OH, CH_3) $\longrightarrow$ (OH, $COCH_3$, CH_3) $\longrightarrow$ (OH, OH, CH_3)

(4) $H_2C{=}CH{-}CH{=}CH_2 \longrightarrow$ 2-hydroxycyclohexanone (OH, O)

4. Write down the products.

(1) $2\,Ph_2C{=}O \xrightarrow[NH_3(\text{liq.})]{Na}$

(2) $2\,H_3C{-}\overset{O}{\overset{\|}{C}}{-}CH_3 \xrightarrow[\text{benzene}]{Mg}$

(3) $H_3C{-}\underset{OH}{\overset{CH_3}{C}}{-}\underset{OH}{CH}{-}\underset{OH}{\overset{CH_3}{C}}{-}\underset{OH}{CH_2} \xrightarrow{HIO_4}$

(4) cyclohexyl$-\overset{O}{\overset{\|}{C}}CH_3 \xrightarrow{RCO_3H}$

(5) $C_2H_5O\overset{O}{\overset{\|}{C}}CH_2CH_2\overset{O}{\overset{\|}{C}}H \xrightarrow{NaBH_4}$

(6) $\begin{array}{l}CH_2(CH_2)_2CO_2C_2H_5\\ |\\ CH_2(CH_2)_2CO_2C_2H_5\end{array} \xrightarrow[2)\,H^+]{1)\,Na}$

(7) cyclodecanone (O) $+ H_2N{-}NH_2 + NaOH \longrightarrow$

(8) (Fischer projection: $COOC_2H_5$; H—OH; HO—H; $COOC_2H_5$) $\xrightarrow{Pb(OAc)_4}$

(9) $\underset{H}{\overset{H_3C}{}}\!\!>C{=}C<\!\!\underset{H}{\overset{CH_3}{}}$ $\xrightarrow[\text{dil. }OH^-]{KMnO_4/H_2O}$

(10) $\underset{H}{\overset{H_3C}{}}\!\!>C{=}C<\!\!\underset{CH_3}{\overset{H}{}}$ $\xrightarrow[\text{dil. }OH^-]{KMnO_4/H_2O}$

(11) $PhCH_2\underset{OH}{\underset{|}{C}H}CH_2Ph$ + cyclohexanone (O) (excess) $\xrightarrow{Al(OBu\text{-}t)_3}$

(12) Ph–C(=O)–Ph + $CH_3CH(OH)CH_3$ (excess) $\xrightarrow{Al(OPr\text{-}i)_3}$

Vocabulary

oxidation [ˌɒksɪ'deɪʃn] 氧化
reduction [rɪ'dʌkʃn] 还原
transfer [træns'fɜ:(r)] 使转移
electronegative [ɪˌlektrəʊ'negətɪv] 电负性的
oxidant ['ɒksɪdənt] 氧化剂
oxirane [ɒksɪ'rɑ:n] 环氧乙烷
epoxidation [epɒksɪ'deɪʃən] 环氧化
ozone ['əʊzəʊn] 臭氧
ozonide ['əʊzəʊnaɪd] 臭氧化物
electron-withdrawing substituent 吸电子基
electron-donating substituent 给电子基
potassium permanganate 高锰酸钾
chromic acid 铬酸
oxalyl chloride 草酰氯
vicinal ['vɪsɪnl] 邻近的
osmium tetroxide 四氧化锇
saturated ketone 饱和酮
stereoselectivity [stɪərɪəsɪlek'tɪvɪtɪ] 立体选择性
Fischer projection 费歇尔投影式
steric hindrance 立体位阻
alkali metal 碱金属
liquid ammonia 液氨
protonate ['prəʊtəneɪt] 使质子化
non-conjugated 非共轭的
delocalize [di:'ləʊkəlaɪz] 使离域
irreversible [ˌɪrɪ'vɜ:səbl] 不可逆的
dissociate [dɪ'səʊʃɪeɪt] 使分离
diborane [daɪ'bɒreɪn] 乙硼烷
tautomerization [tɔ:təməraɪ'zeɪʃən] 互变
hydrazine ['haɪdrəzi:n] 肼
hydrazone ['haɪdrəzəʊn] 腙
disproportionation reaction 歧化反应
benzilic acid 二苯乙醇酸

Chapter 10

Pericyclic Reactions (周环反应)

In a pericyclic reaction, there is concerted bond reorganization and the essential bonding changes occur within a cyclic array of the participating atomic centers. Different from ionic or free radical reactions in a number of respects, *concerted* reactions have three characteristics:

(1) They are relatively unaffected by solvent changes, the presence of radical initiators or scavenging reagents, or (with some exceptions) by electrophilic or nucleophilic catalysts.

(2) They proceed by a simultaneous (concerted) series of bond breaking and bond making events in a single kinetic step, often with high stereospecificity.

(3) These reactions occur without any intermediates including ionic, free radical or other discernible intermediates.

Since such reactions often proceed by nearly simultaneous reorganization of bonding electron pairs by way of *cyclic transition states*, they have been termed as *pericyclic reactions*. We will discuss three principal categories of concerted pericyclic reactions including *electrocyclic reaction*s, cycloaddition reactions, and *sigmatropic rearrangement*s. The common feature is a concerted mechanism involving a cyclic transition state with continuous electronic reorganization. All these reactions are potentially reversible. The reverse of a cycloaddition is called *cycloreversion* and proceeds by a ring cleavage and conversion of two σ-bonds to two π-bonds. The electrocyclic reaction is a ring forming process. The reverse electerocyclic ring opening reaction proceeds by converting a σ-bond to a π-bond. Sigmatropic bond shifts may involve a simple *migrating group*, or may take place between two π-electron systems (e.g. the Cope rearrangement). Pericyclic reactions can however, be influenced only thermally or photochemically. The fundamental aspects of these reactions can be analyzed in terms of orbital symmetry characteristics associated with the transition state.

Although some pericyclic reactions occur spontaneously, most of them require the introduction of energy in the form of heat or light, with the stereoselective structural change of a product dependence on the source of energy used.

周环反应：在加热或光照下，反应物通过一个环状过渡态，原有化学键断裂和新的化学键产生协同完成。

反应特征：①旧键的断裂和新键的形成一步完成；②反应不受溶剂极性、自由基引发剂或抑制剂、酸碱性、催化剂的影响，只需在光照或加热下就能发生，但两种条件下的结果正好相反；③反应有高度的立体专一性。

周环反应包括电环化反应、环加成反应、σ-迁移反应。

10.1 Electrocyclic ring opening/closure reactions（电环化开环/关环反应）

An electrocyclic reaction is the formation of a σ-bond between the *terminal* atoms of a linear conjugated system of π electrons. The reverse reaction may be called electrocyclic ring opening. The number of curved arrows that describe the bond reorganization is half the total number of electrons involved in the process.

In the fist case, thermal electrocyclic opening of cyclobutenes to conjugated butadienes is displayed. *cis*-3,4-Dimethylcyclobutene is converted to *E,Z*-2,4-hexadiene with heat, whereas *trans*-3,4-dimethylcyclobutene yields the *E, E*-isomer. The level of sterospecificity is very high. And the groups bonded to the breaking bond rotate in the same sense during the ring-opening process. Such motion, in which all the substituents rotate either clockwise or counterclockwise, is called the *conrotatory* mode. This mode of reaction is favored by relief of the ring strain, and the reverse ring closure is not normally observed. *The principle of microscopic reversibility* indicates that the ring closure of 1,3-*diene*s to a cyclobutene is also *conrotatory.* Photochemical ring closure can be happened, but the stereospecificity is opposite to that of thermal ring closure.

In the second case, *trans,cis,trans*-2,4,6-octatriene undergoes thermal ring closure to *cis*-5,6-dimethyl-1,3-cyclohexadiene. The sterospecificity of this reaction is demonstrated by closure of the isomeric *trans,cis,cis*-triene to *trans*-5,6-dimethyl-1,3-cyclohexadiene. The ring closure is normally the favored direction of reaction for conjugated trienes because of the greater thermodynamic stability of the cyclic compound. The closure of *Z*-1,3,5-hexatriene to cyclohexa-1,3-diene is exothermic by 16.4 kcal/mol. The ring closure reactions exhibit a high degree of sterospecificity. The groups at the termini of the triene system rotate in the opposite sense during the cyclization process, a mode of electrocyclic reaction known as *disrotatory. The principle of microscopic reversibility* indicates that the ring opening of a 1,3-cyclohexadiene to a 1,3,5-trienes is also *disrotatory.*

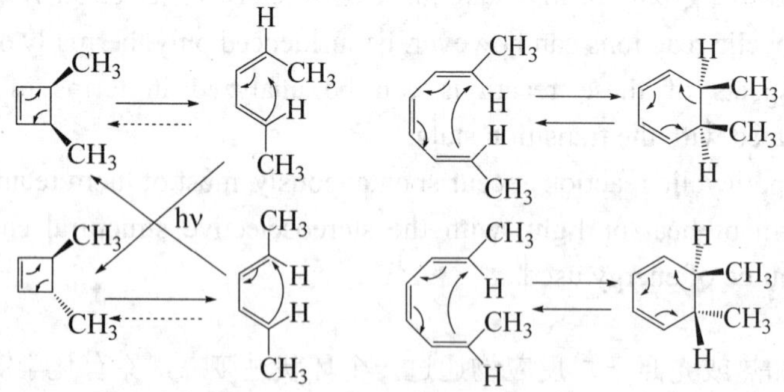

Woodward and Hoffmann proposed that the stereochemistry of the eletrocyclic reactions is controlled by the symmetry properties of the highest occupied molecular orbital (HOMO) of the reacting system, which is an example of *frontier molecular orbital theory (*FMO*)*. The stereochemical features of the reaction are the same in both the forward or reverse directions. For 1,3-dienes, the HOMO is ψ_2 for bonding to form between the terminal carbon atoms, C1 and C4. The positive *lobe* on C1 must overlap with the positive lobe on C4. This overlap of lobes of the same sign can be achieved only by a conrotatory motion. However, disrotatory motion results in overlap of orbitals of opposite sign, and leads to an antibonding overlap that is not allowed to form a bond. So the conrotatory process is preferred for all thermal eletrocyclic reactions of 1,3-dienes (Fig 10.1).

Under photochemical reaction conditions, For 1,3-dienes, the HOMO is ψ_3 for bonding to form between the terminal carbon atoms, C1 and C4. The positive lobe on C1 must overlap with the positive lobe on C4. This overlap of lobes of the same sign can be achieved only by a disrotatory motion. Conrotatory motion results in overlap of orbitals of same sign, and leads to an antibonding overlap that is not allowed to form a bond. So the disrotatory process is preferred for all photochemical eletrocyclic reactions of 1,3-dienes (Fig 10.1).

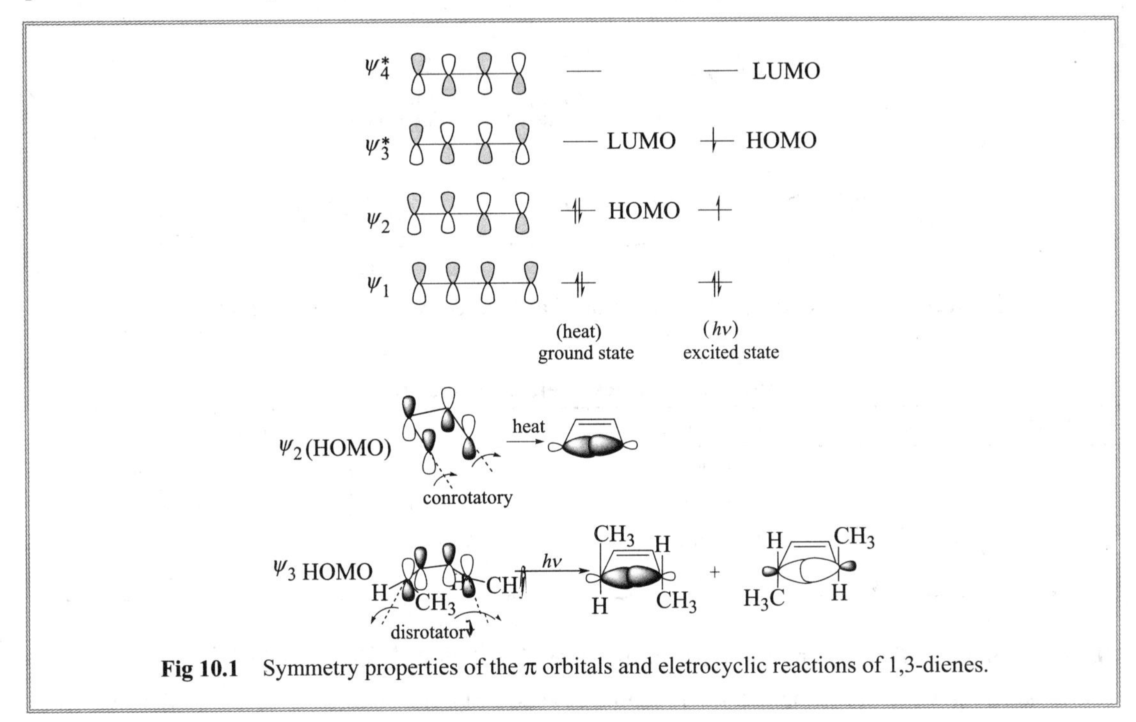

Fig 10.1 Symmetry properties of the π orbitals and eletrocyclic reactions of 1,3-dienes.

For 1,3,5-trienes, the HOMO is ψ_3, which has positive lobes on the same face of the π system. The overlap of lobes of the same sign can be achieved only by a disrotatory motion to form a bond. So disrotatory ring closure (or opening) is general for 1,3,5-trienes, Under photochemical reaction conditions, For 1,3,5-trienes, the HOMO is ψ_4 for bonding to form between the terminal carbon atoms. The overlap of lobes of the same sign can be achieved only by a conrotatory motion to form a bond (Fig 10. 2).

Based on the symmetry patterns of *polyene* HOMO, we can see that systems with $4n\pi$ electrons will undergo electrocyclic reactions by conrotatory motion under thermal reaction conditions, whereas systems with $4n+2$ π electrons will react by the disrotatory motion, as the HOMOs of the $4n\pi$ systems are like those of 1,3-dienes in having opposite phases at the terminal atoms, and the HOMOs of other $4n+2$ π systems are like those of 1,3,5-trienes in having the same phases at the terminal atoms. Rules of electrocyclization under photochemical reaction conditions are opposite to those under thermal reaction conditions. Rules of electrocyclization were shown in Table 10.1.

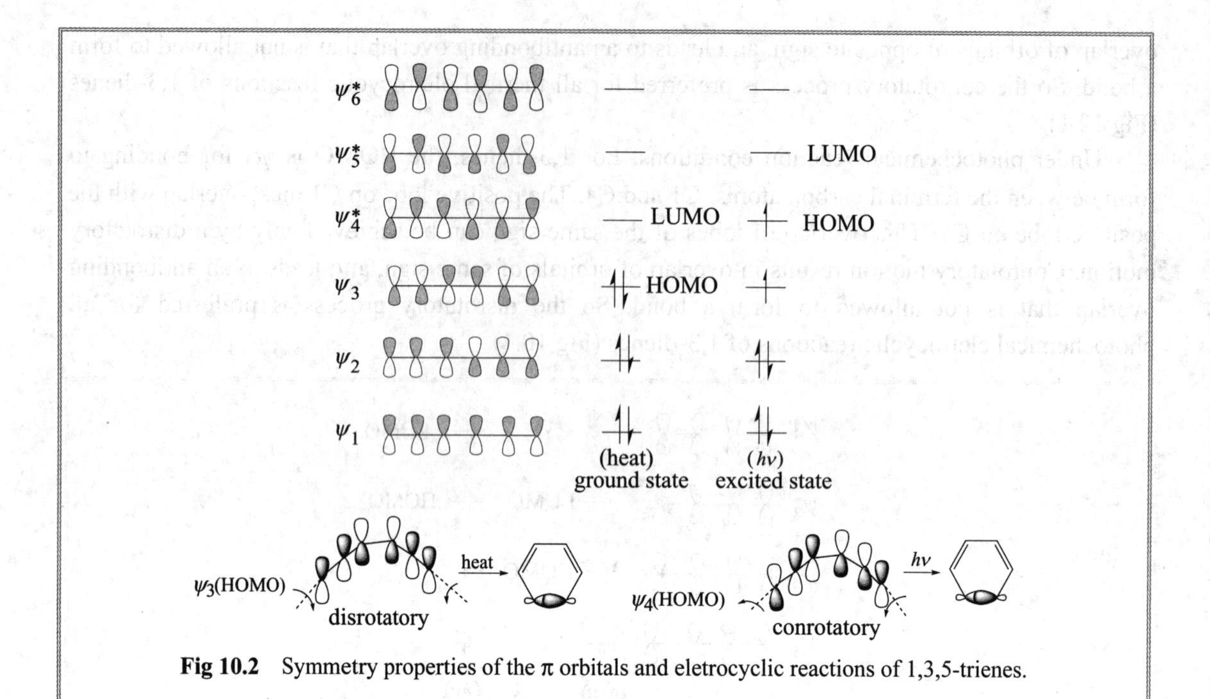

Fig 10.2 Symmetry properties of the π orbitals and eletrocyclic reactions of 1,3,5-trienes.

Table 10.1 Rules of electrocyclization

Number of π electrons	Rotatory mode	Thermal reactions	Photochemical reactions
4*n*	Conrotation	Allowed	Forbidden
	Disrotation	Forbidden	Allowed
4*n*+2	Disrotation	Allowed	Forbidden
	Conrotation	Forbidden	Allowed

10.1.1 Electrocyclic ring opening of cyclobutenes（环丁烯的电环化开环反应）

In keeping with the principle microscopic reversibility the reverse process of thermal ring opening takes exactly the same path as that of thermal ring opening. Due to conrotatory motion, a σ bond will open so as to give the resulting p orbitals which will have the symmetry of the highest occupied π orbital of the product. Since in the case of cyclobutenes the HOMO of the product (i.e., a butadiene) in the thermal reaction is ψ_2, the cyclobutene must open so that on one side the positive lobe lies above the plane, while on the other side it is below it. This process also forces the stereochemistry in the product from substituted cyclobutenes.

Cl, Cl —outward conrotation→ Cl, Cl   Cl, Cl —conrotation→ Cl, Cl

C_5H_{11}, CHO → OHC, C_5H_{11} CH_2OCH_2Ar, CHO → OHC, CH_2OCH_2Ar

10.1.2 Electrocyclization of 1,3,5-hexatrienes (1,3,5-己三烯的电环化反应)

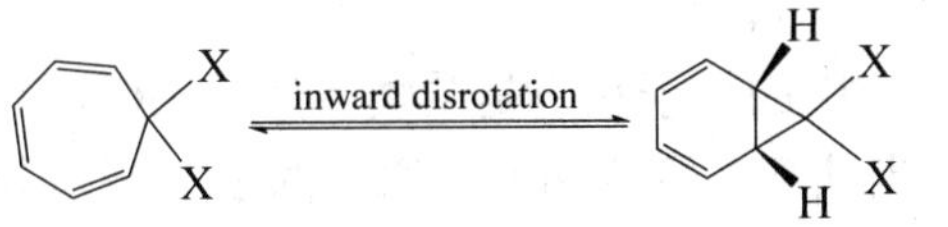

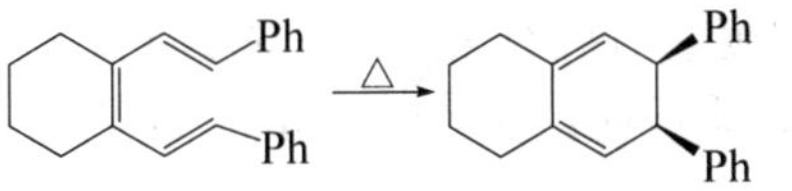

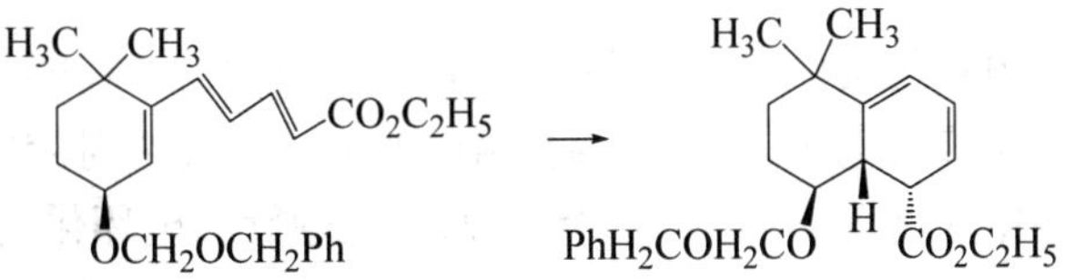

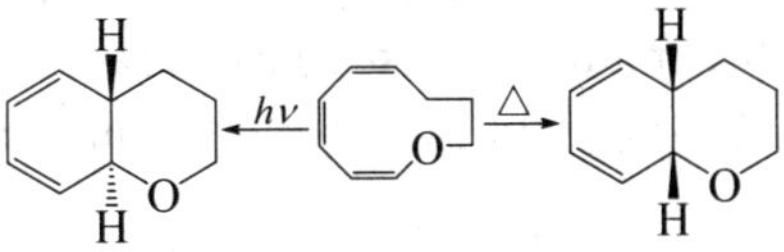

X X inward disrotation H X H X

Ph Ph $\triangle$ Ph Ph

H_3C CH_3 $CO_2C_2H_5$ OCH_2OCH_2Ph → H_3C CH_3 PhH_2COH_2CO H $CO_2C_2H_5$

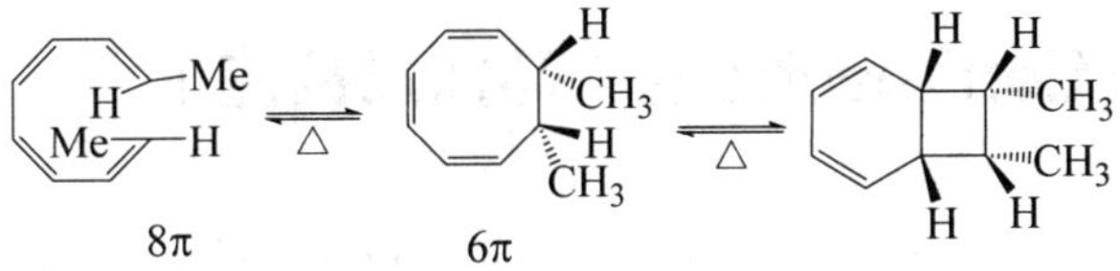

H Me Me H $\triangle$ H CH_3 H CH_3 $\triangle$ H H CH_3 CH_3 H H

8π 6π

H O H hν O $\triangle$ H O H

10.1.3 Electrocyclic reactions of charged species （带电荷物质的电环化反应）

Just as the Hückel aromaticity rule can be applied to charged ring systems, the Woodward-Hoffmann orbital symmetry rules also applied to charged systems. The conversion of cyclopropyl cation to an *allyl cation* involes only two π electrons. Pentadienyl cations can undergo electrocyclization to cyclopentenyl cations, occurring by conrotation because this is a four π-electron system.

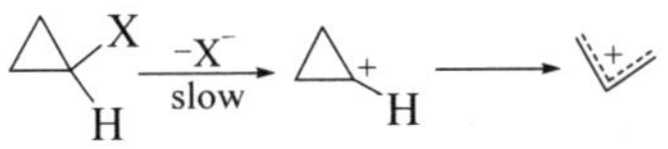

X H $-X^-$ slow + H → + → +

Protonation of *divinyl ketones* give 3-hydroxy-1,4-pentadienyl cations, which proceed through *ketonization* and deprotonation to give the final product.

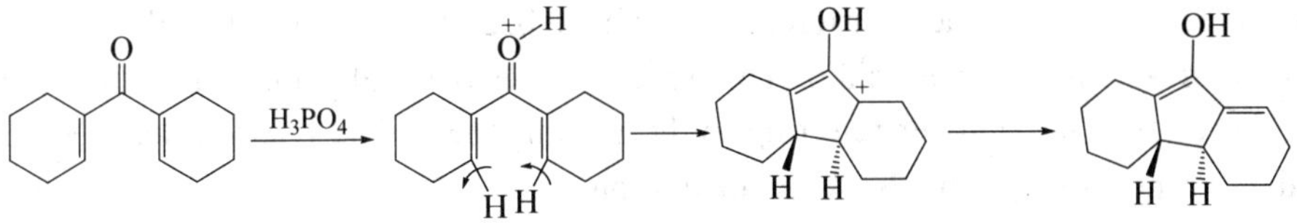

O H_3PO_4 $\overset{+}{O}$—H H H → OH + H H → OH H H

10.1.4 Electrocyclization of *heteroatomic* trienes（杂芳三烯的电环化反应）

When heteroatoms are incorporated into diene, triene, or polyene systems, Electrocyclization can also occur. Electrocyclization of 1-*azatriene* and 1-*oxatriene*s can lead to dihydropyridines and pyrans.

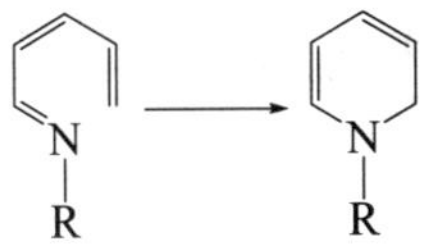

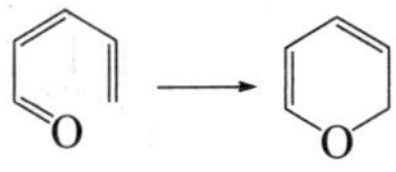

N R → N R O → O

电环化反应指的是共轭多烯末端两个碳原子的 π 电子环合成一个 σ 键，形成比原来分子少一个双键的环烯的反应及其逆反应。

电环化反应中，起决定作用的分子轨道是共轭多烯的 HOMO，反应的立体选择规则主要取决于 HOMO 的对称性。当共轭多烯两端的碳原子的 p 轨道旋转关环生成 σ 键时，必须发生同位相的重叠。

电环化反应产物的立体选择性与电环化反应过程的顺旋和对旋有关：加热时，$4n\pi$ 体系发生顺旋的电环化反应，而（$4n+2$）π 体系发生对旋的电环化反应；光照时，$4n\pi$ 体系发生对旋的电环化反应，（$4n+2$）π 体系发生顺旋的电环化反应。

10.2 Cycloaddition reactions（环加成反应）

A concerted combination of two π-electron systems to form a ring of atoms having two new σ bonds and two fewer π bonds is called a cycloaddition reaction. The number of participating π-electrons in each component is given in brackets preceding the name, and the reorganization of electrons may be depicted by a cycle of curved arrows, each representing the movement of a pair of electrons.

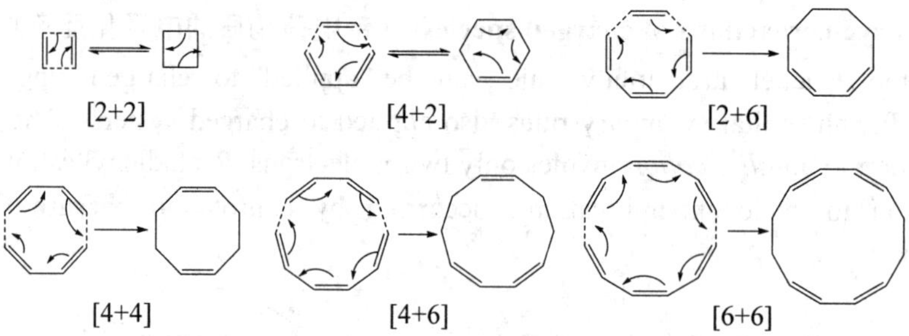

[2+2] [4+2] [2+6]

[4+4] [4+6] [6+6]

Since in a typical cycloaddition reaction, there is addition of two systems containing double bonds, it is logical to expect the addition to occur on the same or the opposite side of the system. Furthermore, as both the π systems are undergoing addition, it is necessary to specify these modes of addition on each of them. These different modes have been termed *suprafacial* (on the same side) and *antarafacial* (on the opposite side). This specification is usually made by placing a suitable subscript (s or a) after the number referring to the π component. The Diels-Alder reaction may be considered as a process involving 2s+4s cycloaddition.

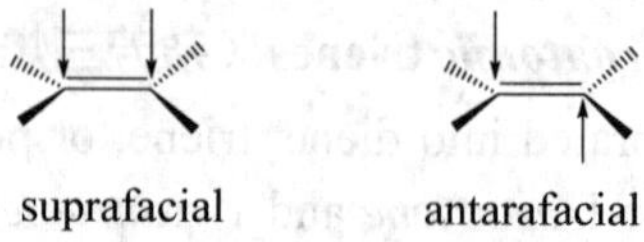

suprafacial antarafacial

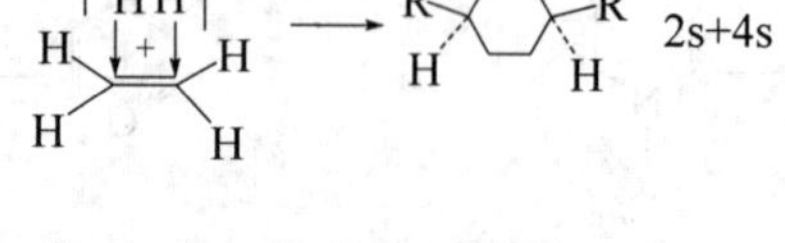

2s+4s

Woodward and Hoffmann formulated the *orbital symmetry principles* for cycloaddition reactions in terms of the frontier orbitals. This frontier-orbital approach is based on the assumption that bonds are formed by a flow of electrons from the highest occupied molecular orbital (HOMO) of one reactant or participating bond to the lowest unoccupied molecular orbital (LUMO) of another reactant or bond. An energetically accessible transition state requires overlap of the frontier orbitals to permit smooth formation of σ bonds. The [2+2] cycloaddition requires the HOMO of one of the ethylenes to overlap with the LUMO of the other. An antibonding interaction results from this overlap, raising the activation energy. For a cyclobutane molecule to result, one of the MOs would have to change its symmetry. Orbital symmetry would not be conserved, so the reaction is symmetry-forbidden. However, the [2+2] cycloaddition of two ethylenes is photochemically "allowed". When a photon with the suitable energy strikes ethylene, one of the π electrons is excited to the next higher molecular orbital. This higher molecular orbital, formerly the LUMO, is now occupied: it is the new HOMO of the excited molecule. The HOMO of the excited ethylene molecule has the same symmetry as the LUMO of the ground-state ethylene. An excited molecule can react with a ground-state molecule to give cyclobutane. The [2+2] cycloaddition is therefore photochemically allowed but thermally forbidden. In most cases, photochemically allowed reactions are thermally forbidden, and thermally allowed reactions are photochemically forbidden.

The requirement for bonding interactions between the HOMO and LUMO are met for [2+4] under thermal reaction conditions but not for [2+2] or [4+4] systems. The HOMO of butadiene has the correct symmetry to overlap in phase with the LUMO of the ethylene, or the LUMO of butadiene has the correct symmetry to overlap in phase with the HOMO of the ethylene. These bonding interactions stabilize the transition state and promote the concerted reaction. This favorable result shows the reaction is symmetry-allowed.

More generally, systems involving $4n+2$ π electrons are favorable (allowed), whereas systems with $4n\pi$ electrons are unfavorable under thermal reaction conditions. On the contrary, systems involving $4n\pi$ electrons are favorable (allowed), whereas systems with $(4n+2)\pi$ electrons are unfavorable under photochemical reaction conditions (Fig 10.3).

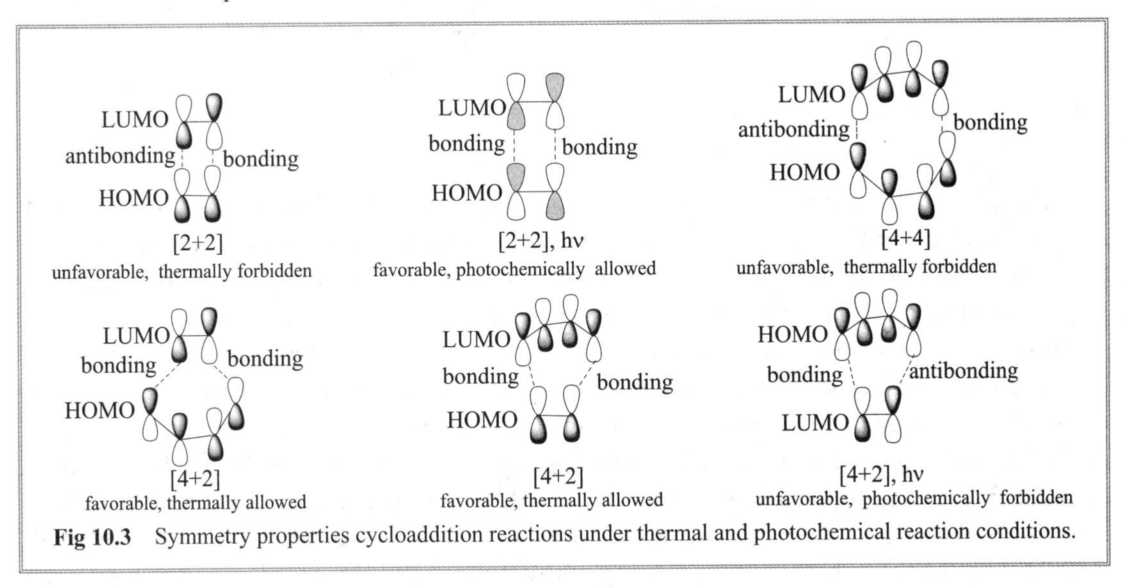

Fig 10.3 Symmetry properties cycloaddition reactions under thermal and photochemical reaction conditions.

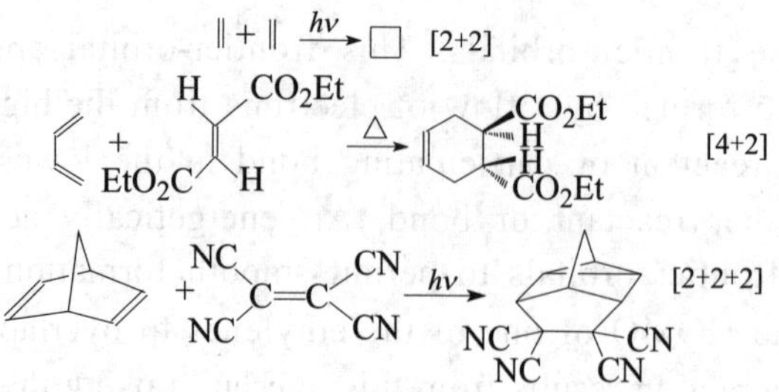

10.2.1 Diels-Alder Reactions （Diels-Alder 反应）

The most common cycloaddition reaction is the [4+2] cyclization known as the Diels-Alder reaction. The two reactants are referred to as the diene and the *dienophile*. The transition structure for a concerted reaction requires that the diene adopt *s-cis* conformation. The diene and dienophile approach each other in approximately parallel planes. For a substituted dienophile, there are two possible stereochemical orientations with respect to the diene. In the *endo* transition state, the reference substitutent on the dienophile is oriented toward the π orbitals of the diene. In the *exo* transition state, the substituent is oriented away from the π system. For many substituted butadiene derivatives, the two transition states lead to two different *stereoisomeric* products. The *endo* mode of addition is usually preferred when an electron withdrawing group is present on the dienophile, even though the *endo* product is the sterically more congested product. This is the *Alder rule*.

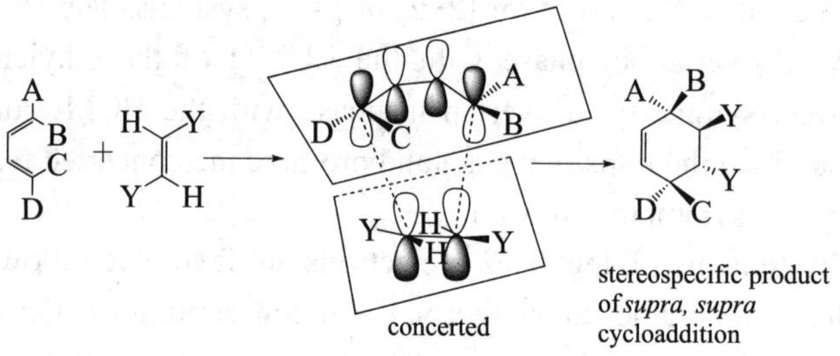

concerted

stereospecific product of *supra, supra* cycloaddition

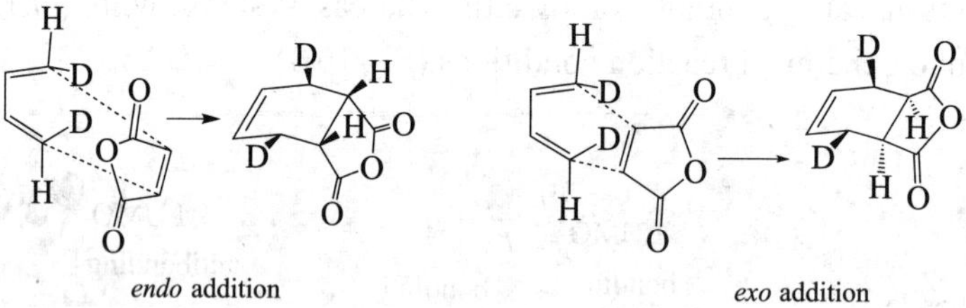

endo addition

exo addition

The following diagram shows examples of [4+2] cycloaddition, and in the second equation a subsequent light induced [2+2] cycloaddition. The stereospecificity of these reactions should be evident. In the first example, the *acetoxy substituents* on the diene have identical *E*-configurations, and they remain *cis* to each other in the cyclic adduct. Likewise, the ester substituents on the dienophile have a *trans*-configuration which is maintained in the adduct. The reactants in the second equation are both monocyclic, so the cycloaddition adduct has three rings. The orientation of the quinone six-membered ring with respect to the bicycloheptane system is *endo*, since the dienophile (quinone) has two activated double bonds, a second cycloaddition reaction is possible, provided sufficient diene is supplied. The second cycloaddition is slower than the first, so the monoadduct shown here is easily prepared in good yield. Although this [4+2] product is stable to further heating, it undergoes a [2+2] cycloaddition when exposed to sunlight.

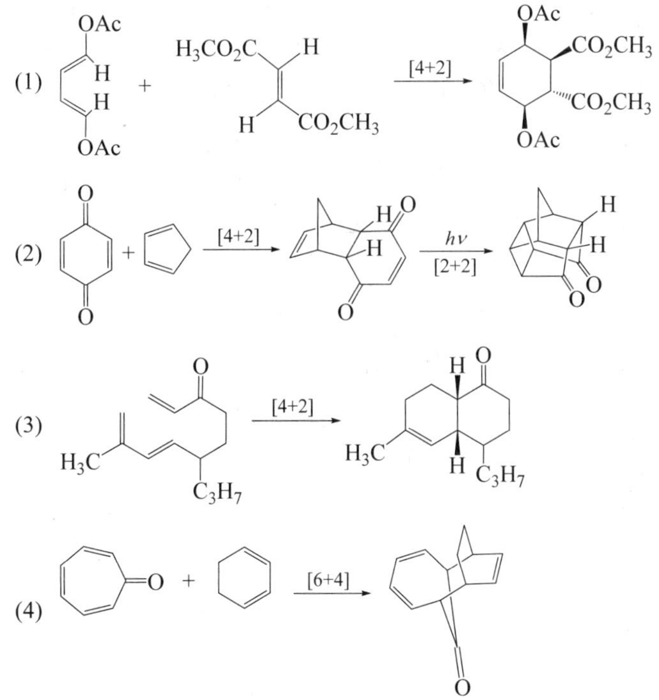

Reaction (3) is an intramolecular Diels-Alder reaction. Since the diene and dienophile are joined by a chain of atoms, the resulting [4+2] cycloaddition actually forms two new rings, one from the cycloaddition and the other from the linking chain. Once again the addition is stereospecific, ignoring the isopropyl substituent, the ring fusion being *cis* and *endo*. The fourth reaction is a [6+4] cycloaddition.

The diene reacts in the *s-cis* conformation, which allows the ends of the conjugated system to reach the doubly bonded carbons of the dienophile. That the *s-cis* geometry of the diene is essential is shown by the unreactive nature of the fixed transoid dienes (Ⅰ and Ⅱ). Moreover, as expected the substituents in the dienes may also affect the cycloaddition sterically. The substituents affect the equilibrium proportion of the diene in the required cisoid form. Consequently Z alkyl or aryl substituents in the 1-position (Ⅲ) of the diene slow down the reaction by sterically hindering formation of the cisoid conformation, while bulky 2-substituents (Ⅳ) make it fast.

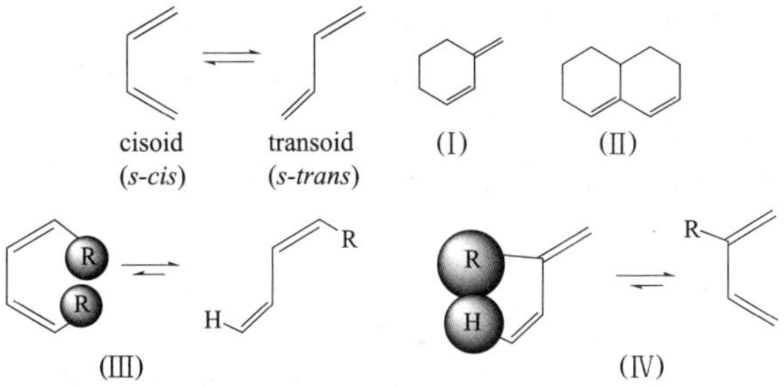

cisoid (*s-cis*)　transoid (*s-trans*)　(Ⅰ)　(Ⅱ)

(Ⅲ)　(Ⅳ)

Electronic substituent affect strongly on the Diels-Alder reaction. The reaction is particularly efficient when the dienophile contains one or more electron withdrawing groups and the diene also contains an electron donating group. The most reactive dienophiles are *quinones*, *maleic anhydrides*, *nitroalkenes*. α,β-Unsaturated ketones, nitriles and esters are also reactive dienophiles. The reactivity relationships can be readily understood in terms of frontier orbital theory. Electron-rich dienes have high-energy HOMOs that interact strongly with the LUMOs of electron-poor dienophiles. When the substituent pattern is reversed, the strongest interaction is between the dienophile HOMO and the diene LUMO, as indicated in Fig 10. 4 and Table 10.2.

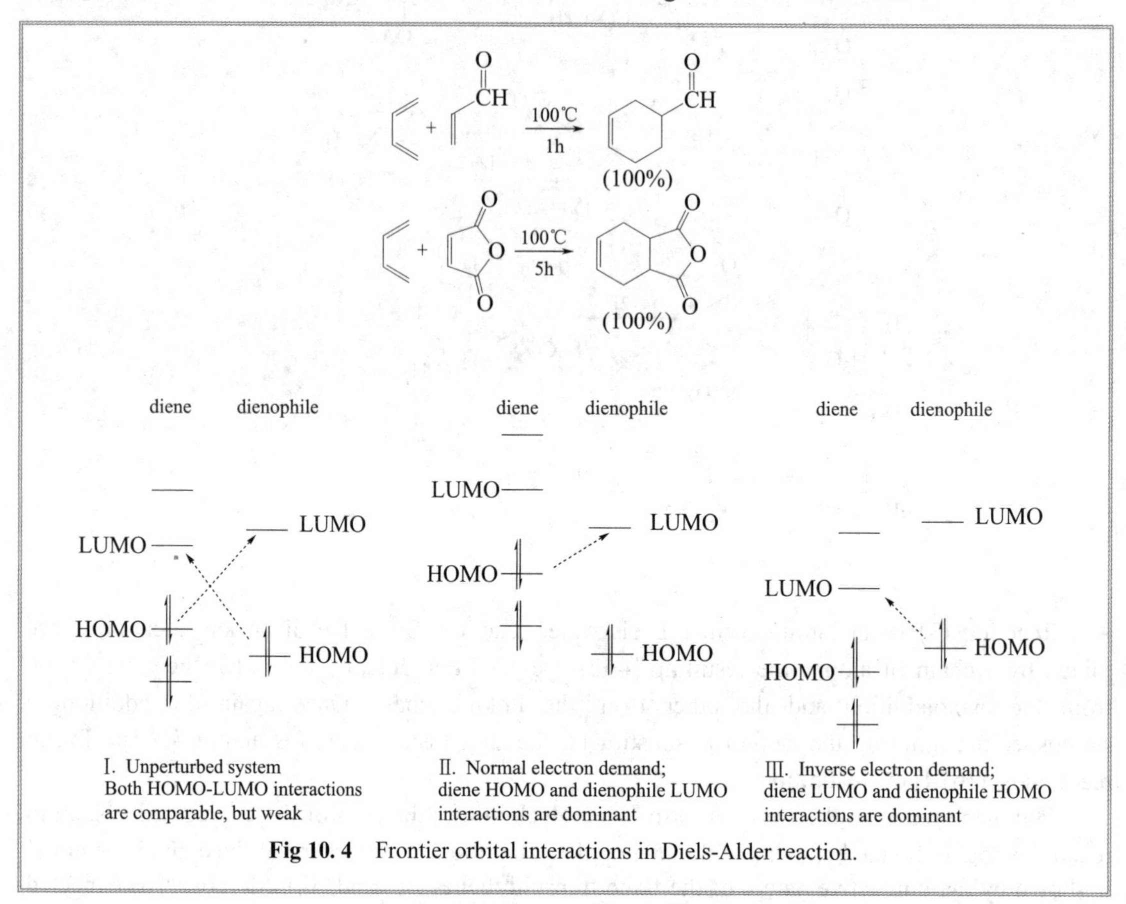

Fig 10. 4 Frontier orbital interactions in Diels-Alder reaction.

Table 10.2 Rules of cycloaddition

Number of π electrons	Thermal reactions	Photochemical reactions
$4n$ (*e.g.* [2+2], [4+4])	Symmetry Forbidden	Symmetry Allowed
$4n+2$ (*e.g.* [4+2],[4+6])	Symmetry Allowed	Symmetry Forbidden

The *regioselectivity* of the Diels-Alder reaction is generalized in Fig 10.5. Wherein we can see that the regioselectivity of the Diels-Alder reaction is determined by the nature and position of the groups on the diene and dienophile.

The essential characteristics of the Diels-Alder cycloaddition reaction may be summarized as follows:

(i) The reaction always creates a new six-membered ring. When intramolecular, another ring may also be formed.

(ii) The diene component must be able to assume a *s-cis* conformation.

(iii) Electron withdrawing groups on the dienophile facilitate reaction.

(iv) Electron donating groups on the diene facilitate reaction.

(v) Steric hindrance at the bonding sites may inhibit or prevent reaction.

(vi) The reaction is stereospecific with respect to substituent configuration in both the dienophile and the diene.

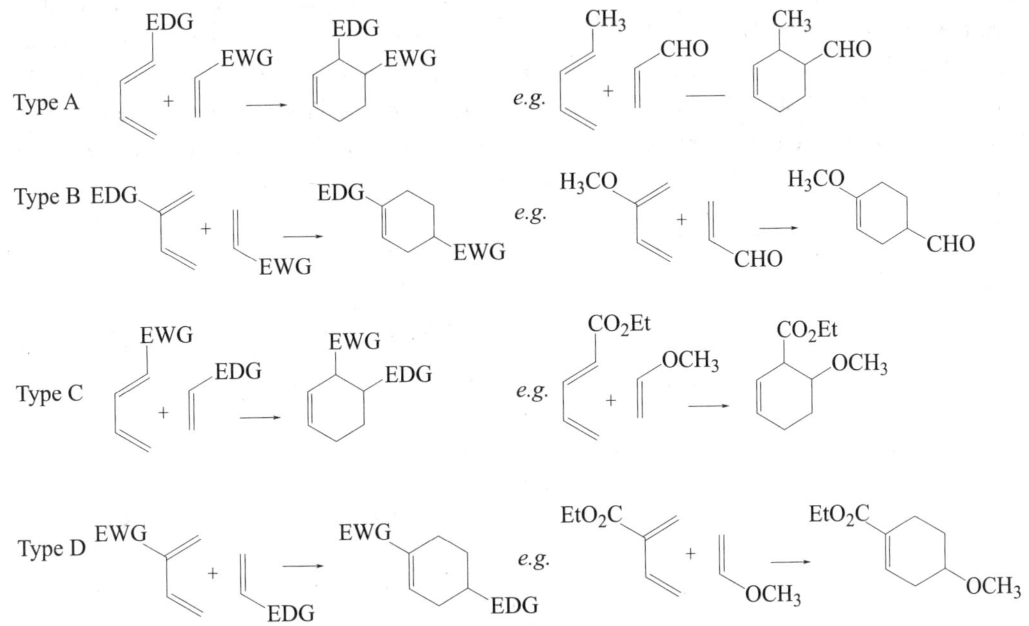

Fig 10.5 The regioselectivity of the Diels-Alder reaction

10.2.2 1,3-Dipolar cycloaddition reactions （1,3-偶极环加成反应）

1,3-Dipolar cycloaddition reactions (1,3-DPCA reactions) are analogous to the Diels-Alder reaction because they are concerted [4+2] cycloadditions. 1,3-Dipolar cycloaddition reactions take place between unsaturated hetero atom compounds, such as diazoalkanes, alkyl and aryl *azide*s, *nitrile oxide*s and nitrones, and alkene or alkyne function group. Although the former reactants are neutral, their Lewis structures have formal charges, and may be written as 1,3-*dipoles*. The 1,3-dipoles have a π-electron system consisting of two filled and one empty orbital and are similar to allyl or propargyl anion. The alkene and alkyne function group to which the dipoles add are called *dipolarophiles*. In general, four resonance *canonical* structures may be written for each dipolar reactant (Fig 10.6). Two have adjacent or 1,2-charge separation, and two have the 1,3-dipolar charge separation. The 1,2-dipolar structures retain valence shell *octet*s for all heavy atoms, suffer less charge separation, and have one more covalent bond than do the 1,3-dipolar structures. Therefore, the most representative Lewis structures for these compounds are 1,2-dipoles, not 1,3-dipoles. Another factor in identifying the best structure for a given compound is electronegativity. Negative charge is best on the most electronegative atom, and positive charge on the least electronegative atom. In the examples drawn for the nitrile oxides, azides and nitrones, the left hand structure is the best 1,2-dipole that can be written. Finally, the general equation (Fig 10.7) demonstrates the danger of thinking about these reactions as a simple addition of a 1,3-dipole to an unsaturated function. Movement of electron pairs out of the dipolarophile to one end of the dipole, with a second electron pair going from the dipole back to the dipolarophile accounts for only four electrons. As shown by the curved arrows in the following reaction, the cycloaddition actually proceeds by a six π-electron transition state, and is suprafacial.

1,3-Dipolar cycloaddition reactions are very useful reactions for the construction of five-membered heterocyclic rings. The bond changes for these reactions involve four π electrons from the 1,3-dipole and

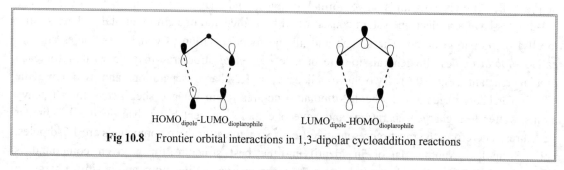

two from the dipolarophile. So these reactions are concerted [4+2] cycloadditions. There is an interaction between the complementary HOMO-LUMO combinations, and depending on the combination, either reactant can be electrophilic or the nucleophilic component. In terms of FMO theory, Sustmann and Trill pointed out the left type should be accelerated by electron donating group in the dipole and electron withdrawing group in the dipolarophile and the right type should be facilitated by an electron withdrawing group in the dipole and an electron donating group in the dipolarophile (Fig 10.8).

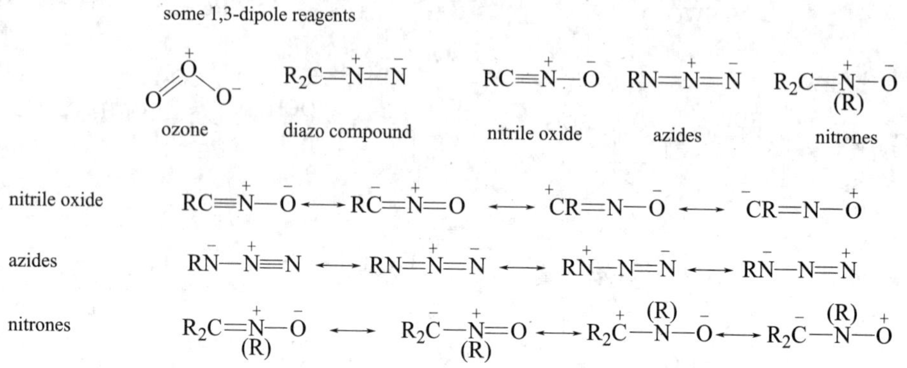

some 1,3-dipole reagents
ozone
diazo compound
nitrile oxide
azides
nitrones
nitrile oxide
azides
nitrones

Fig 10.6 Four resonance canonical structures for some dipolar reactants

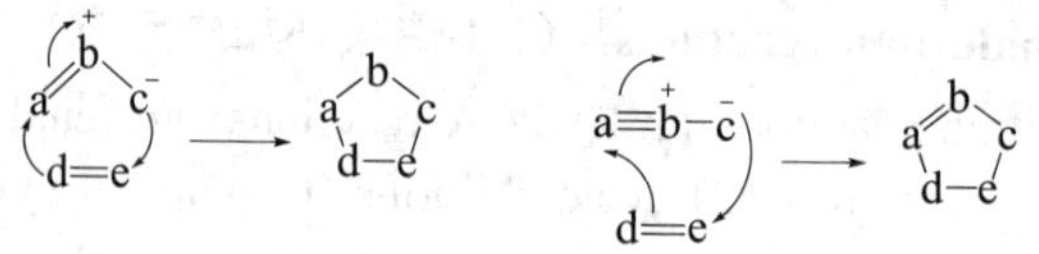

Fig 10.7 1,3-Dipolar cycloaddition reactions

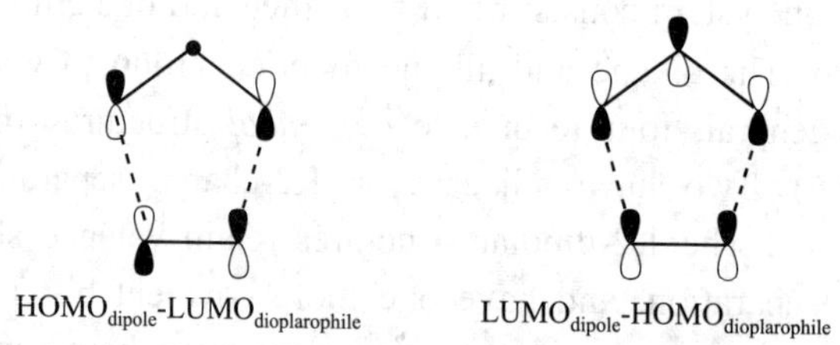

HOMOdipole-LUMOdioplarophile
LUMOdipole-HOMOdioplarophile

Fig 10.8 Frontier orbital interactions in 1,3-dipolar cycloaddition reactions

As with the Diels-Alder reactions, most 1,3-DPCA reactions are highly stereospecific with respect to the dipolarophile.

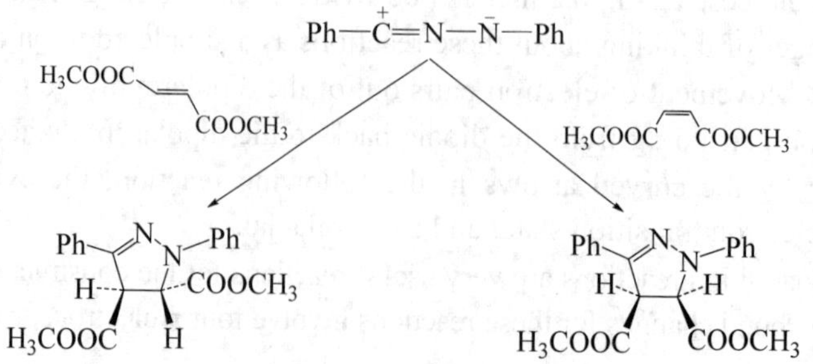

Two possible stereoisomers can be formed if the dipolarophile react with some 1,3-dipoles from two different orientations. For example, substituted diazoalkanes can add to unsymmetrical dipolarophiles to give two *diastereomer*s.

$$Ph-\overset{H}{\underset{-}{C}}-\overset{+}{N}\equiv N + \underset{H_3COOC}{\overset{H}{\diagdown}}C=C\underset{COOCH_3}{\overset{CH_3}{\diagup}} \longrightarrow \text{(Ph, H; N=N; } CH_3\text{; H; } H_3COOC\text{, } COOCH_3\text{)} + \text{(Ph, H; N=N; } CH_3\text{; H; } H_3COOC\text{, } COOCH_3\text{)}$$

环加成反应指的是两个或两个以上的 π 电子体系经由一个环状过渡态得到产物的反应或其逆反应叫环加成反应。

两个分子发生环加成反应时，起决定作用的轨道是一个分子的 HOMO 和另一个分子的 LUMO，轨道必须发生同位相重叠。

光照时，$4n\pi$ 体系环加成反应允许，加热时 $4n\pi$ 体系环加成反应禁阻。光照时，$(4n+2)\pi$ 体系环加成反应禁阻，加热时，$(4n+2)\pi$ 体系环加成反应允许。

Diels-Alder 反应是可逆反应。*s-cis* 型双烯体是发生 Diels-Alder 反应的先决条件。双烯体上有给电子基时亲双烯体上有吸电基时加速反应。遵循最大不饱和度规则：内式加成产物为主；遵循顺式加成规则：取代基的立体关系在反应前后不变；Diels-Alder 反应的区域选择性：以邻、对位加成产物为主。

1,3-偶极加成：反应物为 1,3-偶极分子和亲偶极体，属 $(4n+2)\pi$ 体系环加成反应，与 Diels-Alder 反应相似，用于合成五元杂环化合物。

10.3 Sigmatropic rearrangements（σ重排反应）

Molecular rearrangements in which a σ-bonded atom or group, attached by one or more π-electron systems, shifts to a new location with a corresponding reorganization of the π-bonds are called sigmatropic reactions. The total number of σ-bonds and π-bonds remain unchanged. These rearrangements are described by two numbers set in brackets, which refer to the relative distance (in atoms) each end of the σ-bond has moved, as illustrated by the first equation in the diagram below. The most common atom to undergo sigmatropic shifts is hydrogen or one of its isotopes. The second equation in the diagram shows a facile [1,5] hydrogen shift which converts a relatively unstable allene system into a conjugated triene. Note that this rearrangement, which involves the relocation of three pairs of bonding electrons, may be described by three curved arrows. There are two topologically distinct processes by which a sigmatropic shifts can occur. If the migrating group remains associated with the same face of the conjugated π system throughout the reaction, the migration is termed *suprafacial.* In the alternative mode, called *antarafacial,* the migrating group moves to the opposite face of the conjugated π system throughout the reaction. There is another important element of stereochemistry for the migration of alkyl groups. The migrating group can retain its original configuration (retention) or undergo inversion. The stereochemical features and the number of electrons involved determine whether a reaction is allowed or forbidden.

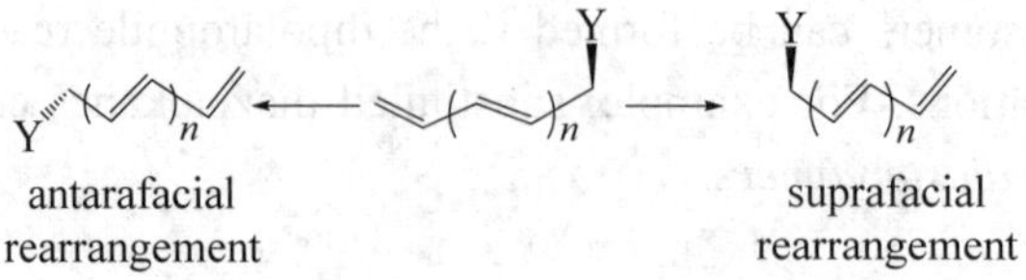

antarafacial rearrangement

suprafacial rearrangement

10.3.1 [1,3]- and [1,5]- sigmatropic shifts of hydrogen and alkyl groups（氢、烷基的[1,3]、[1,5]-σ迁移反应）

The orbital symmetry requirements of sigmatropic reactions can be analyzed by considering the interactions between the frontier orbitals of the π system and those of the migrating fragment. An FMO analysis of 1,3-sigmatropic shift of a hydrogen treats the system as a hydrogen atom interacting with an allyl radical. The frontier orbitals are the H 1s and the allyl ψ_2 orbitals. These interactions are depicted below for both the suprafacial and antarafacial modes. The 1,3-suprafacial shift of hydrogen is forbidden by orbital symmetry considerations. The antarafacial process is symmetry allowed, but it involves a contorted geometry so that this shift is energetically difficult.

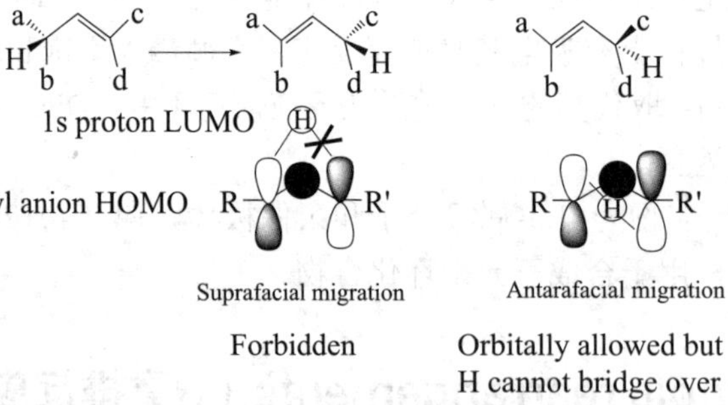

1s proton LUMO

Ψ_2 allyl anion HOMO

Suprafacial migration

Forbidden

Antarafacial migration

Orbitally allowed but H cannot bridge over

The relevant frontier orbitals of 1,5-sigmatropic shift of a hydrogen are the hydrogen 1s and ψ_3 of the pentadienyl radical. The suprafacial mode is symmetry allowed, whereas the antarafacial process is forbidden. The suprafacial shift corresponds to a geometrically favorable six-membered ring.

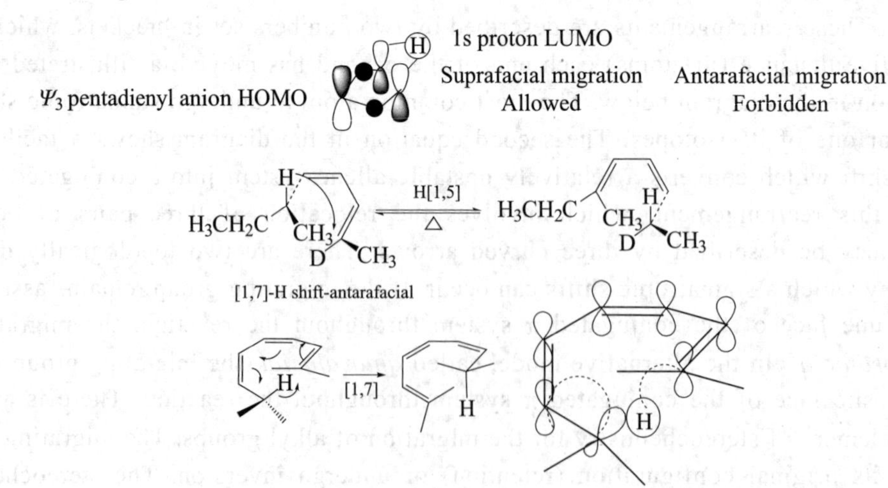

A $[1_s,7_a]$ sigmatropic hydrogen shift may then take place, as described by the four curved arrows. With reference to the approximate plane of this π-electron system, the hydrogen atom departs from the bottom face and bonds to the top face, so the transfer is antarafacial (Table 10.3).

Table 10.3 H [1+*j*] sigmatropic hydrogen shift selection rules

Number of π electrons	Reaction conditions	Selection rules
4*n*+2 [1, 5]	Thermal reactions	Suprafacial allowed
	Photochemical reactions	Antarafacial allowed
4*n* [1, 3], [1,7]	Photochemical reactions	Suprafacial allowed
	Thermal reactions	Antarafacial allowed

Sigmatropic shifts of alkyl groups can also occur. The shift can occur with retention or inversion at the migrating center. The allowed migration includes the suprafacial 1,3-shift with inversion and the suprafacial 1,5-shift with retention.

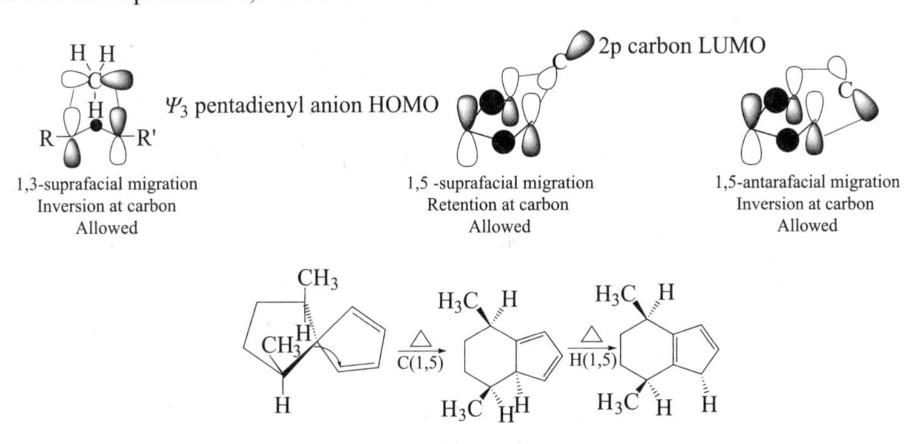

10.3.2 [3, 3]-Sigmatropic rearrangements（[3, 3]-σ重排反应）

The [3,3]-sigmatropic rearrangement of 1,5-dienes or allyl vinyl ethers, known as the Cope and Claisen rearrangements respectively, are among the most commonly used sigmatropic reactions. The transition state for [3,3]-sigmatropic rearrangement can be considered to be two interacting allyl fragments. When the process is suprafacial in both allyl groups (Table 10.4), FMO analysis indicated that both a chairlike and boatlike transition state is possible. But the stereochemistry of the reaction can be predicted and analyzed on the basis of a chair transition state that minimizes steric interactions between substituents. The reaction is both stereospecific and stereoselective. When a chair transition state is favored, the *E, E*- and *Z, Z*- dienes lead to *anti*-3,4-diastereomers, whereas the *E,Z*- and *Z,E*-isomers give the 3,4-*syn* product. The reaction is stereoselective with respect to the configuration of the newly formed double bond. The product ratio reflects product (and transition state) stability. An *E*-arrangement is normally favored for the newly formed double bonds. The stereochemical aspects of the Cope rearrangements for relatively simple reactants are consistent with a chairlike transition state in which the larger substituent at C3 or C4 adopts an equatorial-like conformation.

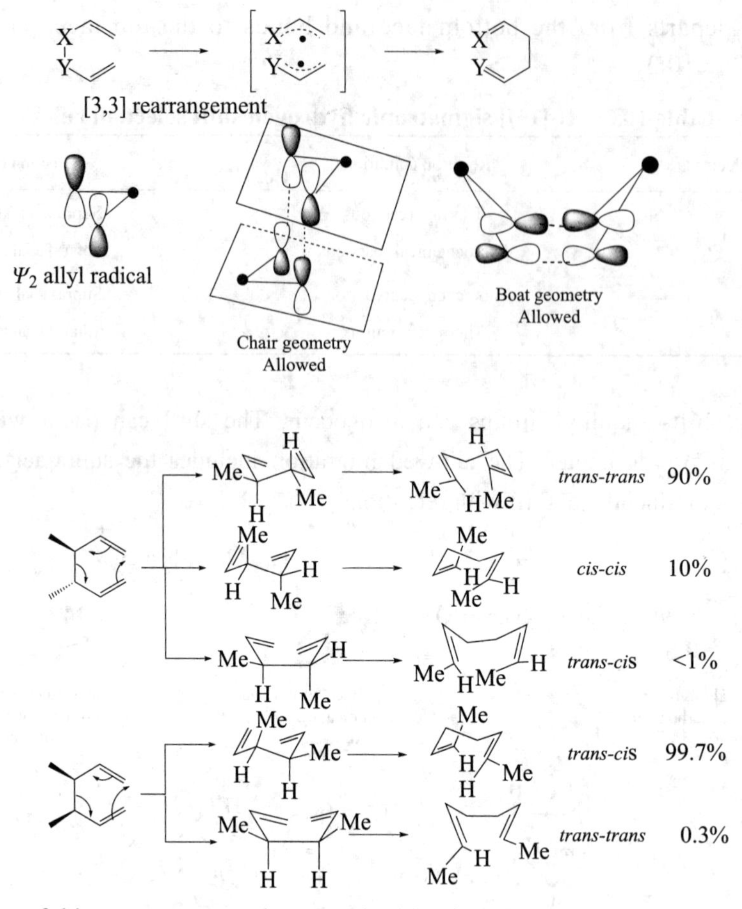

The product of this rearrangement is an enol which immediately *tautomerize*s to its keto form. Such variants are termed as the oxy-Cope rearrangement, and are useful because the reverse rearrangement is blocked by rapid ketonization.

Table 10.4 The orbital symmetry selection rules for [*i*+*j*] sigmatropic processes

Number of π electrons	Supra/Supra	Supra/Antara	Antara/Antara
4*n* [3,5]	Forbidden	Allowed	Forbidden
4*n*+2 [3,3]	Allowed	Forbidden	Allowed

σ键迁移：经过环状过渡态，σ键在分子中发生位置上的变化，同时伴随着双键的位移。

在[1, j]-σ 迁移反应中，起决定作用的分子轨道是奇碳共轭体系中含有单电子的前线轨道。其立体选择规则完全取决于该轨道的对称性。新 σ 键形成时必须发生同位相重叠。

[1，j]-σ 迁移：一个由 σ 键连排的基团通过重排到达 π 共轭体系的末端，同时双键移位。

[1,j]氢迁移: $1+j=4n$，同面迁移热禁阻、光允许，异面迁移热允许、光禁阻；$1+j=4n+2$，同面迁移热允许，异面迁移热禁阻。

[1,j]碳迁移: $1+j=4n$，同面迁移构型翻转；$1+j=4n+2$，同面迁移构型保持。

[3,3]迁移是经过类椅式过渡态按同面-同面迁移的。经过 σ 键的断裂和形成，两端的构型保持不变。典型的反应有 Claisen 重排和 Cope 重排。

Problems

1. Write down the products of the following electrocyclic reactions.

(1) 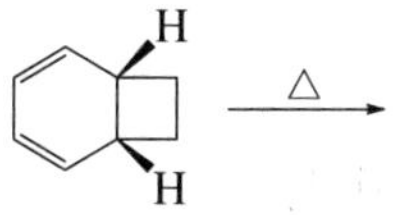 H, H $\xrightarrow{\triangle}$

(2) 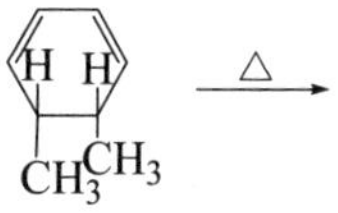 H H, CH_3 CH_3 $\xrightarrow{\triangle}$

(3) H H, CH_3 CH_3 $\xrightarrow{h\nu}$

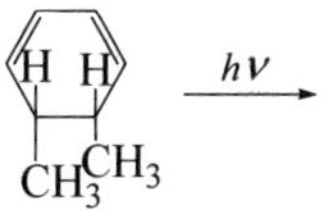

(4) 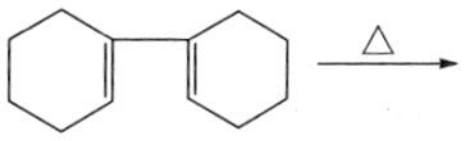 $\xrightarrow{\triangle}$

(5) 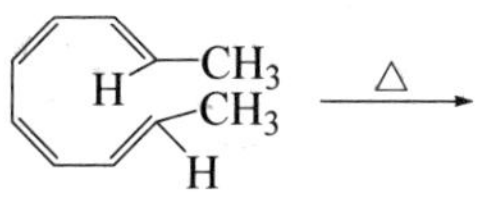 H, CH_3, CH_3, H $\xrightarrow{\triangle}$

(6) 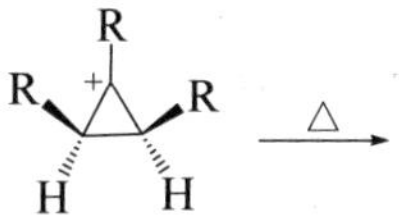 R, R, R, +, H, H $\xrightarrow{\triangle}$

(7) 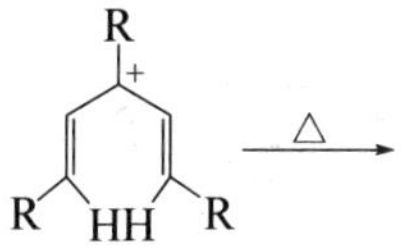 R, +, R, HH, R $\xrightarrow{\triangle}$

(8) 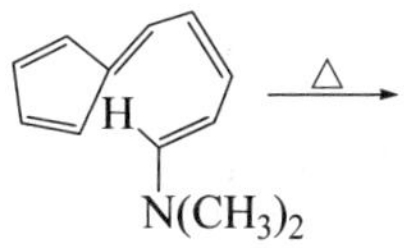 H, $N(CH_3)_2$ $\xrightarrow{\triangle}$

2. Write down the products of the following cycloaddition reactions.

(1) 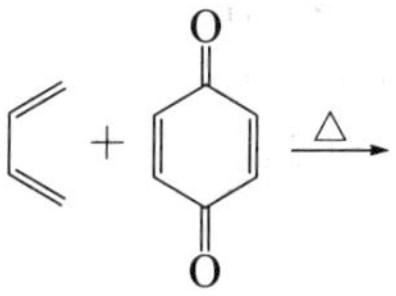 + O, O $\xrightarrow{\triangle}$

(2) 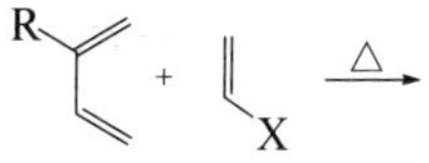 R + X $\xrightarrow{\triangle}$

(3) 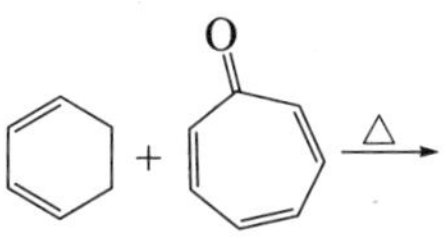 + O $\xrightarrow{\triangle}$

(4) 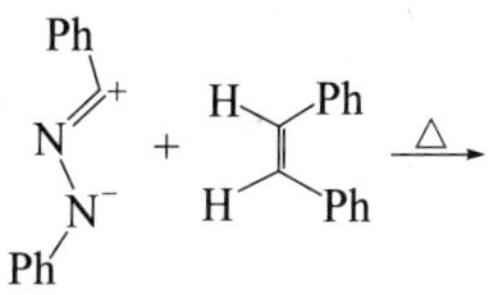 Ph, +, N, N^-, Ph + H, Ph, H, Ph $\xrightarrow{\triangle}$

(5) 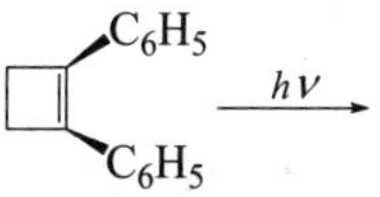 C_6H_5, C_6H_5 $\xrightarrow{h\nu}$

(6) 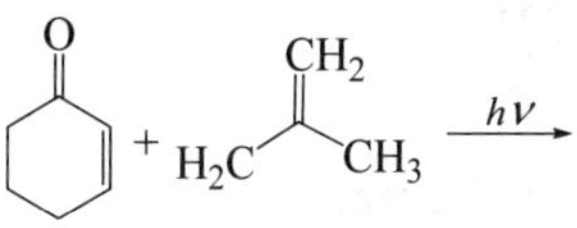 O + CH_2, H_2C, CH_3 $\xrightarrow{h\nu}$

(7) 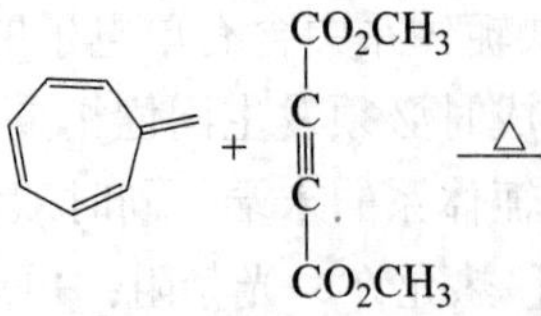 + CO_2CH_3 C ≡ C CO_2CH_3 △ (8) 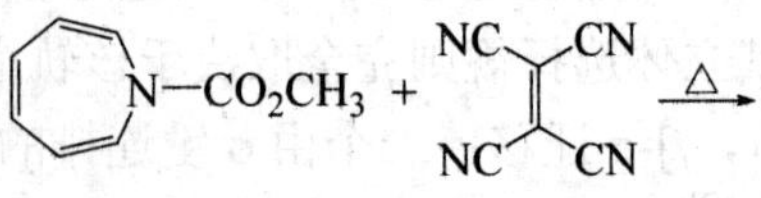 N−CO_2CH_3 + NC CN NC CN △

3. Write down the products of the following reactions upon heating.

(1) 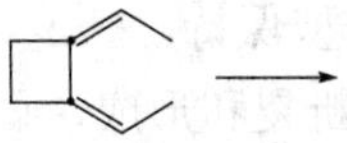 (2) 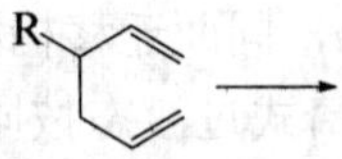 R

(3) 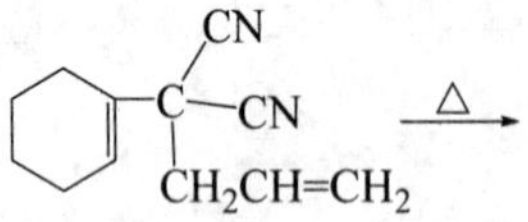 CN CN $CH_2CH{=}CH_2$ △ (4) 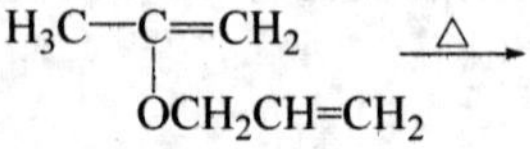 $H_3C-C{=}CH_2$ $OCH_2CH{=}CH_2$ △

(5) 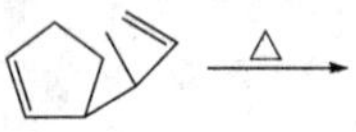 △ (6) 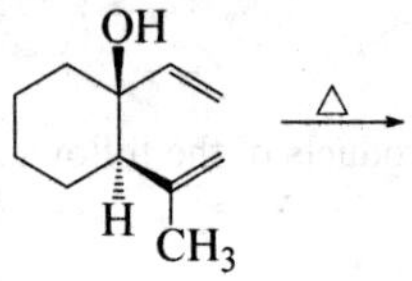 OH H CH_3 △

(7) 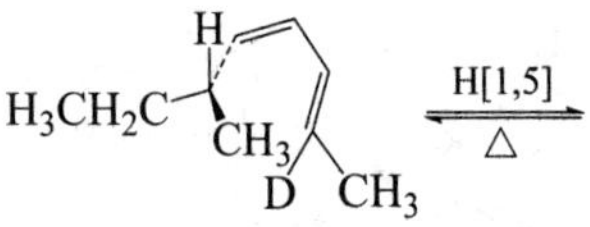 H_3CH_2C H CH_3 D CH_3 H[1,5] △ (8) 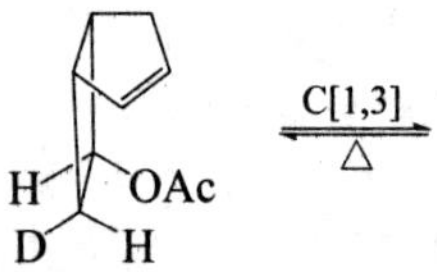 H OAc D H C[1,3] △

4. Give the reaction conditions.

(1) 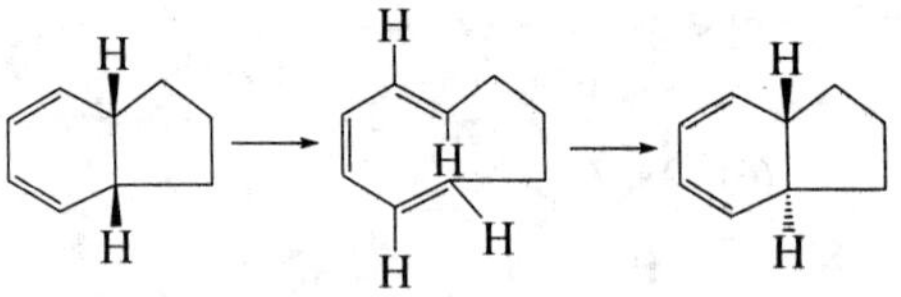 H H H H H H H (2) 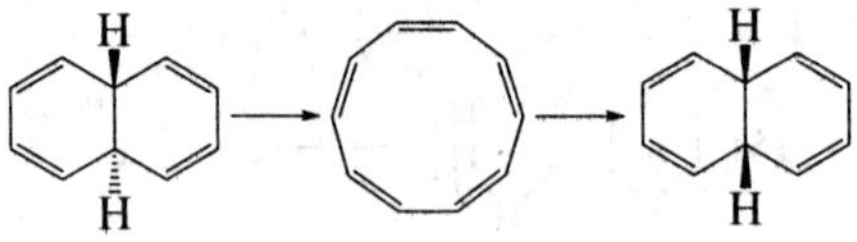 H H H H

(3) 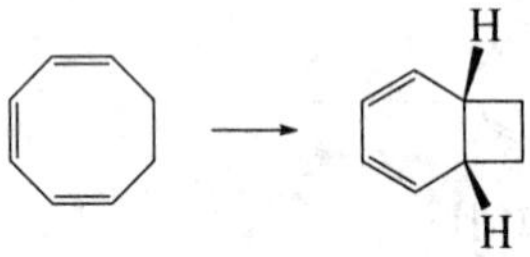 H H (4)

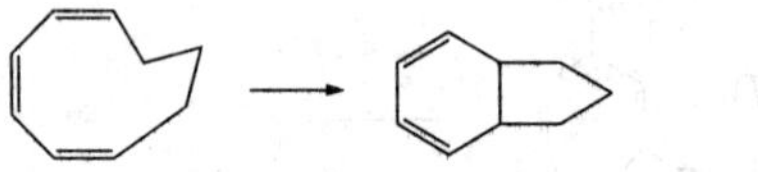

(5) 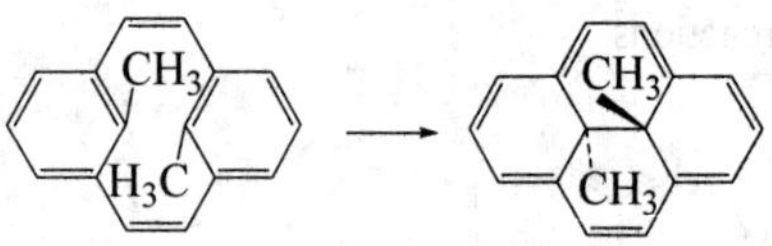 CH_3 H_3C CH_3 CH_3 (6) 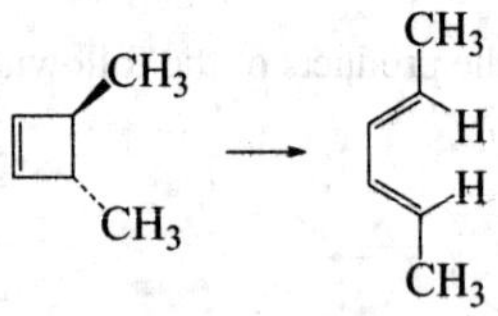 CH_3 CH_3 CH_3 H H CH_3

(7) 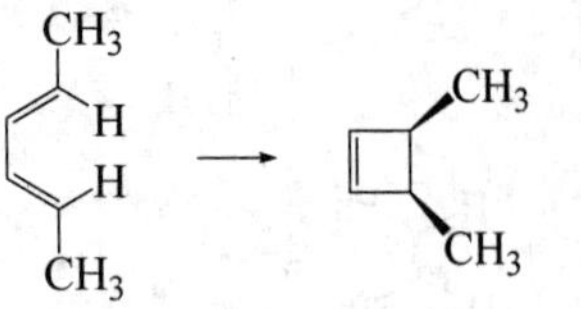 CH_3 H H CH_3 CH_3 CH_3

Vocabulary

concerted [kən'sɜːtɪd] 协同的

cyclic transition state 环状过渡态

pericyclic reaction 周环反应

electrocyclic reaction 电环化反应

cycloaddition reaction 环加成反应

sigmatropic rearrangement σ重排反应

cycloreversion[saɪklərɪ'vɜːʃən] 开（裂）环作用

conrotatory [kənrəʊ'teɪtərɪ] 顺旋

the principle of microscopic reversibility 微观可逆原理
disrotatory [dɪs'rɒtətərɪ] 对旋
frontier molecular orbital theory 前线分子轨道理论
lobe [loʊb] 部分
polyene ['pɒliːˌiːn] 多烯
terminal ['tɜːrmɪnl] 终端
allyl cation 烯丙基阳离子
divinyl ketones 二乙烯基酮
ketonization [kiːtoʊnaɪ'zeɪʃən] 酮化作用
heteroatomic[hetiərəʊ'tɒmɪk] 杂原子的
1-azatriene 1-氮杂三烯
1-oxatriene 1-氧杂三烯
orbital symmetry principles 轨道对称性原理
diene 双烯
dienophile [daɪ'enəfaɪl] 亲双烯体
endo['endəʊ] 内型
exo 外型
stereoisomeric [steriːəʊaɪ'səʊmerɪk] 立体异构的
acetoxy substituent 酰氧取代基
quinones [kwɪ'nəʊnz] 醌类
maleic anhydride 马来酸酐
nitroalkenes 硝基烯烃
regioselectivity [riːdʒiːəʊsɪlek'tɪvɪtɪ] 区域选择性
dipoles 偶极体
dipolarophiles 亲偶极体
canonical [kə'nɑːnɪkl] structures 规范结构式
octet [ɑːk'tet] 八电子
nitrile oxides 氧化腈
azides 叠氮化合物
suprafacial [sjuːprə'feɪʃəl] 同面的
diastereomer [daɪə'stɪərɪoʊmə] 非对映体的
antarafacial [æn'tærɑːfeɪʃl] 异面的
migrating group 迁移基团
tautomerize [tɔː'tɒməraɪz] 发生互变异构现象

Appendix Ⅰ

《Advanced Organic Chemistry》 Final Test(1)

1. Select or fill in blanks (2 points each, 40 points total)

(1) Which one doesn't have aromaticity (　　).

A. 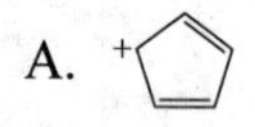　B.　C.　D.

(2) Which one has the biggest dipole moment (　　).

A. Cl—⟨benzene⟩—CH_3　B. Cl, —CH_3　C. Cl, —CH_3

(3) Which one is chiral(　　).

A. CH_2OH; HO—H; HO—H; CH_2OH　B. COOHH; H COOH　C. H, CH_3　D. Br, Cl, Cl, Br

(4) The important organic intermediates are (　　), (　　), (　　), (　　), (　　) and (　　).

(5) The chemical shift of methyl groups H_3C (α), CH_3 (β) in α or β positions, the smaller δ is in (　　), due to (　　).

(6) Which one is the easiest leaving group (　　).

A. Br—⟨benzene⟩—O^-　B. CH_3—⟨benzene⟩—O^-　C. ⟨benzene⟩—O^-　D. O_2N—⟨benzene⟩—O^-

(7) To bombard the following compounds to form carbcation, the highest energy needed is (　　).

A. CH_3—C(Br)(CH_3)(CH_3)　B. Br　C. Br

(8) Which one is the strongest acid (　　).

A. $CH_3CH_2CO_2H$　B. Cl_2CHCO_2H　C. $ClCH_2CO_2H$　D. $Cl_2CHCH_2CO_2H$

(9) Which one is the strongest base (　　).

A. $CH_3CH_2O^-$　B. $(CH_3)_3CO^-$　C. $CH_3CH_2CH_2O^-$　D. $(CH_3)_2CHO^-$

(10) The relationship of the following is(　　).

COOH; H—NH_2; HO—H; CH_3　　H, COOH, OH, NH_2, H_3C, H

A. the same　B. racemate　C. enantiomer　D. diastereomer

(11) The fastest solvolysis in actic acid is(　　).

A. 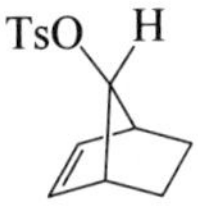 TsO H　　B. 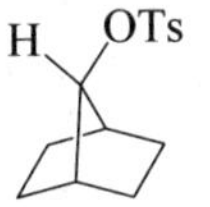 H OTs　　C. 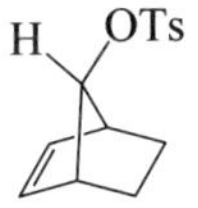 H OTs

(12) For $R—X + Nu^- \xrightarrow{S_N2} R—Nu + X^-$, when R is (　　), the reaction is fastest.

A. CH_3　　B. CH_3CH_2　　C. $(CH_3)_2CH$　　D. $(CH_3)_3C$

(13) The largest reaction rate for S_N1 is(　　).

A. $C_2H_5OCH_2CH_2Cl$　　B. $CH_3CH_2CH_2CH_2Cl$　　C. $C_2H_5OCH_2Cl$

(14) The major product for the intermolecular elimination reaction of AcO H Ph Ph D H is (　　).

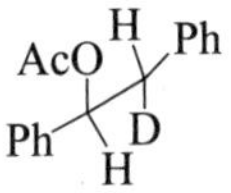

A. 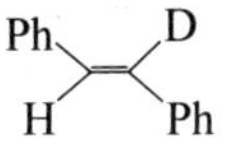 Ph D H Ph　　B. 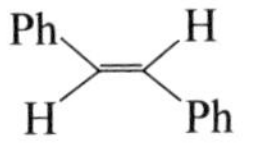 Ph H H Ph

(15) 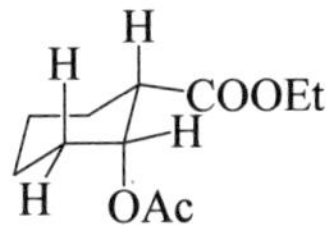 H H COOEt H H OAc $\xrightarrow{heat}$ (　　).

A. 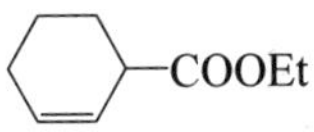 COOEt　　B. 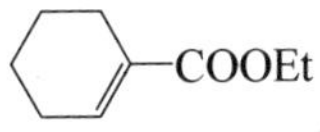 COOEt

(16) The strongest nucleophile is(　　).

A. OH^-　　B. CH_3^-　　C. NH_2^-　　D. F^-

(17) The strongest reactivity in its nucleophilic addition reaction(　　).

A. $C_6H_5COCH_3$　　B. HCHO　　C. CH_3COCH_3　　D. CH_3CHO

(18) Using $LiAlH_4$ as a reducing reagent of Ph Et H $-COCH_3$, the major product is (　　).

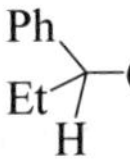

A. 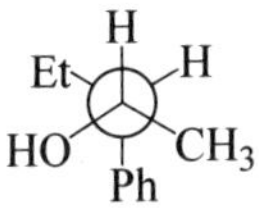 H Et H HO CH_3 Ph　　B. 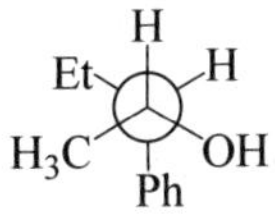 H Et H H_3C OH Ph

(19) When HCN add to the following compounds, the largest equilibrium constant K is (　　).

A. 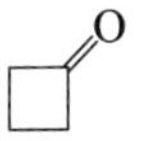 O　　B. 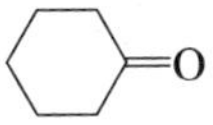 =O　　C.  O

(20) In KOH/alcohol solutions, the faster dehydrohalogenation is (　　).

A. 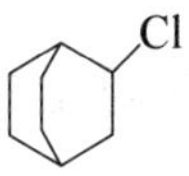 Cl　　B. 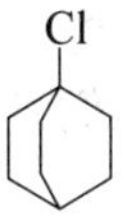 Cl

2. Complete the following reactions (write down the configuration when necessary, 2 points each, 40 points total)

(1) 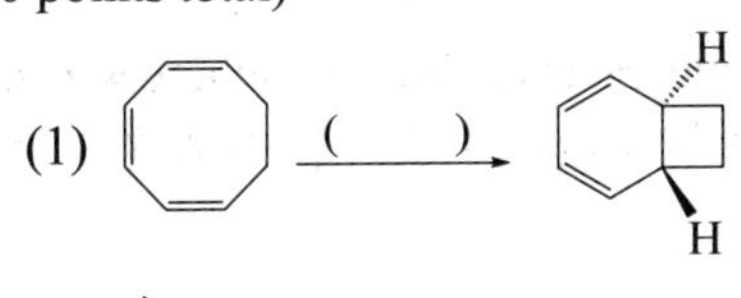

(2) 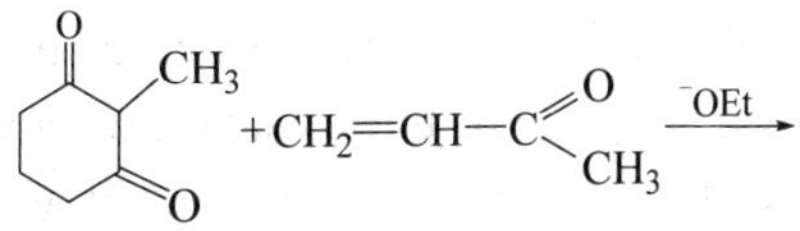 O CH_3 O $+CH_2{=}CH{-}C(=O)CH_3$ $\xrightarrow{^-OEt}$

(3) O $\xrightarrow{LiAlH_4}$ $\xrightarrow{H_3O^+}$

(4) CH_3 H $\xrightarrow{HBr}$

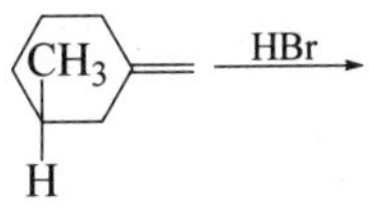

(5) 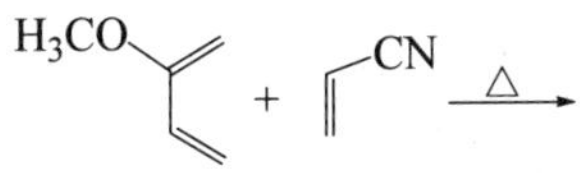 H_3CO + CN $\xrightarrow{\triangle}$

(6) 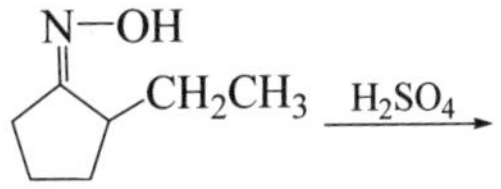 N—OH CH_2CH_3 $\xrightarrow{H_2SO_4}$

(7) 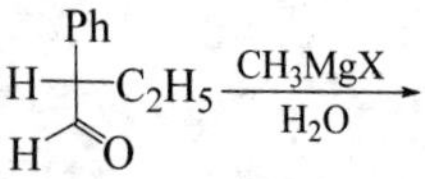 Ph; H—C—C_2H_5; H—C=O $\xrightarrow[H_2O]{CH_3MgX}$

(8) 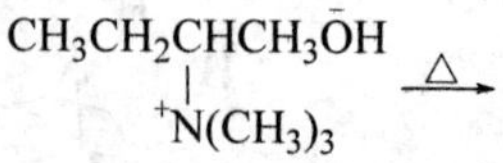 $CH_3CH_2CHCH_2OH$ with $^+N(CH_3)_3$ $\xrightarrow{\triangle}$

(9) 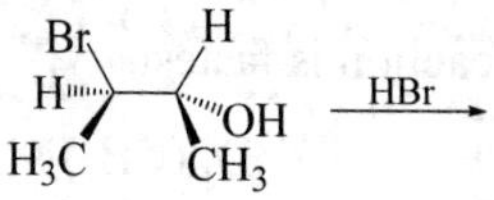 Br, H, H, OH, H_3C, CH_3 $\xrightarrow{HBr}$

(10) 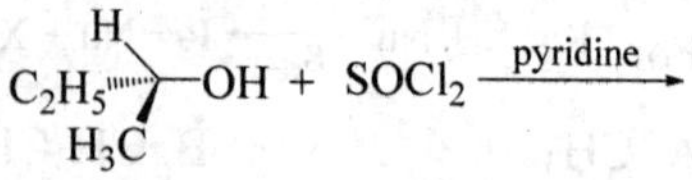 H, C_2H_5—C—OH, H_3C + $SOCl_2$ $\xrightarrow{pyridine}$

(11) 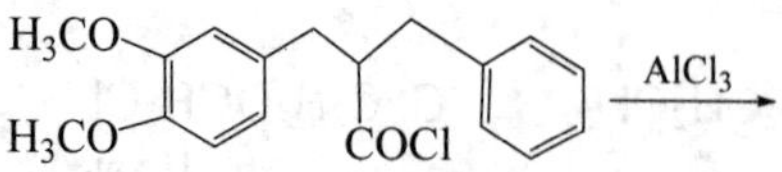 H_3CO, H_3CO, COCl $\xrightarrow{AlCl_3}$

(12) 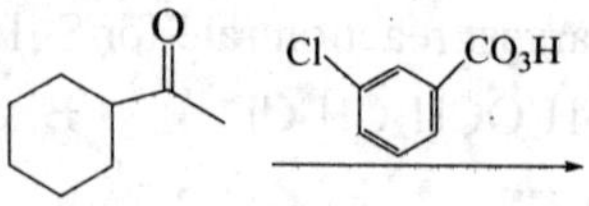 O; Cl, CO_3H $\longrightarrow$

(13) 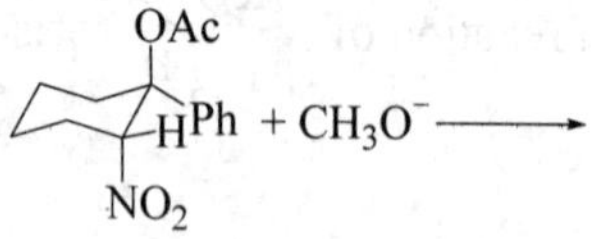 OAc, Ph, H, NO_2 + CH_3O^- $\longrightarrow$

(14) 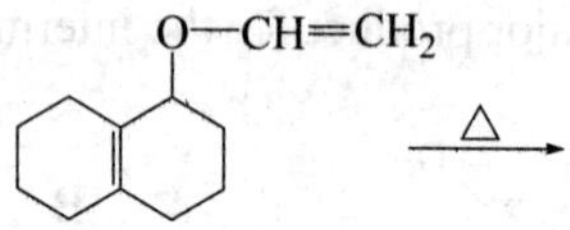 O—CH=CH_2 $\xrightarrow{\triangle}$

(15) 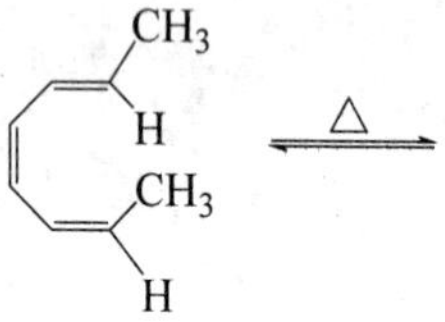 CH_3, H, CH_3, H $\xrightleftharpoons{\triangle}$

(16) 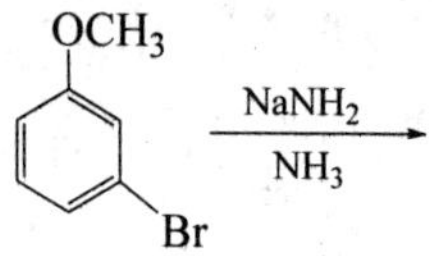 OCH_3, Br $\xrightarrow[NH_3]{NaNH_2}$

(17) 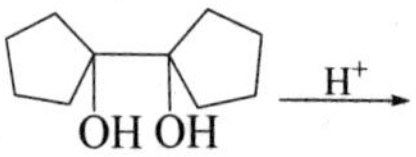 OH OH $\xrightarrow{H^+}$

(18) 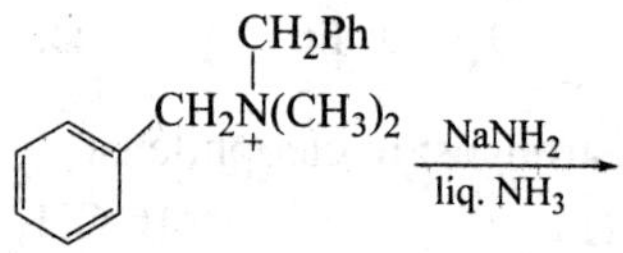 CH_2Ph; $CH_2\overset{+}{N}(CH_3)_2$ $\xrightarrow[liq.\ NH_3]{NaNH_2}$

(19) 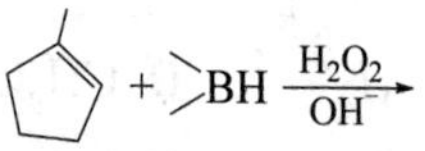 + >BH $\xrightarrow[OH^-]{H_2O_2}$

(20) 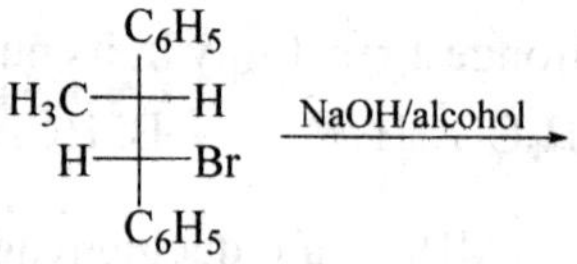 C_6H_5; H_3C—C—H; H—C—Br; C_6H_5 $\xrightarrow{NaOH/alcohol}$

3. Write down the reaction mechanisms (5 points each, 10 points total)

(1) 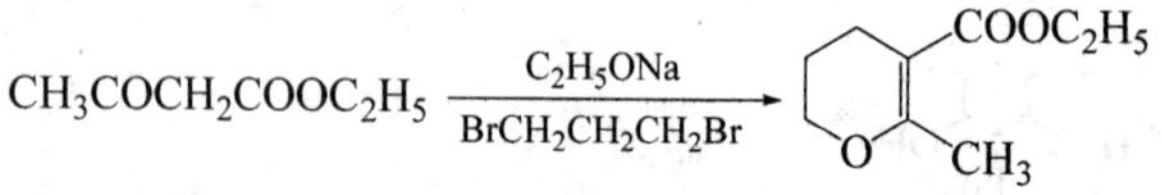 $CH_3COCH_2COOC_2H_5$ $\xrightarrow[BrCH_2CH_2CH_2Br]{C_2H_5ONa}$ $COOC_2H_5$, O, CH_3

(2) 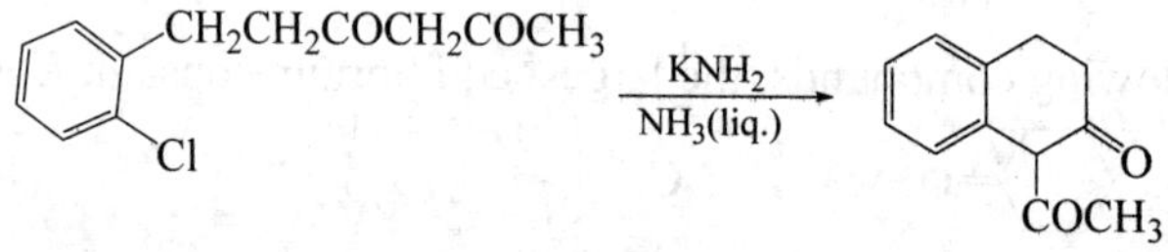 $CH_2CH_2COCH_2COCH_3$, Cl $\xrightarrow[NH_3(liq.)]{KNH_2}$ O, $COCH_3$

4. Explain the following results (5 points each, 10 points total)

(1) What kind of concerted reactions in this reaction?

(下列二步转变，都包括一个协同反应，试说明它们发生了什么样的协同反应。)

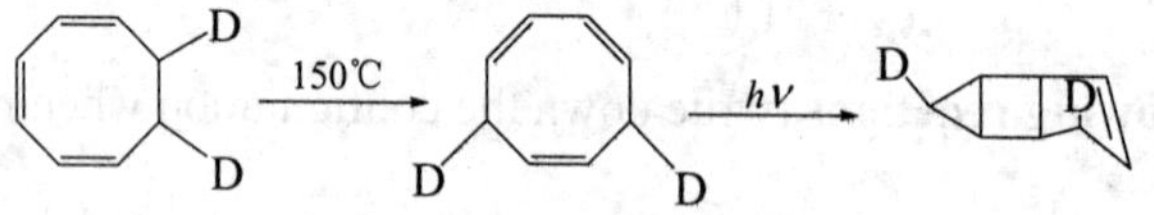

D, D $\xrightarrow{150℃}$ D, D $\xrightarrow{h\nu}$ D, D

(2) Racematic products can be resulted when cyclobutene reacts with Br_2, whereas a meso product can only be resulted when it react with H_2/Pt. (环丁烯与溴反应生成一种外消旋产物，然而环丁烯与重氢在铂催化下反应却生成一种内消旋产物。)

《高等有机化学》测试题（1）参考答案

1. 选择或填空题 (每小题 2 分，总分 40 分)

(1) A (2) A (3) B (4) 碳正离子，碳负离子，碳烯，自由基，苯炔，鎓内盐 (5) α-CH_3 上的氢。该氢位于双键π电子的屏蔽区内。(6) D (7) C (8) B (9) B (10) D (11) C (12) A (13) C (14) A (15) A (16) B (17) B (18) A (19) C (20) A

2. 完成下列反应 (涉及立体构型时，写出立体构型，每小题 2 分，总分 40 分)

(1) hν

(2) 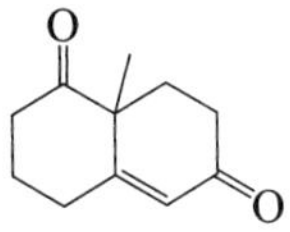 O, O

(3) 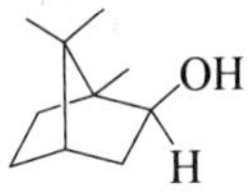 OH, H

(4) 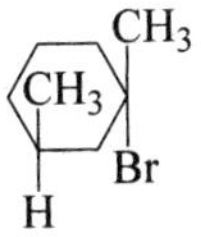 CH_3, CH_3, Br, H

(5) 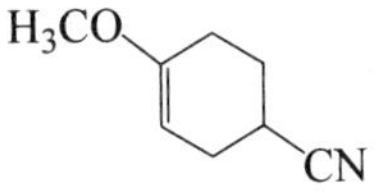 H_3CO, CN

(6) 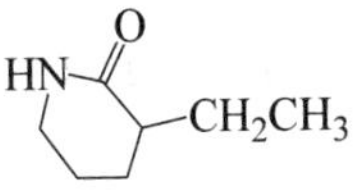 HN, O, CH_2CH_3

(7) 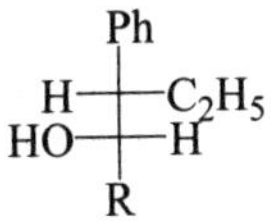 Ph, H, C_2H_5, HO, H, R

(8) CH_2═$CHCH_2CH_3$

(9) 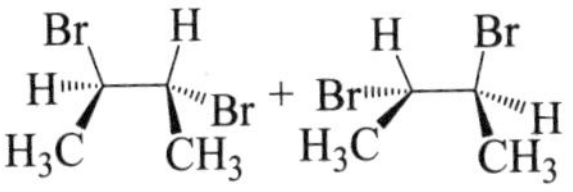 Br, H, H, Br, H_3C, CH_3 + Br, H, H, Br, H_3C, CH_3

(10) 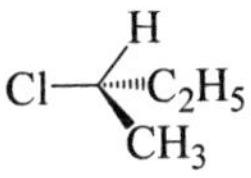 Cl, H, C_2H_5, CH_3

(11) 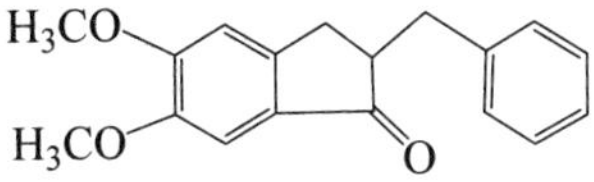 H_3CO, H_3CO, O

(12) 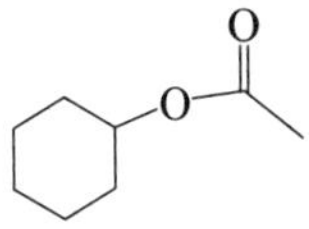 O, O

(13)  Ph, NO_2

(14) 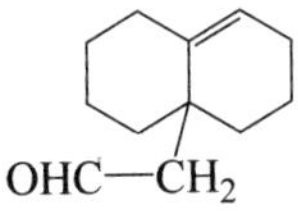 OHC—CH_2

(15) 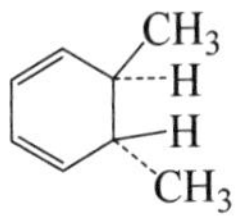 CH_3, H, H, CH_3

(16) 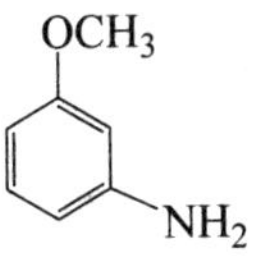 OCH_3, NH_2

(17) 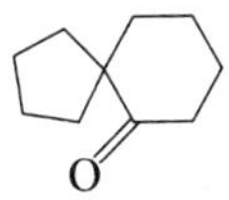 O

(18) 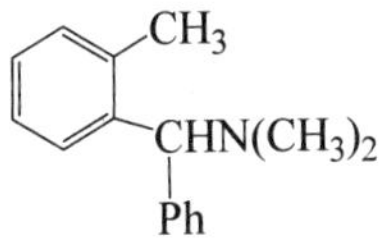 CH_3, $CHN(CH_3)_2$, Ph

(19) 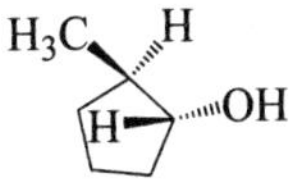 H_3C, H, H, OH

(20) 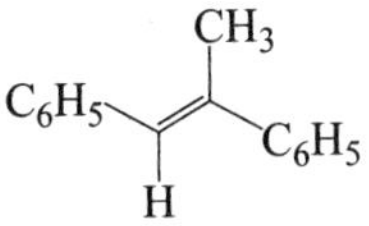 CH_3, C_6H_5, C_6H_5, H

3. 写出下列反应的机理 (每小题 5 分，总分 10 分)

(1)

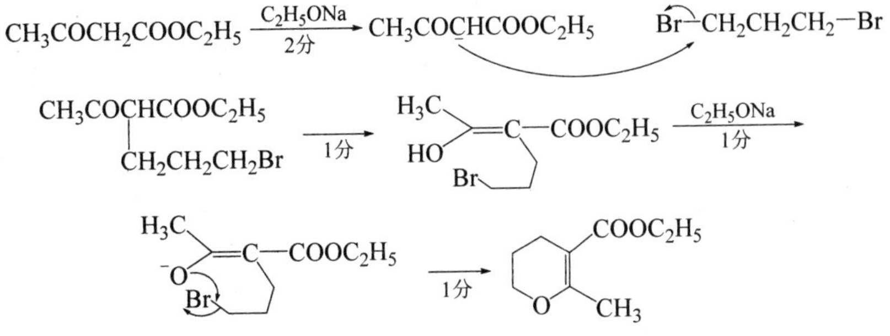

$CH_3COCH_2COOC_2H_5 \xrightarrow[2分]{C_2H_5ONa} CH_3CO\bar{C}HCOOC_2H_5$ Br—$CH_2CH_2CH_2$—Br

$CH_3COCHCOOC_2H_5$ ($CH_2CH_2CH_2Br$) $\xrightarrow[1分]{}$ H_3C(HO)C═C—$COOC_2H_5$ (Br) $\xrightarrow[1分]{C_2H_5ONa}$

H_3C($^-$O)C═C—$COOC_2H_5$ (Br) $\xrightarrow[1分]{}$ $COOC_2H_5$, O, CH_3

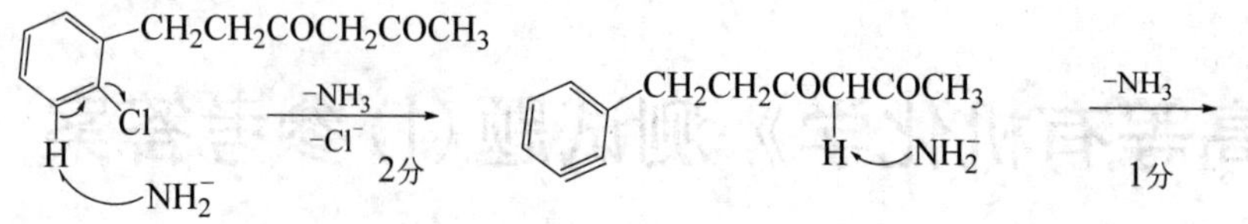

(2)

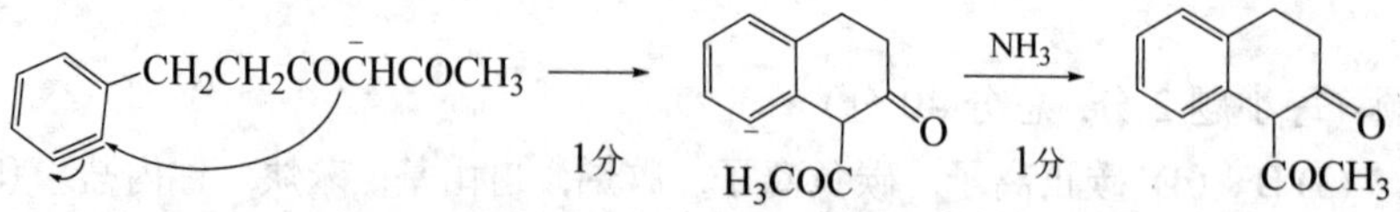

$-NH_3$, $-Cl^-$ 2分　　$-NH_3$ 1分

1分　　NH_3 1分

4. 简要说明下列问题 (每小题 5 分, 总分 10 分)

(1) 答：第一步是[1, 5]氢迁移(2.5 分), 第二步是[2+2]环加成(2.5 分)。

(2) 答：环丁烯与溴的加成是通过溴鎓离子中间体进行的, Br^-从反面进攻溴鎓离子, 两个碳进攻机会均等, 从而得到外消旋体(3 分)。

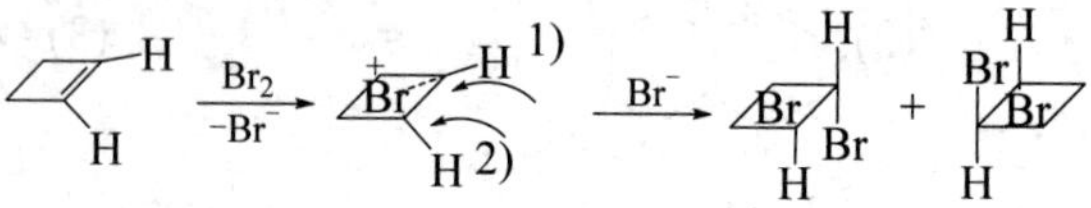

Br_2, $-Br^-$　　Br^-

而烯烃双键与氢的加成为顺式加成, 加重氢时, 两个氘原子加到环的同一边, 成为内消旋体 (2 分)。

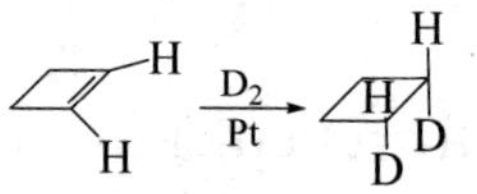

D_2, Pt

Appendix Ⅱ

《Advanced Organic Chemistry》Final Test(2)

1. Selection (2 points each, 40 points total)

(1) The one with the biggest dipole moment (偶极矩) is (　　).

A. 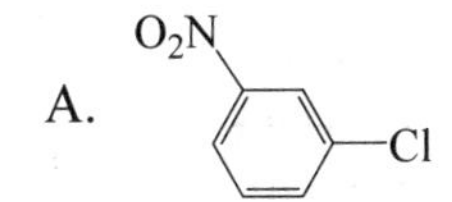

B. NO_2 (nitrobenzene)　　C. O_2N–C_6H_4–CH_3

(2) Which one is chiral (手性的) (　　).

A. CH_2Cl; H_3C–H; H_3C–H; CH_2Cl　　B. H, CH_3　　C. Br, CH_3, Br, CH_3　　D. Cl Br, Br Cl

(3) Which one is the easiest leaving group (离去基) (　　).

A. Cl–C_6H_4–SO_3^-　　B. O_2N–C_6H_4–SO_3^-　　C. C_6H_5–SO_3^-　　D. H_3CO–C_6H_4–SO_3^-

(4) The strongest base(碱)is (　　).

A. 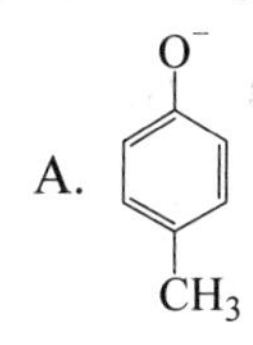

B. O^- (phenoxide)　　C. O^-, Cl　　D. O^-, NO_2

(5) The strongest nucleophile(亲核试剂)is (　　).

A. CH_3O^-　　B. $CH(CH_3)_2O^-$　　C. $CH_3CH_2CH_2O^-$　　D. $C(CH_3)_3O^-$

(6) The correct order of reactivity in its nucleophilic addition (亲核加成) reaction (　　).

a. 乙醛　　b.丙酮　　c. 苯乙酮　　d. 二苯甲酮

A. d>c>b>a　　B.a>b>c>d　　C. a>c>b>d　　D. c>d>b>a

(7) The most reactive one for electrophile substitution (亲电取代) reaction is (　　).

A. OCH_3　　B. Cl　　C. CH_3　　D. NO_2

(8) The relationship between [Newman projection: H, H, Et, HO, CH_3, Ph] and [Newman projection: H, H, Et, H_3C, OH, Ph] is (　　).

A.构象异构体　　B. 对映体　　C. 非对映体　　D. 相同化合物

(9) H_3C [cyclohexane] $CHMe_2$, Cl $\xrightarrow{OH^-}$ (　　).

A. 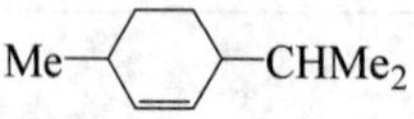 Me—(cyclohexene)—$CHMe_2$ B. 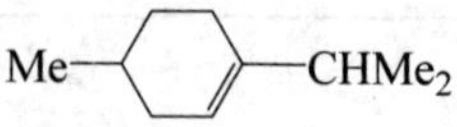 Me—(cyclohexene)—$CHMe_2$

(10) After the reaction of MeMgBr and Ph, Et, H—C—$COCH_3$, the major product is( ).

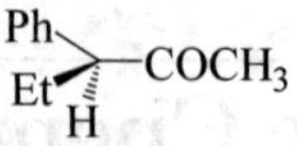

A. 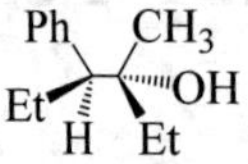 Ph, Et, H / CH_3, OH, Et B. 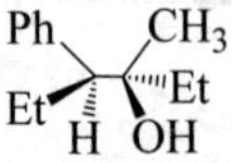 Ph, Et, H / CH_3, Et, OH

(11) Which one has no aromaticity(芳香性)( ).

A. 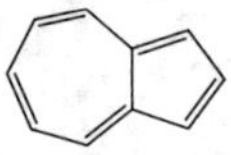 B. + C. 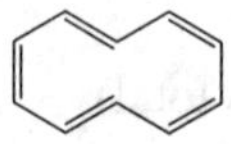 D. 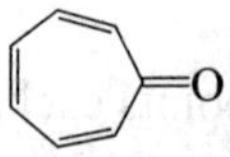 =O

(12) 顺-2-戊烯与溴反应的产物为( ).

A. 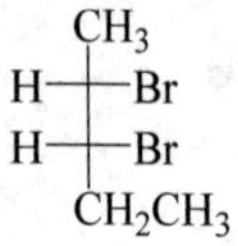 CH_3; H—Br; H—Br; CH_2CH_3 B. 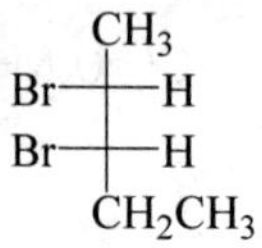 CH_3; Br—H; Br—H; CH_2CH_3 C. 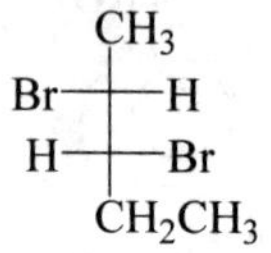 CH_3; Br—H; H—Br; CH_2CH_3 D. 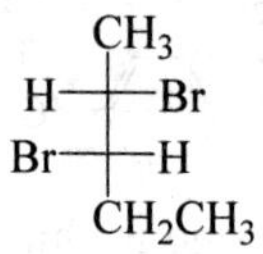 CH_3; H—Br; Br—H; CH_2CH_3 E. C and D

(13) 下列化合物中 pK_a 最小的是( ).

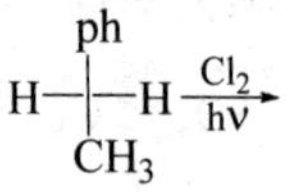

A.  B.  C.

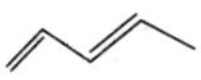

(14) 下列哪个反应的产物能被拆分为对映体？( ).

A. 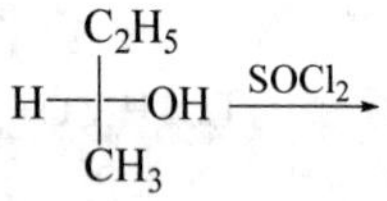 H—C(C_2H_5)(CH_3)—OH $\xrightarrow{SOCl_2}$ B. H—C(ph)(CH_3)—H $\xrightarrow[h\nu]{Cl_2}$ C. H—C(CH_2CH_3)(CH_2CH_3)—Cl $\xrightarrow{AgNO_3}$

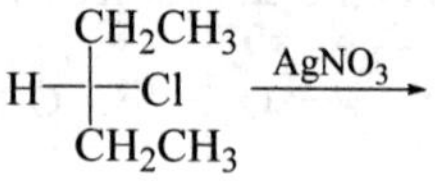

(15) 下列试剂碱性最强的是( ).

A. F^- B. OH^- C. NH_2^- D. CH_3^-

(16) The major product for the following reaction (furan) + (maleic anhydride) $\xrightarrow{\triangle}$ is( ).

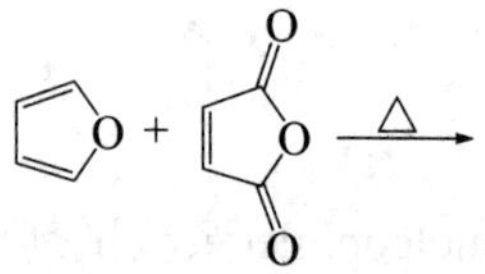

A. 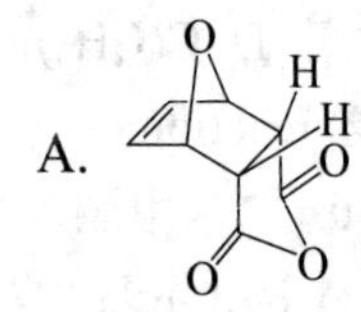

B.

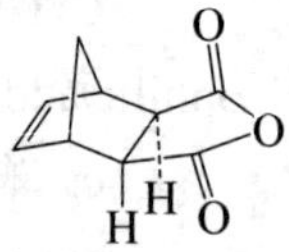

(17) The most stable anion (负离子) is( ).

A. $(CF_3)_3C^-$ B. $(CF_3)_2C^-H$ C. $(CF_3)C^-H_2$ D. C^-F_3

(18) The hardest one to proceed in(进行)both S_N1 and S_N2 reactions is ( ).

A. $PhCH_2Br$ B. 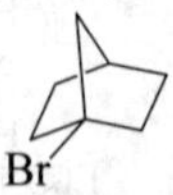 (1-bromonorbornane) Br C. $PhCOCH_2Br$ D. Ph_2CHBr

(19) 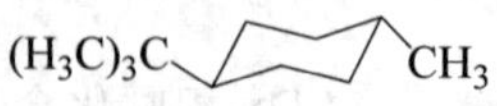 $(H_3C)_3C$—(cyclohexane)—CH_3 和 $C(CH_3)_3$—(cyclohexane)—CH_3 互为 ( ).

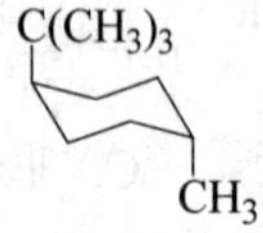

A. 构型异构体 B. 对映体 C. 构象异构体 D. 互变异构体

(20) 与 HCl 反应速度最快的是(　　).

A. $CF_3CH=CH_2$　　B. $BrCH=CH_2$　　C. $CH_3OCH=CHCH_3$　　D. $CH_3CH=CHCH_3$

2. Complete the following reactions (write down the configuration when necessary, 必要时写出立体构型, 2 points each, 40 points total)

(1) 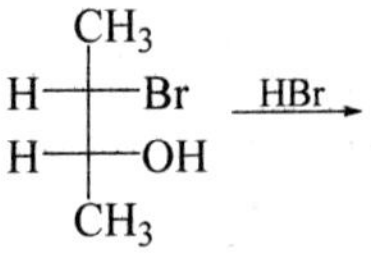 CH_3 / H—Br / H—OH / CH_3 $\xrightarrow{HBr}$

(2) H_3CH_2C—C(H)(CH_3)—OH + $SOCl_2$ 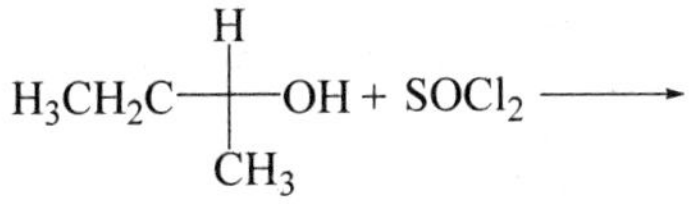 $\longrightarrow$

(3) 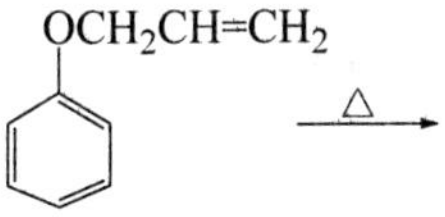 $OCH_2CH{=}CH_2$ $\xrightarrow{\triangle}$

(4) 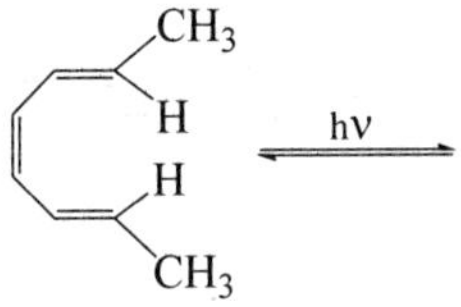 CH_3 H H CH_3 $\overset{h\nu}{\rightleftharpoons}$

(5) 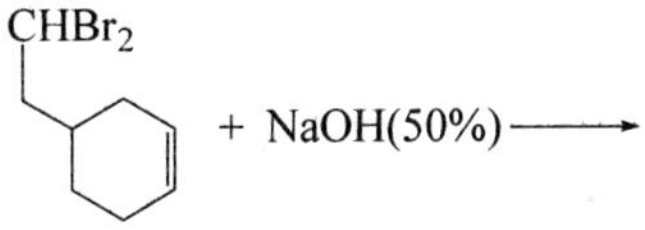 $CHBr_2$ + NaOH(50%) $\longrightarrow$

(6) 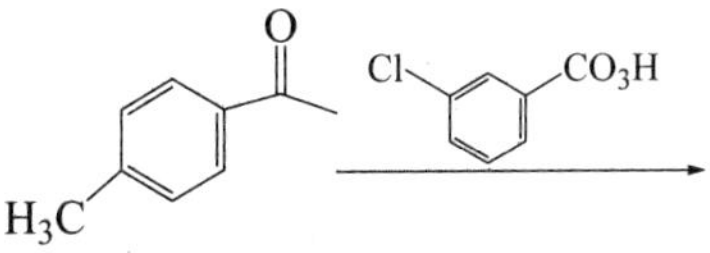 O H_3C $\xrightarrow{Cl\text{—}C_6H_4\text{—}CO_3H}$

(7) 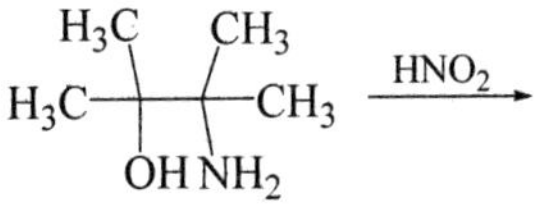 H_3C CH_3 H_3C— —CH_3 OH NH_2 $\xrightarrow{HNO_2}$

(8) 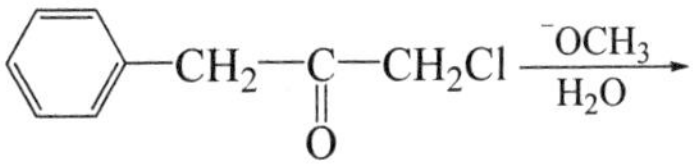 C_6H_5—CH_2—C(=O)—CH_2Cl $\xrightarrow[H_2O]{^-OCH_3}$

(9) 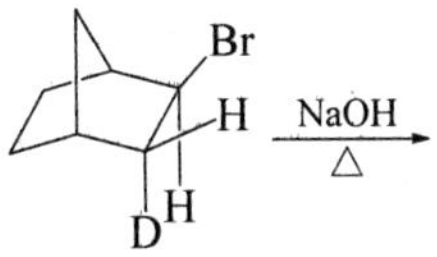 Br H D H $\xrightarrow[\triangle]{NaOH}$

(10) 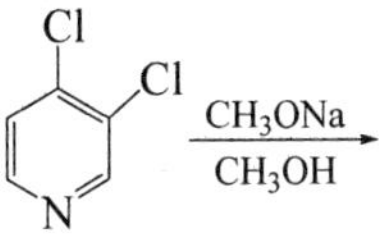 Cl Cl N $\xrightarrow[CH_3OH]{CH_3ONa}$

(11) 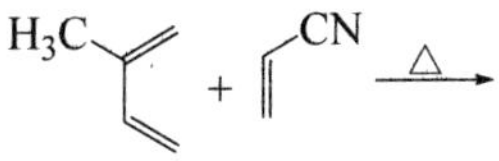 H_3C + CN $\xrightarrow{\triangle}$

(12) 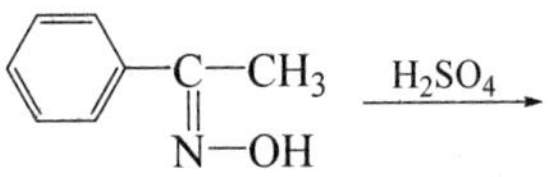 C_6H_5—C(=N—OH)—CH_3 $\xrightarrow{H_2SO_4}$

(13) 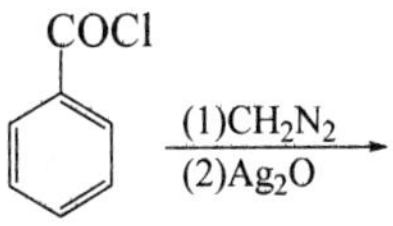 COCl $\xrightarrow[(2)Ag_2O]{(1)CH_2N_2}$

(14) 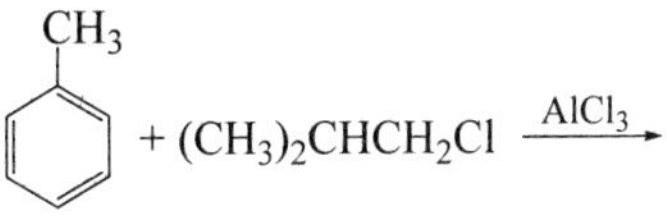 CH_3 + $(CH_3)_2CHCH_2Cl$ $\xrightarrow{AlCl_3}$

(15) 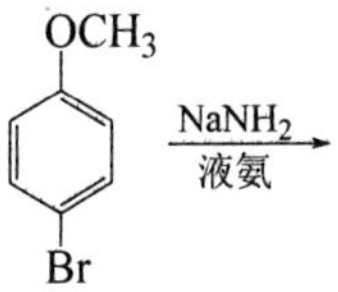 OCH_3 Br $\xrightarrow[\text{液氨}]{NaNH_2}$

(16) 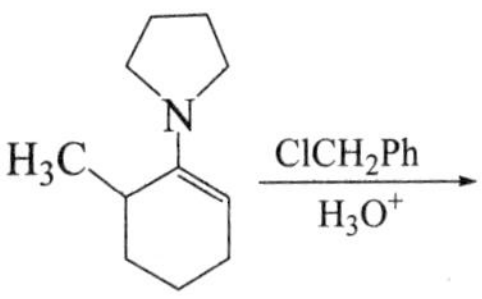 N H_3C $\xrightarrow[H_3O^+]{ClCH_2Ph}$

(17) 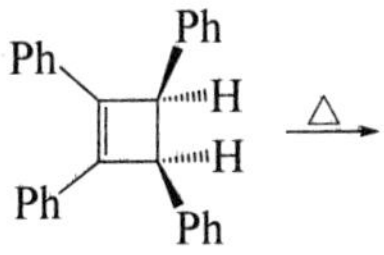 Ph Ph H H Ph Ph $\xrightarrow{\triangle}$

(18) 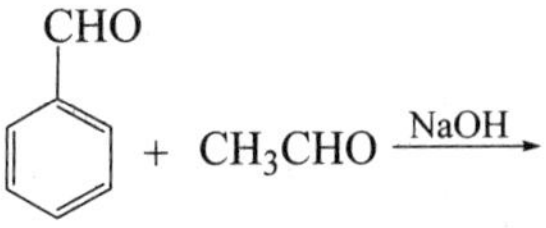 CHO + CH_3CHO $\xrightarrow{NaOH}$

(19) 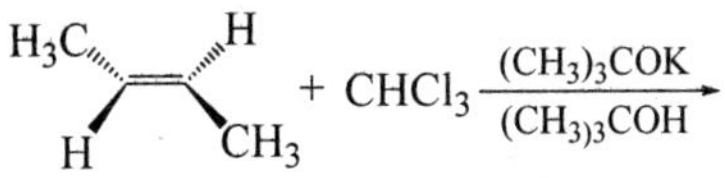 H_3C H H CH_3 + $CHCl_3$ $\xrightarrow[(CH_3)_3COH]{(CH_3)_3COK}$

(20) 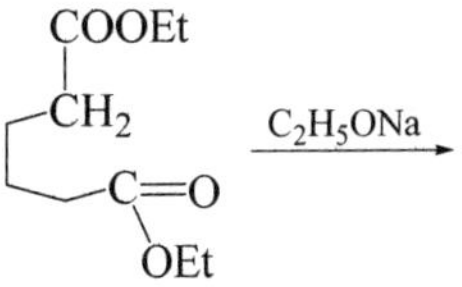 COOEt CH_2 C=O OEt $\xrightarrow{C_2H_5ONa}$

3. Write down the reaction mechanisms (20 points total)

(1) CCl_3CHO + (chlorobenzene) $\xrightarrow{H_2SO_4}$ Cl–C_6H_4–$CH(CCl_3)$–C_6H_4–Cl (7 points)

(2) EtOOC, O, O $\xrightarrow{Me_2CuLi}$ O, O, O (6 points)

(3) O, O $\xrightarrow{Br_2}$ $\xrightarrow{KOH}$ $\xrightarrow{H^+}$ O, COOH (7 points)

《高等有机化学》测试题(2)参考答案

1. 选择题 (每小题 2 分，总分 40 分)

(1) C (2) D (3) B (4) A (5) A (6) B (7) A (8) C (9) B (10) B (11) C (12) E (13) A (14) B (15) D (16) A (17) A (18) B (19) C (20) C

2. 完成下列反应 (涉及立体构型时，写出立体构型。每小题 2 分，总分 40 分)

(1) 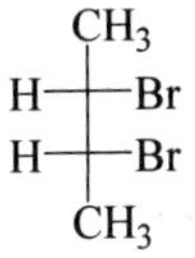 CH_3 / H—Br / H—Br / CH_3

(2) H / H_3CH_2C—Cl / CH_3

(3) OH / $CH_2CH{=}CH_2$

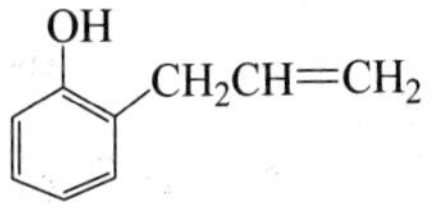

(4) 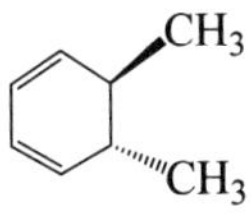 CH_3 / CH_3

(5)

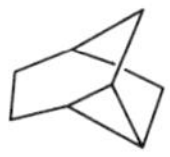

(6) 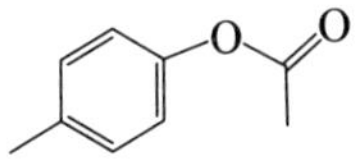 O / O

(7) 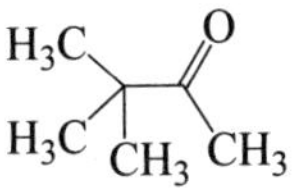 H_3C / O / H_3C / CH_3 / CH_3

(8) 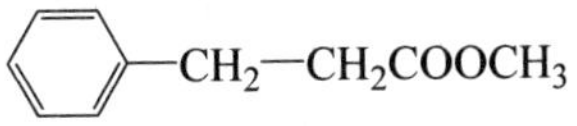 —CH_2—CH_2COOCH_3

(9) 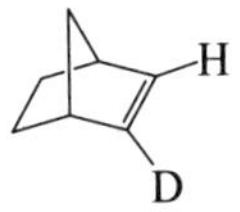 H / D

(10) 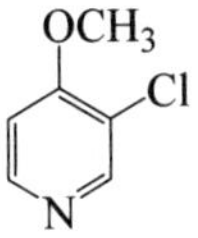 OCH_3 / Cl / N

(11) 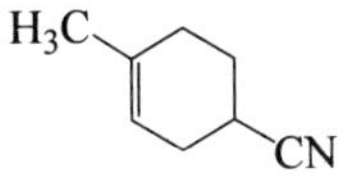 H_3C / CN

(12) 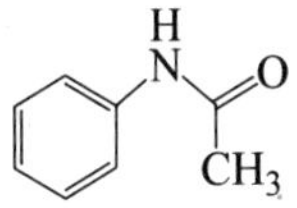 H / N / O / CH_3

(13) 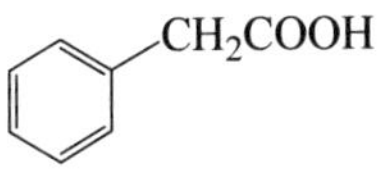 CH_2COOH

(14)

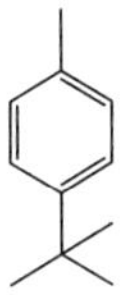

(15) 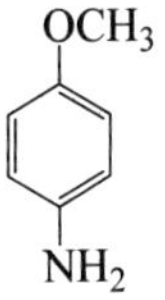 OCH_3 / NH_2

(16) 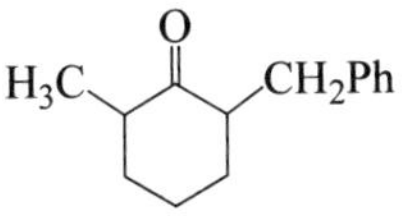 H_3C / O / CH_2Ph

(17) 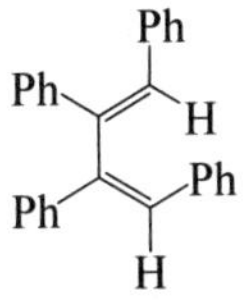 Ph / Ph / H / Ph / Ph / H

(18) 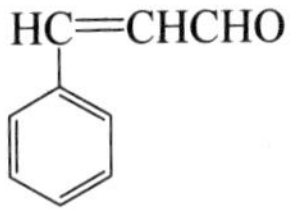 HC$=$CHCHO

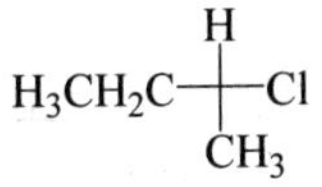

(19) 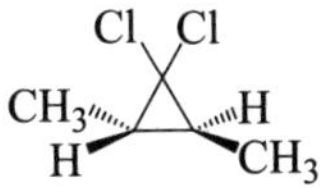 Cl Cl / CH_3 / H / H / CH_3

(20) 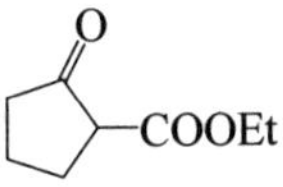 O / COOEt

3. 写机理 (共 20 分)

(1) 共 7 分

$CCl_3CHO \xrightarrow[2分]{H^+}$ HO—$\overset{+}{C}H$—CCl_3 $\xrightarrow{-H^+}$ (2分) Cl—(苯环)—CH(CCl_3)—OH $\xrightarrow[-H_2O]{H^+}$ (2分)

Cl—(苯环)—$\overset{+}{C}H$(CCl_3) $\xrightarrow{-H^+}$ (1分) Cl—(苯环)—CH(CCl_3)—(苯环)—Cl

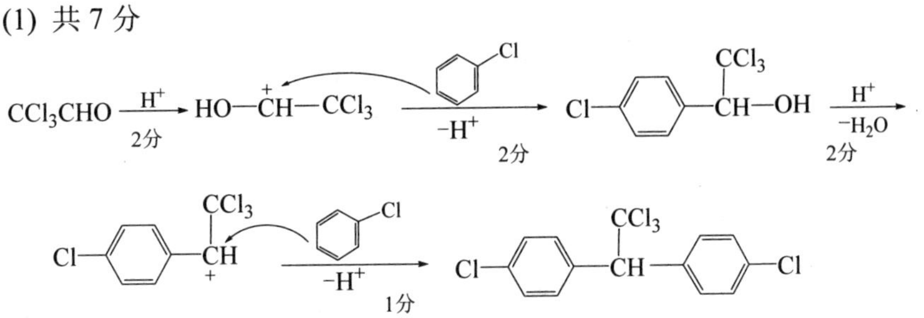

(2) 共 6 分

EtOOC O Me_2CuLi 2分 O EtO O 2分 O ^-O OEt 2分 O O

(3) 共 7 分

Br O O Br_2 1分 Br Br O O ^-OH 1分 Br Br O OH O 1分 ^-OH 1分 Br Br O^- COO^-

$-Br^-$ 1分 Br O COO^- ^-OH $-HBr$ 1分 O COO^- H^+ 1分 O COOH

Appendix Ⅲ

《Advanced Organic Chemistry》Final Test（3）

1. Selection (2 points each, 32 points total)

(1) A. 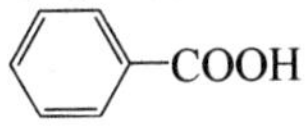 C_6H_5—COOH B. Br—C_6H_4—COOH C. C_6H_5—SO_3H D. 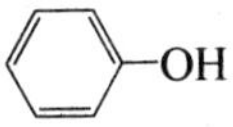 C_6H_5—OH

上述化合物的酸性由强到弱的排列顺序为（ ）.

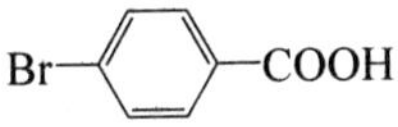

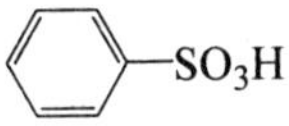

A.a>b>c>d B. c>d>b>a C. c>b>a>d D. b>a>c>d

(2) 试剂的亲核活性由强到弱的顺序为（ ）.

A. $CH_3CH_2^- > CH_3O^- > (CH_3)_2N^- > F^-$ B. $CH_3CH_2^- > (CH_3)_2N^- > CH_3O^- > F^-$

C. $F^- > CH_3O^- > (CH_3)_2N^- > CH_3CH_2^-$ D. $(CH_3)_2N^- > CH_3CH_2^- > CH_3O^- > F^-$

(3) 发生 S_N1 反应时的相对速度由大到小的顺序为（ ）.

a. t-BuBr b.i-PrBr c.$PhCH_2Br$ d. 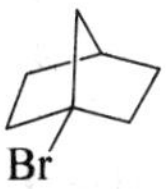 (1-bromobicyclo[2.2.1]heptane, Br)

A. a>b>c>d B. d>c>b>a C. c> a > b >d D. b>a>c>d

(4) 如果亲核取代反应发生了重排反应，则该反应的历程为（ ）.

A. S_N1 B. S_N2 C.不确定 D. 可以为 S_N1，也可以为 S_N2

(5) 发生 Diels-Alder 反应的立体选择性为（ ）.

A. 加热时，同面/同面允许 B. 光照，为同面/同面允许

C. 加热时，同面/异面允许 D. 加热和光照时都没有立体选择性

(6) 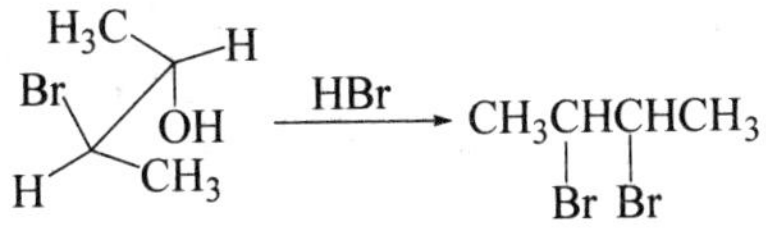 (H₃C, H, OH / Br, H, CH₃ structure) $\xrightarrow{HBr}$ $CH_3CH(Br)CH(Br)CH_3$ 反应产物为（ ）.

A. 外消旋体 B. 左旋体 C. 右旋体 D. 内消旋体

(7) 1, 3-丁二烯的 HOMO 轨道为（ ）.

A. Ψ_1 B. Ψ_2 C. Ψ^*_3 D. Ψ^*_4

(8) The most stable carbocation is（ ）.

A. 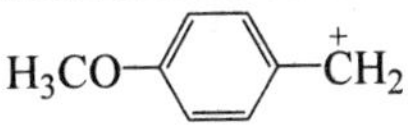 H_3CO—C_6H_4—$\overset{+}{C}H_2$ B. O_2N—C_6H_4—$\overset{+}{C}H_2$ C. C_6H_5—$\overset{+}{C}H_2$ D. 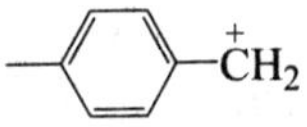 CH_3—C_6H_4—$\overset{+}{C}H_2$

(9) Which one doesn't have aromaticity (芳香性)（ ）.

A. 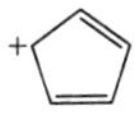 (cyclopentadienyl cation, +) B. (cyclopropenyl cation, +) C.  (cyclopentadienyl anion, −) D. 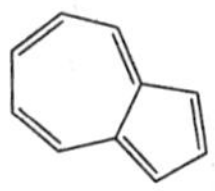 (azulene)

(10) Which one is chiral（ ）.

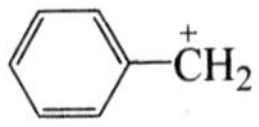

A. 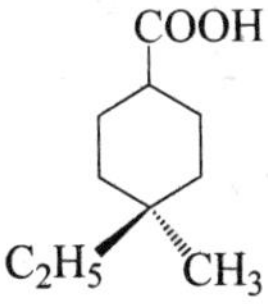 (COOH; cyclohexane; C_2H_5, CH_3) B. (CH_2Br; H—OH; H—OH; CH_2Br) C. (H_3C, Cl)C=C=C(CH_3, Cl) D.  (H_3C, H)cyclobutane(H, COOH)

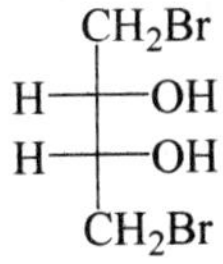

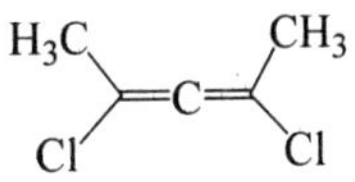

(11) The lowest energy of the following σ-complex is (　　).

A. 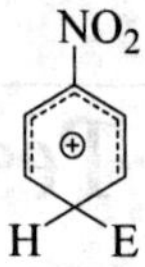 (σ-complex with NO_2, H, E)　B.  (σ-complex with Cl, H, E)　C. 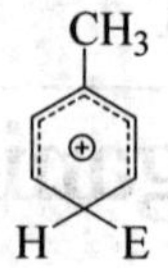 (σ-complex with CH_3, H, E)　D. 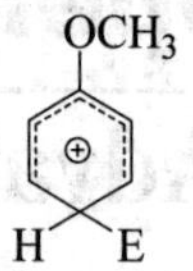 (σ-complex with OCH_3, H, E)

(12) $C_6H_5CHICH_2CH_3$ 在丙酮-水溶液中放置时，变为相应的醇,则醇的构型为(　　).

A. 内消旋体　B.构型翻转　C. 外消旋体　D.构型不变

(13) Which one can't be hydrolyzed by dilute acid (　　).

A. 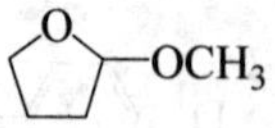 2-methoxytetrahydrofuran ($-OCH_3$)　B. 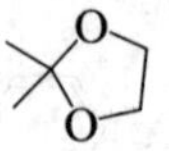 2,2-dimethyl-1,3-dioxolane　C. 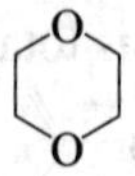 1,4-dioxane　D. 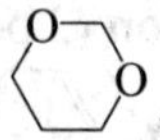 1,3-dioxane

(14) The fastest reaction for S_N1 is (　　).

A. $C_2H_5SCH_2CH_2Cl$　B. $C_2H_5SCH_2Cl$　C. $CH_3CH_2CH_2CH_2Cl$

(15) The major product for the following reaction furan + $CH_2=CHCOOEt$ $\xrightarrow{\triangle}$ is (　　).

A. 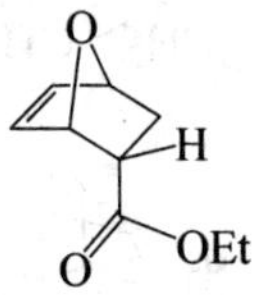 (oxabicyclic adduct, H, $COOEt$)　B. 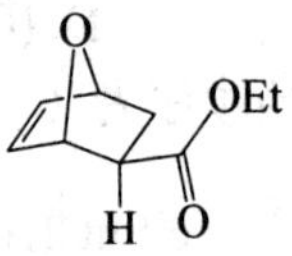 (oxabicyclic adduct, H, $COOEt$)

(16) 与 HBr 反应速度最慢的是(　　).

A. $CF_3CH{=}CH_2$　B. $ClCH{=}CH_2$　C. $CH_3OCH{=}CHCH_3$　D. $CH_3CH{=}CHCH_3$

2. Complete the following reactions (write down the configuration when necessary, 必要时写出立体构型, 2 points each, 40 points total)

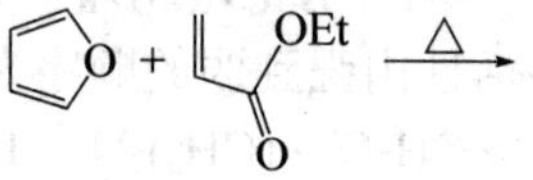

(1) 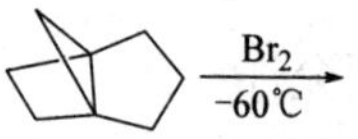 (tricyclic alkane) $\xrightarrow[-60℃]{Br_2}$

(2) 1-methylcyclohexene (CH_3) $\xrightarrow[OH^-]{KMnO_4(稀)}$

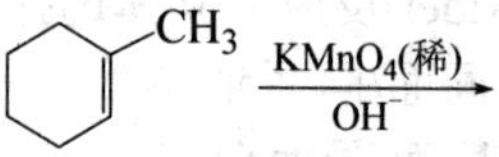

(3) $C_6H_5CH_2OCH_3 \xrightarrow[H_2O]{(CH_3)_3COK}$

(4) 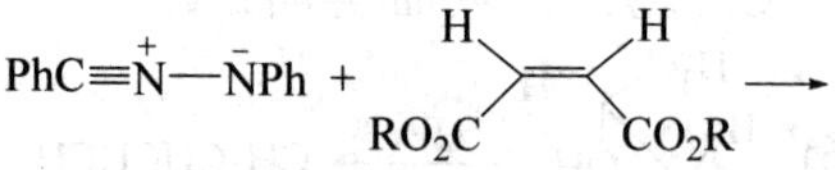 $PhC{\equiv}\overset{+}{N}{-}\overset{-}{N}Ph$ + (H)(H)$C{=}C$(RO_2C)(CO_2R) $\longrightarrow$

(5) 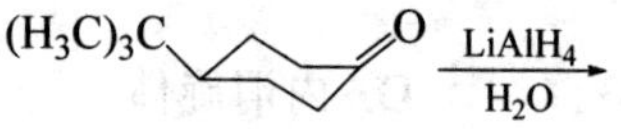 $(H_3C)_3C$-cyclohexanone $\xrightarrow[H_2O]{LiAlH_4}$

(6) 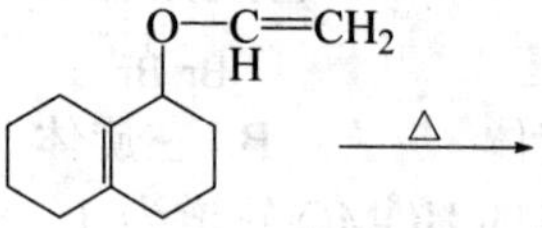 (octahydronaphthalene with $O{-}\underset{H}{C}{=}CH_2$) $\xrightarrow{\triangle}$

(7) 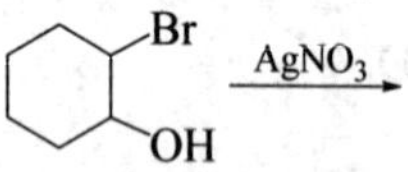 2-bromocyclohexanol (Br, OH) $\xrightarrow{AgNO_3}$

(8) 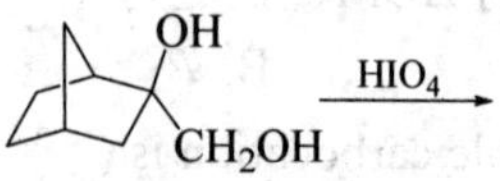 (bicyclic with OH, CH_2OH) $\xrightarrow{HIO_4}$

(9) 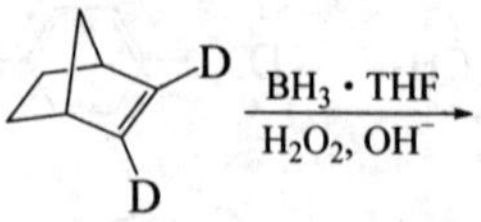 (bicyclic alkene with D, D) $\xrightarrow[H_2O_2,\ OH^-]{BH_3 \cdot THF}$

(10) 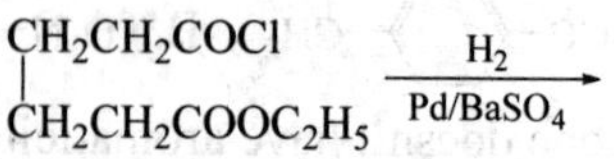 CH_2CH_2COCl / $CH_2CH_2COOC_2H_5$ $\xrightarrow[Pd/BaSO_4]{H_2}$

(11) 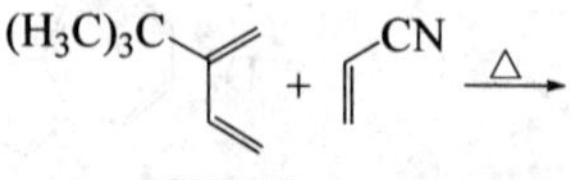 $(H_3C)_3C$-diene + $CH_2{=}CHCN$ $\xrightarrow{\triangle}$

(12) 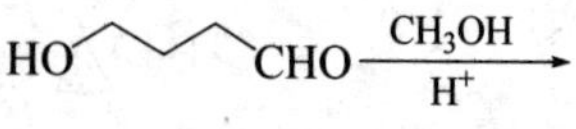 $HO\diagup\diagdown\diagup CHO \xrightarrow[H^+]{CH_3OH}$

(13) 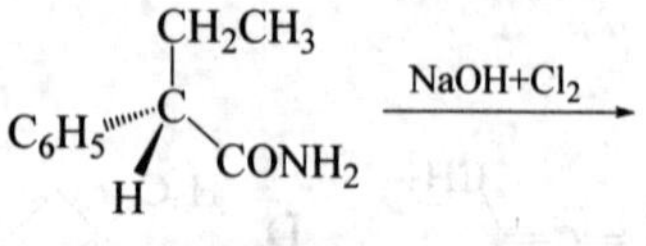 C_6H_5, H, CH_2CH_3, $CONH_2$ (chiral C) $\xrightarrow{NaOH+Cl_2}$

(14) 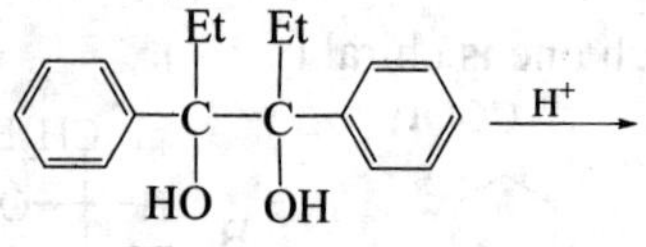 Ph-C(Et)(HO)-C(Et)(OH)-Ph $\xrightarrow{H^+}$

(15) 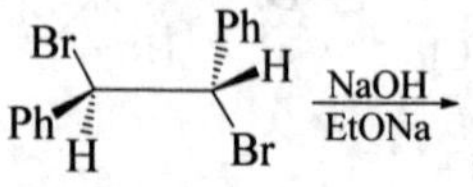 Br, Ph, H / Ph, H, Br $\xrightarrow[EtONa]{NaOH}$

(16) 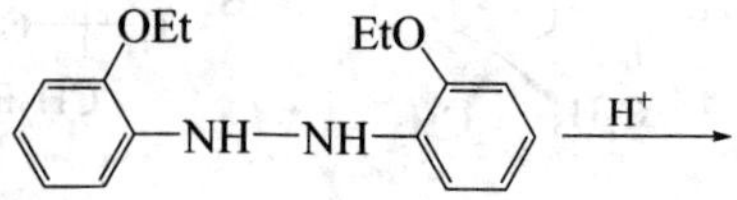 (2-OEt-phenyl)-NH—NH-(2-EtO-phenyl) $\xrightarrow{H^+}$

(17) 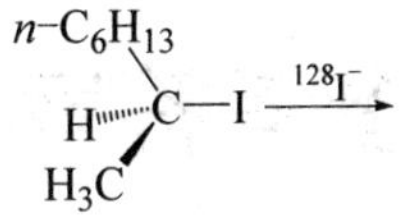 n-C_6H_{13} / H···C—I / H_3C $\xrightarrow{^{128}I^-}$

(18) 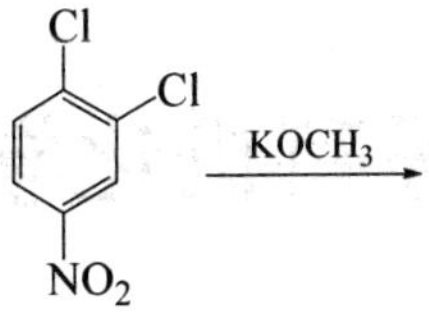 3,4-dichloronitrobenzene (Cl, Cl, NO_2) $\xrightarrow{KOCH_3}$

(19) 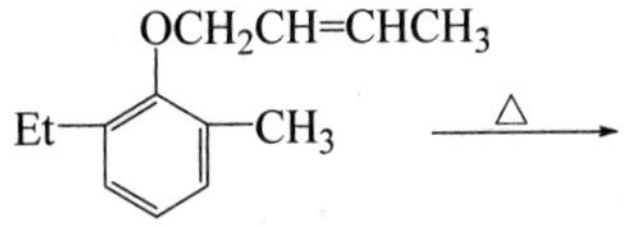 Et— benzene ring —CH_3 with $OCH_2CH{=}CHCH_3$ $\xrightarrow{\triangle}$

(20) $CH_2{=}CH{-}COOCH_3 + CH_2I_2 + Zn(Cu) \longrightarrow$

3. Write down the reaction mechanisms (28 points total)

(1) 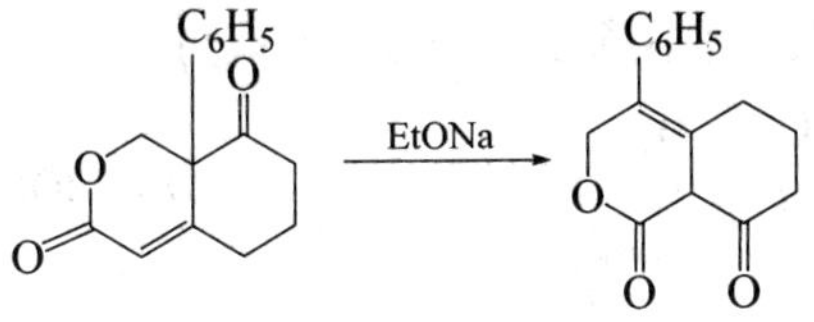 [structure with C_6H_5, O, O, O] $\xrightarrow{EtONa}$ [structure with C_6H_5, O, O, O] (7 points)

(2) 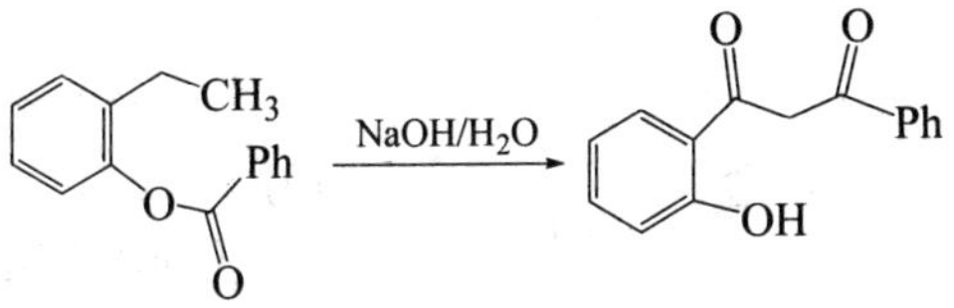 [structure with CH_3, Ph, O, O] $\xrightarrow{NaOH/H_2O}$ [structure with O, O, Ph, OH] (7 points)

(3) 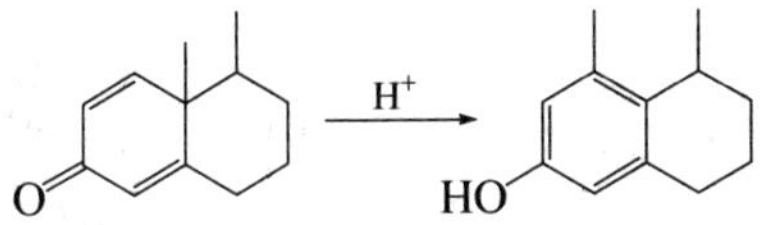 [structure with O] $\xrightarrow{H^+}$ [structure with HO] (7 points)

(4) 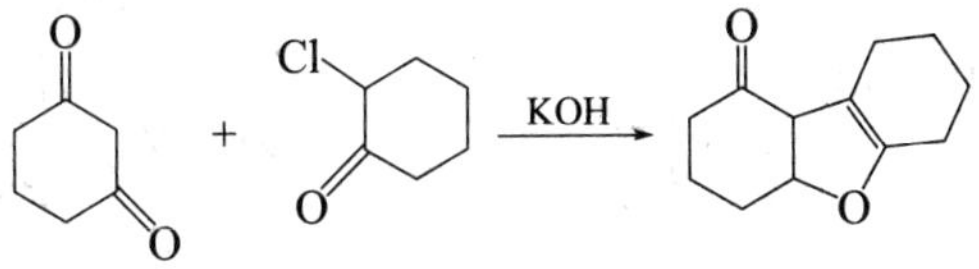 [structure with O, O] + [structure with Cl, O] $\xrightarrow{KOH}$ [structure with O, O] (7 points)

《高等有机化学》测试题(3)参考答案

1. 选择题 (每小题 2 分, 总分 32 分)

(1) C (2) B (3) C (4) A (5) A (6) D (7) B (8) A

(9) A (10) C (11) D (12) B (13) C (14) B (15) A (16) A

2. 完成下列反应 (涉及立体构型时, 写出立体构型。每小题 2 分, 总分 40 分)

(1) 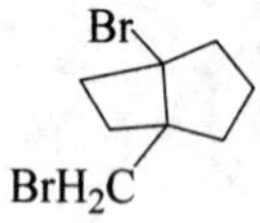 Br; BrH_2C

(2) 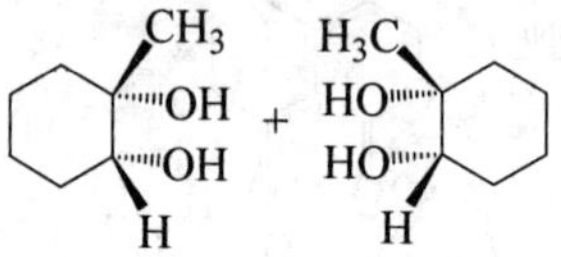 CH_3, OH, OH, H + H_3C, HO, HO, H

(3) C_6H_5CHOH / CH_3

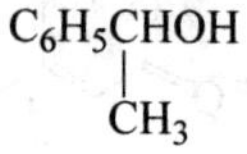

(4) 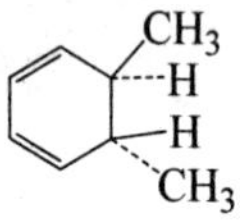 CH_3, H, H, CH_3

(5) 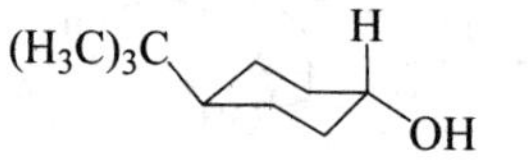 $(H_3C)_3C$, H, OH

(6) 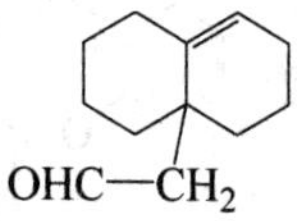 $OHC—CH_2$

(7) 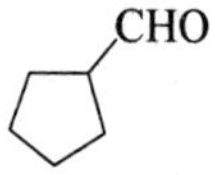 CHO

(8) 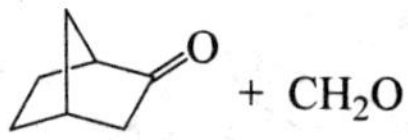 O + CH_2O

(9) 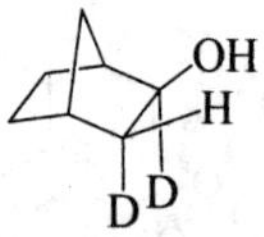 OH, H, D, D

(10) 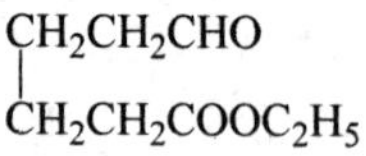 CH_2CH_2CHO / $CH_2CH_2COOC_2H_5$

(11) 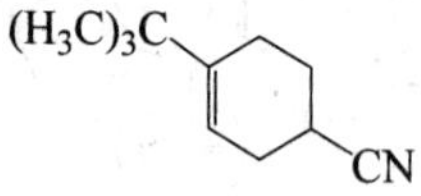 $(H_3C)_3C$, CN

(12) 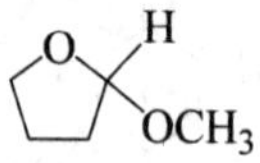 O, H, OCH_3

(13) 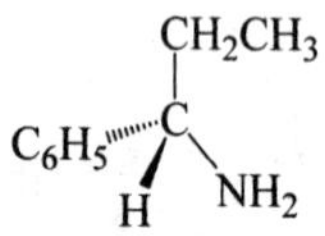 CH_2CH_3, C_6H_5, H, NH_2

(14) 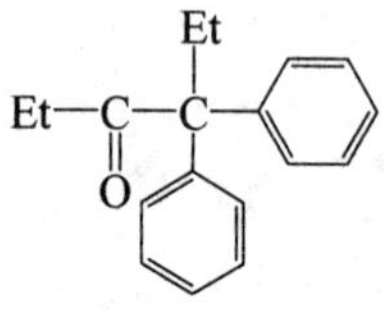 Et, Et—C(=O)—C

(15) 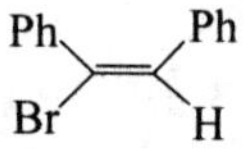 Ph, Ph, Br, H

(16) 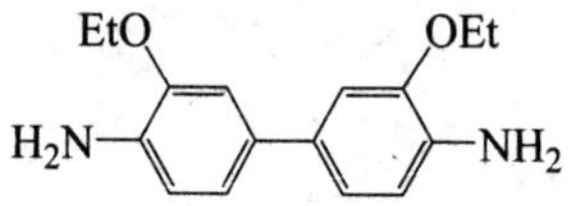 EtO, OEt, H_2N, NH_2

(17) 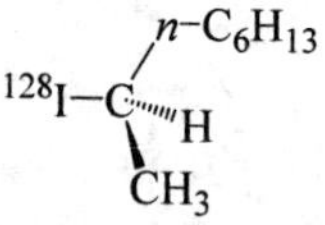 $n\text{-}C_6H_{13}$, $^{128}I—C$, H, CH_3

(18) 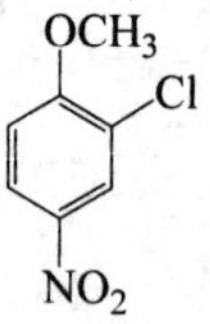 OCH_3, Cl, NO_2

(19) 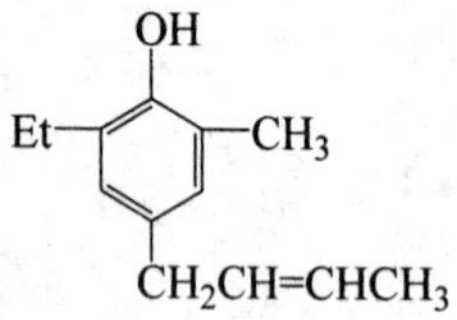 OH, Et, CH_3, $CH_2CH{=}CHCH_3$

(20) 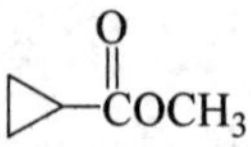 O, $COCH_3$

3. 写机理 (共 28 分)

(1) 共 7 分

C_6H_5 … $\xrightarrow[2分]{EtO^-}$ C_6H_5, O^-, OEt $\xrightarrow{2分}$ C_6H_5, EtO, O $\xleftrightarrow{1分}$ C_6H_5, EtO, O $\longrightarrow$ C_6H_5, ^-O, EtO, 1分 $\xrightarrow{1分}$ C_6H_5, O, O

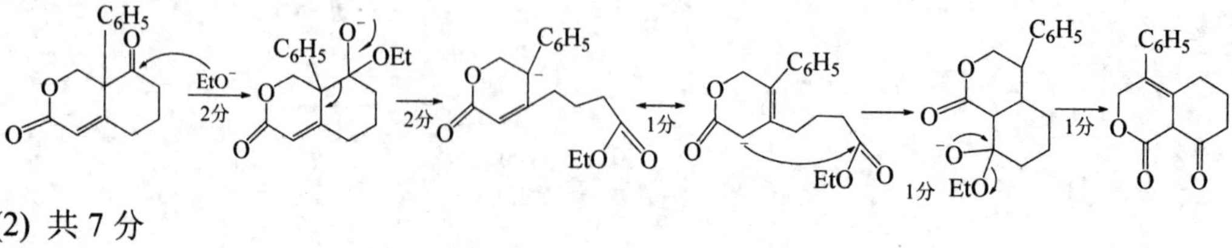

(2) 共 7 分

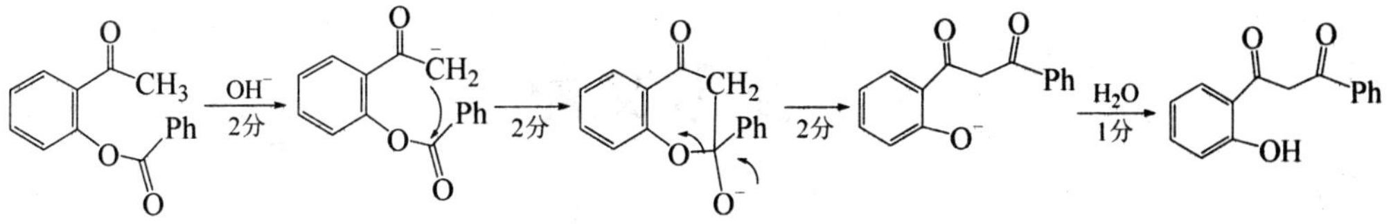

CH_3 O Ph O O $\xrightarrow[2分]{OH^-}$ O $\bar{C}H_2$ Ph O O $\xrightarrow{2分}$ O CH_2 Ph O O^- $\xrightarrow{2分}$ O O Ph O^- $\xrightarrow[1分]{H_2O}$ O O Ph OH

(3) 共 7 分

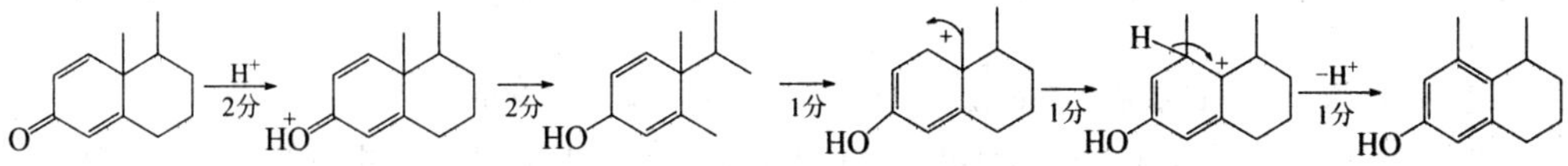

O $\xrightarrow[2分]{H^+}$ $H\overset{+}{O}$ $\xrightarrow{2分}$ HO $\xrightarrow{1分}$ + HO $\xrightarrow{1分}$ H + HO $\xrightarrow[1分]{-H^+}$ HO

(4) 共 7 分

O O $\xrightarrow[1分]{KOH}$ O O Cl O 1分 $\xrightarrow{1分}$ O O O^- Cl $\xrightarrow{1分}$ O O OH Cl

$\xrightarrow{1分}$ O OH Cl O^- $\xrightarrow[1分]{-Cl^-}$ O HO O $\xrightarrow[1分]{-H_2O}$ O O

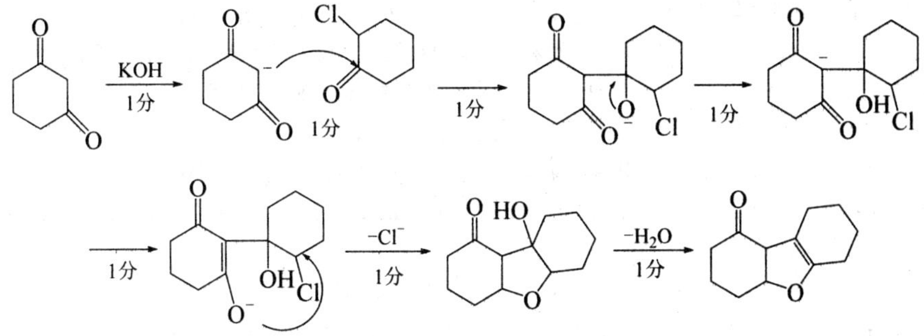

参考文献

[1] Michael B Smith, Jerry March. March's Advanced Organic Chemistry. 6th Ed. New York: John Wiley & Sons Inc, 2007.

[2] [美]韦德. 有机化学改编版. 第 6 版. 北京：高等教育出版社, 2009.

[3] [美]约翰・麦克默里. Fundamentals of Organic Chemistry. 第 4 版. 北京：机械工业出版社, 2003.

[4] 傅相锴等. 高等有机化学. 北京：高等教育出版社, 2003.

[5] [美] J. 马奇. 陶慎熹等译. 高等有机化学（上）. 北京：人民教育出版社, 1981.

[6] Bernard Miller 著. 高等有机化学. 吴范宏译, 荣国斌校. 上海：华东理工大学出版社, 2005.

[7] 恽魁宏, 高鸿宾, 任贵忠. 高等有机化学. 北京：高等教育出版社, 1995.

[8] 梁世懿, 成本诚. 高等有机化学. 北京：高等教育出版社, 1993.

[9] 邢其毅, 裴伟伟, 徐瑞秋. 有机化学. 第 2 版. 北京: 北京大学出版社, 2005.

[10] Francis A Carey, Richard J Sundberg. Advanced Organic Chemistry (Part A: Structure and Mechanisms).5th Ed. Berlin: Springer.

[11] 汪秋安. 高等有机化学. 第 2 版. 北京：化学工业出版社, 2007.

[12] L G Wade JR. Organic Chemistry. 5th Ed. New Jersey: Pearson Education, INC.

[13] 王积涛. 高等有机化学. 北京：高等教育出版社, 1982.

[14] 沈世瑜. 高等有机化学习题集. 合肥：中国科技大学出版社, 1992.

[15] Neil S Isaacs. Physical Organic Chemistry. 2nd Ed. London: Addison Wesley Longman Limited, 1995.